ISLAND
LIFE

MAP SHEWING THE DISTRIBUTION OF THE TRUE JAYS.

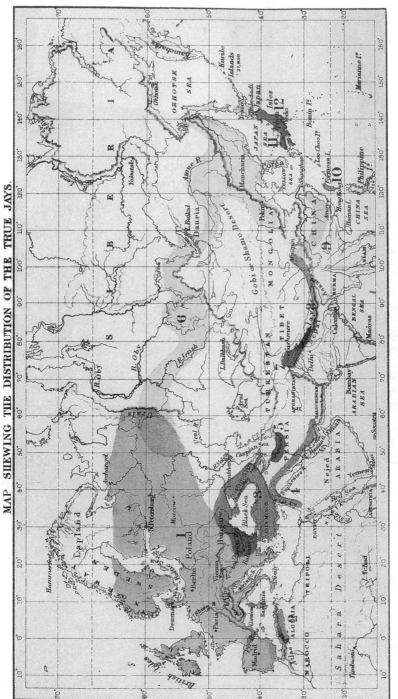

1. *Garrulus glandarius.* 2. *G. cervicalis.* 3. *G. krynicki.* 4. *G. atricapillus.* 5. *G. hyrcanus.* 6. *G. brandti.* 7. *G. lanceolatus.* 8. *G. bispecularis.* 9. *G. sinensis.* 10. *G. taivanus.* 11. *G. japonicus.* 12. *G. lidthi.*

Harper & Brothers New York.

ISLAND LIFE

ALFRED RUSSEL WALLACE

With an introduction by H. James Birx

GREAT MINDS SERIES

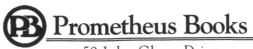

Prometheus Books

59 John Glenn Drive
Amherst, New York 14228-2197

Published 1998 by Prometheus Books

59 John Glenn Drive, Amherst, New York 14228–2197,
716–691–0133. FAX: 716–691–0137.

Library of Congress Cataloging-in-Publication Data

Wallace, Alfred Russel, 1823–1913.
　Island life / Alfred Russel Wallace ; with an introduction by
H. James Birx.
　　　　p.　　cm. — (Great minds series)
　Previously published: New York : Harper & Bros., 1881. With
new introd.
　Includes bibliographical references and index.
　ISBN 1–57392–177–7 (alk. paper)
　1. Island ecology. 2. Biogeography. 3. Glacial epoch.
I. Title. II. Series.
QH541.5.I8W34　　1997
577.5′2—dc21　　　　　　　　　　　　　　　　　97–44046
　　　　　　　　　　　　　　　　　　　　　　　　　　CIP

Printed in the United States of America on acid-free paper.

Also Available in Prometheus's Great Minds Paperback Series

See the back of this volume for a complete list of titles in Prometheus's Great Books in Philosophy and Great Minds series.

CONTENTS

CONTENTS

MAPS AND ILLUSTRATIONS.

Introduction

Alfred Russel Wallace (1823–1913), born into a poor family in Usk, Monmouthshire, Wales, had to work hard in order to support himself. He lacked the economic and educational advantages offered to Charles Robert Darwin (1809–1882), because of the latter's birth into the British upper class and all those opportunities that resulted from this fortunate circumstance. Despite their social differences, however, both Wallace and Darwin were avid naturalists who, through critical reflections on their perceptive observations of organic nature, independently developed the same scientific theory of organic evolution. To support his interests in nature, particularly botany and entomology, the self-educated Wallace spent time surveying, teaching, and collecting zoological specimens (especially butterflies) for museums in England.

In terms of science and reason, Wallace and Darwin offered a conception of natural history that argued for the mutability of all species. Interestingly enough, the major influences on these two insightful biologists had been the same. This resulted in their both first doubting the eternal fixity of species and then accepting the fact of organic evolution; a process which, they held, was primarily the result of natural selection or the "survival of the fittest" (as the contemporary philosopher Herbert Spencer referred to it).

To begin with, Wallace and Darwin were naturalists dedicated to field research, especially in entomology; Wallace rigorously devoted his time and effort to collecting butterflies and plants while Darwin was deeply interested in beetles as well as rock formations. Each had read Alexander von Humboldt's major work, *Personal Narrative* (1805–1834). Considered the father of scientific exploration, Humboldt had traveled to the Western hemisphere as a naturalist interested in understanding and appreciating plants and animals within a holistic perspective of the earth. His writings on the flora and fauna of South America inspired both Wallace and Darwin, and they in turn had an intense desire to travel around the world as naturalists committed to the pursuit of scientific knowledge. Consequently, it is not surprising that they both finally did do extensive research in South America.

Like Darwin, Wallace benefited enormously from having read Charles Lyell's three-volume work, *Principles of Geology* (1830–1833). Understandably, Lyell's sweeping geological perspective of vast periods of time and pervasive global changes due to natural forces greatly influenced the conceptual framework of both naturalists. Changing environments over millions of years implied that plant and animal forms would likewise have to change if they are to adapt to new habitats, surviving long enough to reproduce. In short, historical geology and global biogeography clearly suggested that floral and faunal species are mutable; this was a revolutionary idea in the biology of the nineteenth century.

In 1845, Wallace read Robert Chamber's *Vestiges of the Natural History of Creation* (1844) and became convinced of the fact of evolution. He was now very interested in questions concerning the origin of species and therefore wanted to explore other areas on this planet. In 1848, with the entomologist Henry Walter Bates, Wallace sailed to South America in order to collect and study some of the life forms found in the equatorial regions of the Amazon. During his four-year stay, he developed an abiding interest in the geographical distribution of plants and animals, and subsequently pondered the relationship between species and their habitats. (During his global voyage on the HMS *Beagle,* Darwin had experienced the diversity of life, particularly among insects, in the tropical rain forests of Brazil and later unearthed the fossil remains of giant mammals in Argentina.) Unfortunately, while returning to England in 1852, Wallace lost most of his unique specimens and field notes during a fire aboard ship.

It was the time that Wallace and Darwin spent on oceanic archipelagos that contributed to their doubting the eternal fixity of species and taking more and more seriously the heretical idea in natural philosophy that plant and animal forms are, in fact, mutable. For Darwin, it was his five-week visit to the Galapagos Islands (1835) that would convince him of the mutability of species, although the significance of his stay in this archipelago was understood and appreciated by the great naturalist only after he had returned to England in the following year. Wallace's eight-year stay in Malaysia (1854–1861) offered him the opportunity to study the incredible diversity of plants and animals in that part of the world. His diligent research resulted in several books, including *The Malay Archipelago: The Land of the Orang-Utan and the Bird of Paradise* (1869) and *Island Life: or, The Phenomena and Causes of Insular Faunas and Floras* (1880).

Wallace's keen observations of the biogeographical distribution of life forms resulted in his establishing the "Wallace line"—a natural gap which divides the Malay Archipelago into two separate biological areas of distinct plant and animal species, Indo-Malaysia and Austro-Malaysia (with the line running between the islands of Bali and Lombok).

Wallace had presented an introduction to his theory of evolution in two published works, *The Sarawak Law* (1855) and the *Ternate Essay* (1858). However, he still had no adequate explanatory mechanism to account for the mutability of species throughout organic history.

Like Darwin, Wallace had also read Thomas Malthus's monograph, *An Essay on the Principle of Population* (1798). It argued for the geometrical progression of animals but only the arithmetic growth of plants; this assumed discrepancy resulted in a struggle for existence throughout the animal world for the limited availability of foodstuffs in the biosphere. This description of nature implied that a selective process always pervades the living world.

In February 1858, while he was seriously ill with a fever in the Moluccas islands of Indonesia, Wallace first thought of the explanatory mechanism of natural selection. He then sent a letter and manuscript to Darwin, who had already accepted the fact of organic evolution and formulated an explanation in terms of natural selection—twenty years earlier! Regrettably, Darwin had not published anything on his theory of evolution before Wallace's abstract arrived.

Clearly, both Wallace and Darwin explained organic evolution in terms of the pervasive principle of natural selection. Darwin's major work, *On the Origin of Species* (1859), appeared only after the Darwin/Wallace positions had been read at the Linnean Society meeting in London on the evening of July 1, 1858, with Darwin being given priority for discovering the mechanism of natural selection to account for evolution. Unfortunately, Wallace is usually forgotten as the co-discoverer of biological evolution grounded in the explanatory mechanism of natural selection.

But like Darwin, Wallace referred to neither God nor metaphysical speculations when he first and independently formulated the theory of evolution many years after Darwin had reached the same conclusion about the dynamic history of life on earth. In the beginning, they were both strictly naturalistic in their interpretation of organic evolution. Each rejected a special creation to account for the appearance of the hominid zoological group. Yet the remarkable similarities between Darwin and Wallace were not to last. As the years passed, Wallace became more and more a believer in miracles, occultism, phrenology, mesmerism, and the supernatural (not clearly discriminating between real science and the pseudosciences of his time). He eventually gave a spiritualistic interpretation of human evolution that focused on the alleged uniqueness of our own species that separates it from all other forms of life (including the apes). Oddly enough, he steadfastly maintained that the explanatory mechanism of natural selection is both necessary and sufficient to account for organic evolution, but only up to the emergence of our recent species. That is, natural selection alone was responsible for the evolution of all life forms on earth before the emergence of the human animal.

But for Wallace, both the origin of life and the appearance of humankind require an explanation beyond naturalism, i.e., an interpretation of organic evolution beyond Darwinian mechanistic materialism, in order to explain both the beginning and the end of organic evolution as Wallace saw it.

According to Wallace, our species is a unique animal separated from the biological world of plants and other animals. He claimed that human evolution has always represented a single continuity and essential unity. Actually, the growing fossil hominid record clearly shows that our remote ancestors were, in fact, a very diversified group of species (with many hominid lines of evolution becoming extinct).

Wallace tenaciously held that our species has physical as well as mental traits, moral and spiritual aspects that isolate it from the rest of the living world. He argued that the emergence of these special characteristics cannot be explained merely as the result of natural selection operating throughout the gradual evolution of hominids from apelike forms to our present species. Surprisingly, Wallace firmly maintained that since man's mental abilities are superior to his needs for mere survival, then a spiritualistic explanation is necessary to account for this discrepancy. Elaborating on his argument, he further claimed that man's artistic, musical, mathematical, and metaphysical faculties could not have progressively developed in continuity by degree from earlier animals as a result of natural selection or the ability of the fittest to adapt, survive, and reproduce. In particular (for Wallace), human wit, speech, humor, morality, hairless and sensitive skin, and specialized and perfected brain, hands, and feet could not be the product of natural selection alone. Obviously, Darwin did not agree.

Furthermore, Wallace claimed that our species's exceptional mental capabilities demonstrate the existence of a Supreme Mind or Overruling Intelligence that is guiding human evolution to its assumed spiritual end (as he saw it). Obviously, he believed in the personal immortality of the human soul. And for Wallace, this alleged uniqueness of our species had required a major change in his explanation for human evolution. To account for the assumed special place our species occupies within evolving nature, Wallace had moved from naturalism to spiritualism.

Not surprisingly, Darwin was very disappointed with Wallace's turn to spiritualism. It is a bitter irony that it was Wallace's rigid commitment to natural selection that contributed to his maintaining that Mind and Purpose must manifest themselves in human evolution. Unlike Darwin, Wallace embraced teleology; he saw meaning in and a direction to the further evolution of our species in terms of spirit. Whereas Darwin emphasized chance in evolution, seeing no overall purpose or ultimate end-goal to human existence, Wallace focused on mental and sociocultural (rather than natural) selection and therefore saw both meaning in and a direction to the evolution of our species. In later writings, Wallace placed emphasis on the assumed intelligence throughout nature. Actually, his final view is self-contradictory in terms of accounting for the uniqueness of the human animal if all of living nature manifests purposive intelligence.

Wallace actually attended séances (once with T. H. Huxley, although Darwin refused to go). He believed that the ghosts of deceased persons could materialize through a medium at such contrived meetings. No doubt for emotional reasons (e.g., the untimely death of his beloved brother Herbert due to yellow fever in 1852, while they were both in South America), Wallace continuously clung to his will to believe in both pseudoscience and the ongoing spiritual progress of our species. Darwin, of course, never took such wishful thinking seriously and was understandably both perplexed and disillusioned with Wallace's willingness to be taken in by such obvious trickery.

Naturalists and humanists may rightly respect and admire Darwin's pervasive materialism and unabashed commitment to evolution during the Victorian Age. To his lasting credit, Darwin gave priority to science and reason. On the other hand, Wallace's occultism and spiritualism represent not only his gullibility concerning the paranormal, but also his inability to come to grips with the naturalistic ramifications of organic evolution for our species. Surely, self-deception and a failure of nerve are no substitutes for truth and wisdom.

Wallace even rejected the possibility that life forms and intelligent beings may exist elsewhere in this cosmos. In *Man's Place in the Universe* (1903), he dogmatically claimed, as a spiritualist, that the unique human species, as an intelligent being on earth, must be absolutely alone in the great diversity of this cosmos. He still defended evolutionism but maintained that Darwinism is not sufficient to interpret organic evolution. (At least once, while in a Brazilian rain forest, Darwin himself speculated about the existence of life forms on other worlds.)[1]

In *The World of Life* (1910), Wallace clearly replaced his earlier naturalism with vitalism, teleology, and spiritualism. For him, beyond all the phenomena and causes and laws of nature, there is Mind and Purpose: the ultimate aim of human evolution is, for him, the further development of our species for an enduring spiritual existence in the future. This universe is the best of all possible worlds (so he thought) and it exists to bring about a predestined end-goal for humankind in terms of spirit. Subsequently, Wallace believed that an infinite Deity had designed the whole cosmos and is the acting power within all life; the evolving world of matter is altogether subordinate to an unseen universe of Spirit.[2]

Wallace ultimately allowed himself to be consumed by simplistic spiritualism. His interpretation of evolution is both earth-bound and human-centered in terms of life and intelligence, respectively. Such a myopic view of the natural world sets limits to scientific inquiry in biology, anthropology, and astronomy. But at least Wallace committed himself to the fact of evolution, whereas the religious creationists and biblical fundamentalists of today reject the evolutionary framework outright in spite of the overwhelming empirical evidence to support it.

*   *   *

Island Life is an impressive, detailed work that illustrates Wallace's substantial contributions to the sciences of geography, ecology, botany and zoology. This volume is still relevant in terms of modern evolutionary thought; e.g., it helps support the hypothesis that new species emerge in small isolated populations (conditions found throughout the Galapagos Islands and the Malay Archipelago) and heightens our awareness that environmental differences influence the distribution of life forms over the planet, resulting in the rich diversity of plants and animals necessary for sustaining the complex ecosystem of the earth. It is as a rigorous naturalist and devoted evolutionist that Wallace will always be best remembered in the history of scientific thought and critical thinking.

H. James Birx, Ph.D.
Canisius College, Harvard University

ENDNOTES

1. Darwin wrote, "How great would be the desire in every admirer of nature to behold, if such were possible, the scenery of another planet!" Refer to Charles Darwin, *The Voyage of the Beagle* (Garden City, N.Y.: Doubleday/Anchor Books, 1962), p. 494.

2. This interpretation of evolution is very similar to the spiritualistic worldview of the mystical geopaleontologist and Jesuit priest Pierre Teilhard de Chardin and is best represented in his major work, *The Phenomenon of Man* (1938–1940). Refer to Pierre Teilhard de Chardin, *The Phenomenon of Man,* 2d ed. (New York: Harper Colophon, 1975).

FURTHER READINGS

Birx, H. James. *Interpreting Evolution: Darwin & Teilhard de Chardin.* Amherst, N.Y.: Prometheus Books, 1991, esp. pp. 60–63, 112–65.

Brooks, J. L. *Just Before the Origin: Alfred Russel Wallace's Theory of Evolution.* New York: Columbia University Press, 1984.

Clements, Harry. *Alfred Russel Wallace: Biologist and Social Reformer.* London: Hutchinson, 1983.

Clodd, Edward. *Pioneers of Evolution: From Thales to Huxley.* London: Watts, 1921, esp. pp. 61–69.

Darwin, Charles. *The Descent of Man.* Amherst, N.Y.: Prometheus Books, 1998. Refer to the introduction by H. James Birx, pp. ix–xxviii.

Desmond, Adrian. *Huxley: From Devil's Disciple to Evolution's High Priest.* Reading, Mass.: Addison–Wesley, 1997.

INTRODUCTION

Desmond, Adrian, and James Moore. *Darwin: The Life of a Tormented Evolutionist*. New York: Warner Books, 1991.

Eiseley, Loren C. "Alfred Russel Wallace." *Scientific American* 200, no. 2 (February 1959): 70–82, 84.

Fichman, Martin. *Alfred Russel Wallace*. Boston: Twayne, 1981.

George, Wilma Beryl. *Biologist Philosopher: A Study of the Life and Writings of Alfred Russel Wallace*. New York: Abelard-Schuman, 1964.

Kottler, Malcolm Jay. "Alfred Russel Wallace, the Origin of Man, and Spiritualism." *Isis* 65, no. 227 (June 1974): 144–92.

Marchant, J. *Alfred Russel Wallace: Letters and Reminiscences*. New York: Arno Press, 1975.

Mayer, William V. "Wallace and Darwin." *The American Biology Teacher* 49, no. 8 (November/December 1987): 406–10.

McKinney, H. Lewis. *Wallace and Natural Selection*. New Haven: Yale University Press, 1972.

Raby, Peter. *Bright Paradise: Victorian Scientific Travellers*. Princeton: Princeton University Press, 1997, esp. pp. 148–77.

Shermer, Michael. "A Gentlemanly Arrangement: Alfred Russel Wallace, Charles Darwin, and the Resolution of a Scientific Priority Dispute." *Skeptic* 3, no. 2 (1995): 80–89.

———. "A Heretic-Scientist Among the Spiritualists: Alfred Russel Wallace and 19th-Century Spiritualism." *Skeptic* 3, no. 1 (1995): 70–83.

Turner, Frank Miller. *Between Science and Religion: The Reaction to Scientific Naturalism in Late Victorian England*. New Haven: Yale University Press, 1974, esp. pp. 68–103.

Wallace, Alfred Russel. *Contributions to the Theory of Natural Selection: A Series of Essays*. New York: Macmillan, 1870.

———. *Darwinism: An Exposition on the Theory of Natural Selection, with Some of its Applications*. London: Macmillan, 1889.

———. *Man's Place in the Universe: A Study of the Results of Scientific Research in Relation to the Unity or Plurality of Worlds*. New York: McClure/Phillips, 1903.

———. *My Life: A Record of Events and Opinions*. Rev. ed. London: Chapman & Hall, 1908.

———. *Natural Selection and Tropical Nature: Essays on Descriptive and Theoretical Biology*. London: Macmillan, 1895.

———. *The World of Life: A Manifestation of Creative Power, Directed Mind and Ultimate Purpose*. New York: Moffat/Yard, 1910.

PREFACE.

THE present volume is the result of four years' additional thought and research on the lines laid down in my "Geographical Distribution of Animals," and may be considered as a popular supplement to and completion of that work.

It is, however, at the same time, a complete work in itself; and, from the mode of treatment adopted, it will, I hope, be well calculated to bring before the intelligent reader the wide scope and varied interest of this branch of natural history. Although some of the earlier chapters deal with the same questions as my former volumes, they are here treated from a different point of view; and, as the discussion of them is more elementary and at the same time tolerably full, it is hoped that they will prove both instructive and interesting. The plan of my larger work required that *genera* only should be taken account of; in the present volume I often discuss the distribution of *species*, and this will help to render the work more intelligible to the unscientific reader.

The full statement of the scope and object of the present essay given in the "Introductory" chapter, together with the "Summary" of the whole work and the general view of the more important arguments given in the "Conclusion," render it unnecessary for me to offer any further remarks on these points. I may, however, state generally that, so far as I am able to judge, a real advance has here been made in the mode of treating problems in geographical distribution, owing to the firm establishment of a number of preliminary doctrines or " principles," which in many cases lead to a far simpler and yet more complete solution of such problems than has been hitherto possible. The most important of these doctrines are those which establish and define—(1) The former wide extension of all groups now discontinuous, as being a necessary result of "evolution;" (2) The permanence of the great features of the distribution of land and water on the earth's surface; and (3) The nature and frequency of climatal changes throughout geological time.

I have now only to thank the many friends and correspondents who have given me information or advice. Besides those whose assistance is acknowledged in the body of the work, I am especially indebted to four gentlemen who have been kind enough to read over the proofs of chapters dealing with questions on which they have special knowledge, giving me the benefit of valuable emendations and suggestions. Mr. Edward R. Alston has looked over those parts of the earlier chapters which relate to the mammals of Europe and the north temperate zone; Mr. S. B. J. Skertchley, of the Geological Survey, has read the chapters which discuss the glacial epoch and other geological questions; Professor A. Newton has looked over the passages referring to the birds of the Madagascar group; while Sir Joseph D. Hooker has given me the invaluable benefit of his remarks on my two chapters dealing with the New Zealand flora.

PART I.

THE DISPERSAL OF ORGANISMS: ITS PHENOMENA
LAWS, AND CAUSES

ISLAND LIFE.

CHAPTER I.

INTRODUCTORY.

WHEN an Englishman travels by the nearest sea-route from Great Britain to Northern Japan he passes by countries very unlike his own, both in aspect and natural productions. The sunny isles of the Mediterranean, the sands and date-palms of Egypt, the arid rocks of Aden, the cocoa groves of Ceylon, the tiger-haunted jungles of Malacca and Singapore, the fertile plains and volcanic peaks of Luzon, the forest-clad mountains of Formosa, and the bare hills of China, pass successively in review; till after a circuitous voyage of thirteen thousand miles he finds himself at Hakodadi in Japan. He is now separated from his starting-point by the whole width of Europe and Northern Asia, by an almost endless succession of plains and mountains, arid deserts or icy plateaux, yet when he visits the interior of the country he sees so many familiar natural objects that he can hardly help fancying he is close to his home. He finds the woods and fields tenanted by tits, hedge-sparrows, wrens, wagtails, larks, redbreasts, thrushes, buntings, and house-sparrows, some absolutely identical with our own feathered friends, others so closely resembling them that it requires a practised ornithologist to tell the difference. If he is fond of insects he notices many butterflies and a host of beetles which, though on close examination they are found to be distinct from ours, are yet of the same general aspect, and seem just what might be expected in any part of Europe. There are also of course many birds and insects which are quite new and peculiar, but these are by no means so numerous or conspicuous as to remove the general impression of a wonderful resemblance between the productions of such remote islands as Britain and Yesso.

Now let an inhabitant of Australia sail to New Zealand, a distance of less than thirteen hundred miles, and he will find himself in a country whose productions are totally unlike those of his own. Kangaroos and wombats there are none, the birds are almost all entirely new, insects are very scarce and quite unlike the handsome or strange Australian forms, while even the vegetation is all changed, and no gum-tree, or wattle, or grass-tree meets the traveller's eye.

But there are some more striking cases even than this, of the diversity of the productions of countries not far apart. In the Malay Archipelago there are two islands, named Bali and Lombok, each about as large as Corsica, and separated by a strait only fifteen miles wide at its narrowest part. Yet these islands differ far more from each other in their birds and quadrupeds than do England and Japan. The birds of the one are extremely *unlike* those of the other, the difference being such as to strike even the most ordinary observer. Bali has red and green woodpeckers, barbets, weaver-birds, and black-and-white magpie-robins, none of which are found in Lombok, where, however, we find screaming cockatoos and friar-birds, and the strange mound-building megapodes, which are all equally unknown in Bali. Many of the kingfishers, crow-shrikes, and other birds, though of the same general form, are of very distinct species; and though a considerable number of birds are the same in both islands, the difference is none the less remarkable—as proving that mere distance is one of the least important of the causes which have determined the likeness or unlikeness in the animals of different countries.

In the western hemisphere we find equally striking examples. The eastern United States possess very peculiar and interesting plants and animals, the vegetation becoming more luxuriant as we go south, but not altering in essential character, so that when we reach the southern extremity of Florida we still find ourselves in the midst of oaks, sumachs, magnolias, vines, and other characteristic forms of the temperate flora; while the birds, insects, and land-shells are almost identical with those found farther north. But if we now cross over the narrow strait, about fifty miles wide, which separates Florida from the Bahama Islands, we find ourselves in a totally different country, surrounded by a vegetation which is essentially tropical and generally identical with that of Cuba. The change is most striking, because there is no difference of climate, of soil, or apparently of position, to account for it; and when we find that the birds, the insects, and especially the land-shells are almost all West Indian, while the North American types of plants and animals have almost all completely disappeared, we shall be convinced that such differences and resemblances cannot be due to existing conditions, but must depend upon laws and causes to which mere proximity of position offers no clew.

Hardly less uncertain and irregular are the effects of climate. Hot countries usually differ widely from cold ones in all their organic forms; but the difference is by no means constant, nor does it bear any proportion to difference of temperature. Between frigid Canada and sub-tropical Florida there are less marked differences in the animal productions than between Florida and Cuba or Yucatan, so much more alike in climate and so much nearer together. So the differences between the birds and quadrupeds of temperate Tasmania and tropical North Australia are slight and unimportant as compared with the enormous differences we find when we pass from the latter country to equally tropical Java. If we compare corresponding portions of different continents, we find no indication that the almost perfect similarity of climate and general conditions has any tendency to produce similarity in the animal world. The equatorial parts of Brazil and of the west coast of Africa are almost identical in climate and in luxuriance of vegetation, but their animal life is totally diverse. In the former we have tapirs, sloths, and prehensile-tailed monkeys; in the latter, elephants, antelopes, and man-like apes; while among birds, the toucans, chatterers, and humming-birds of Brazil are replaced by the plantain-eaters, bee-eaters, and sun-birds of Africa. Parts of South-temperate America, South Africa, and South Australia correspond closely in climate; yet the birds and quadrupeds of these three districts are as completely unlike each other as those of any parts of the world that can be named.

If we visit the great islands of the globe, we find that they present similar anomalies in their animal productions, for while some exactly resemble the nearest continents, others are widely different. Thus the quadrupeds, birds, and insects of Borneo correspond very closely to those of the Asiatic continent, while those of Madagascar are extremely unlike African forms, although the distance from the continent is less in the latter case than in the former. And if we compare the three great islands Sumatra, Borneo, and Celebes—lying, as it were, side by side in the same ocean—we find that the two former, although farthest apart, have almost identical productions, while the two latter, though closer together, are more unlike than Britain and Japan, situated in different oceans and separated by the largest of the great continents.

These examples will illustrate the kind of questions it is the object of the present work to deal with. Every continent, every country, and every island on the globe offer similar problems of greater or less complexity and interest, and the time has now arrived when their solution can be attempted with some prospect of success. Many years' study of this class of subjects has convinced me that there is no short and easy method of dealing with them; because they are, in their very nature, the visible outcome and residual product of the whole past history of the earth. If we take the organic productions of a small island, or of any very limited tract of country, such as a moderate-sized country parish, we have, in their relations and affinities—in the fact that they are *there* and others are *not* there, a problem which involves all the migrations of these species and their ancestral forms—all the vicissitudes of climate and all the changes of sea and land which have affected those migrations—the whole series of actions and reactions which have determined the preservation of some forms and the extinction of others—in fact, the whole history of the earth, inorganic and organic, throughout a large portion of geological time.

We shall perhaps better exhibit the scope and complexity of the subject, and show that any intelligent study of it was almost impossible till quite recently, if we concisely enumerate the great mass of facts and the number of scientific theories or principles which are necessary for its elucidation.

We require, then, in the first place, an adequate knowledge of the fauna and flora of the whole world, and even a detailed knowledge of many parts of it, including the islands of more special interest and their adjacent continents. This kind of knowledge is of very slow growth, and is still very imperfect;[1] and in many cases it can never now be obtained, owing to the reckless destruction of forests, and with them of countless species of plants and animals. In the next place, we require a true and natural classification of animals and plants, so that we may know their real affinities; and it is only now that this is being

---

[1] I cannot avoid here referring to the enormous waste of labor and money with comparatively scanty and unimportant results to natural history of most of the great scientific voyages of the various civilized governments during the present century. All these expeditions combined have done far less than private collectors in making known the products of remote lands and islands. They have brought home fragmentary collections, made in widely scattered localities, and these have been usually described in huge folios, whose value is often in inverse proportion to their bulk and cost. The same species have been collected again and again, often described several times over under new names, and not unfrequently stated to be from places they never inhabited. The result of this wretched system is that the productions of some of the most frequently visited and most interesting islands on the globe are still very imperfectly known, while their native plants and animals are being yearly exterminated; and this is the case even with countries under the rule or protection of European governments. Such are the Sandwich Islands, Tahiti, the Marquesas, the Philippine Islands, and a host of smaller ones; while Bourbon and Mauritius, St. Helena, and several others have only been adequately explored after an important portion of their productions has been destroyed by cultivation or the reckless introduction of goats and pigs. The employment in each of our possessions, and those of other European powers, of a resident naturalist at a very small annual expense, would have done more for the advancement of knowledge in this direction than all the expensive expeditions that have again and again circumnavigated the globe.

generally arrived at. We further have to make use of the theory of "descent with modification" as the only possible key to the interpretation of the facts of distribution; and this theory has only been generally accepted within the last twenty years. It is evident that, so long as the belief in "special creations" of each species prevailed, no explanation of the complex facts of distribution *could* be arrived at or even conceived; for if each species was created where it is now found, no further inquiry can take us beyond that fact, and there is an end of the whole matter. Another important factor in our interpretation of the phenomena of distribution is a knowledge of the extinct forms that have inhabited each country during the tertiary and secondary periods of geology. New facts of this kind are daily coming to light, but except as regards Europe, North America, and parts of India, they are extremely scanty; and even in the best-known countries the record itself is often very defective and fragmentary. Yet we have already obtained remarkable evidence of the migrations of many animals and plants in past ages, throwing an often unexpected light on the actual distribution of many groups.[1] By this means alone can we obtain positive evidence of the past migrations of organisms; and when, as too frequently is the case, this is altogether wanting, we have to trust to collateral evidence and more or less probable hypothetical explanations. Hardly less valuable is the evidence of stratigraphical geology; for this often shows us what parts of a country have been submerged at certain epochs, and thus enables us to prove that certain areas have been long isolated, and the fauna and flora allowed time for special development. Here, too, our knowledge is exceedingly imperfect, though the blanks upon the geological map of the world are yearly diminishing in extent. Lastly, as a most valuable supplement to geology, we require to know the exact depth and contour of the ocean-bed, since this affords an important clew to the former existence of now submerged lands, uniting islands to continents, or affording intermediate stations which have aided the migrations of many

---

[1] The general facts of Palæontology, as bearing on the migrations of animal groups, are summarized in my "Geographical Distribution of Animals," Vol. I., Chapters VI., VII., and VIII.

organisms. This kind of information has only begun to be obtained during the last few years; and it will be seen in the latter part of this volume that some of the most recent deep-sea soundings have afforded a basis for an explanation of one of the most difficult and interesting questions in geographical biology—the origin of the fauna and flora of New Zealand.

Such are the various classes of evidence that bear directly on the question of the distribution of organisms; but there are others of even a more fundamental character, and the importance of which is only now beginning to be recognized by students of nature. These are, firstly, the wonderful alterations of climate which have occurred in the temperate and polar zones, as proved by the evidences of glaciation in the one and of luxuriant vegetation in the other; and, secondly, the theory of the permanence of existing continents and oceans. If glacial epochs in temperate lands and mild climates near the poles have, as now believed by men of eminence, occurred several times over in the past history of the earth, the effects of such great and repeated changes, both on the migration, modification, and extinction of species, must have been of overwhelming importance—of more importance, perhaps, than even the geological changes of sea and land. It is therefore necessary to consider the evidence for these climatal changes; and then, by a critical examination of their possible causes, to ascertain whether they were isolated phenomena, were due to recurrent cosmical actions, or were the result of a great system of terrestrial development. The latter is the conclusion we arrive at; and this conclusion brings with it the conviction that, in the theory which accounts for both glacial epochs and warm polar climates, we have the key to explain and harmonize many of the most anomalous biological and geological phenomena, and one which is especially valuable for the light it throws on the dispersal and existing distribution of organisms. The other important theory, or rather corollary from the preceding theory—that of the permanence of oceans and the general stability of continents throughout all geological time—is as yet very imperfectly understood, and seems, in fact, to many persons in the nature of a paradox. The evidence for it, however, appears to me to be conclusive; and it is certainly the

most fundamental question in regard to the subject we have to
deal with; since, if we once admit that continents and oceans
may have changed places over and over again (as many writers
maintain), we lose all power of reasoning on the migrations of
ancestral forms of life, and are at the mercy of every wild theo-
rist who chooses to imagine the former existence of a now submerged continent to explain the existing distribution of a group
of frogs or a genus of beetles.

As already shown by the illustrative examples adduced in this
chapter, some of the most remarkable and interesting facts in
the distribution and affinities of organic forms are presented by
islands in relation to each other and to the surrounding continents. The study of the productions of the Galapagos—so peculiar, and yet so decidedly related to the American continent
—appears to have had a powerful influence in determining the
direction of Mr. Darwin's researches into the origin of species;
and every naturalist who studies them has always been struck
by the unexpected relations or singular anomalies which are so
often found to characterize the fauna and flora of islands. Yet
their full importance in connection with the history of the earth
and its inhabitants has hardly yet been recognized; and it is in
order to direct the attention of naturalists to this most promis-
ing field of research that I restrict myself in this volume to an
elucidation of some of the problems they present to us. By far
the larger part of the islands of the globe are but portions of
continents undergoing some of the various changes to which
they are ever subject; and the correlative statement, that every
part of our continents have again and again passed through insular conditions, has not been sufficiently considered, but is, I
believe, the statement of a great and most suggestive truth, and
one which lies at the foundation of all accurate conception of
the physical and organic changes which have resulted in the
present state of the earth.

The indications now given of the scope and purpose of the
present volume render it evident that, before we can proceed
to the discussion of the remarkable phenomena presented by
insular faunas and floras, and the complex causes which have
produced them, we must go through a series of preliminary

studies, adapted to give us a command of the more important facts and principles on which the solution of such problems depends. The succeeding eight chapters will, therefore, be devoted to the explanation of the mode of distribution, variation, modification, and dispersal of species and groups, illustrated by facts and examples; of the true nature of geological change as affecting continents and islands; of changes of climate, their nature, causes, and effects; of the duration of geological time and the rate of organic development.

CHAPTER II.

THE ELEMENTARY FACTS OF DISTRIBUTION.

Importance of Locality as an Essential Character of Species.—Areas of Distribution.
—Extent and Limitations of Specific Areas.—Specific Range of Birds.—Generic
Areas.—Separate and Overlapping Areas.—The Species of Tits as Illustrating
Areas of Distribution.—The Distribution of the Species of Jays.—Discontinuous
Generic Areas.—Peculiarities of Generic and Family Distribution.—General Feat-
ures of Overlapping and Discontinuous Areas.—Restricted Areas of Families.—
The Distribution of Orders.

So long as it was believed that the several species of animals
and plants were " special creations," and had been formed ex-
pressly to inhabit the countries in which they are now found,
their habitat was an ultimate fact which required no explanation. It was assumed that every animal was *exactly* adapted
to the climate and surroundings amid which it lived, and that
the only, or, at all events, the chief, reason why it did not inhabit another country was, that the climate or general conditions of that country were not suitable to it, but in what the
unsuitability consisted we could rarely hope to discover. Hence
the exact locality of any species was not thought of much importance from a scientific point of view, and the idea that anything could be learned by a comparative study of different floras
and faunas never entered the minds of the older naturalists.

But so soon as the theory of evolution came to be generally
adopted, and it was seen that each animal could only have come
into existence in some area where ancestral forms closely allied
to it already lived, a real and important relation was established
between an animal and its native country, and a new set of
problems at once sprang into existence. From the old point of
view, the *diversities* of animal life in the separate continents,
even where physical conditions were almost identical, was the
fact that excited astonishment; but seen by the light of the

evolution theory, it is the *resemblances* rather than the diversities in these distant continents and islands that are most difficult to explain. It thus comes to be admitted that a knowledge of the exact area occupied by a species or a group is a real portion of its natural history, of as much importance as its habits, its structure, or its affinities; and that we can never arrive at any trustworthy conclusions as to how the present state of the organic world was brought about until we have ascertained with some accuracy the general laws of the distribution of living things over the earth's surface.

Areas of Distribution.—Every species of animal has a certain area of distribution to which, as a rule, it is permanently confined, although, no doubt, the limits of its range fluctuate somewhat from year to year, and in some exceptional cases may be considerably altered in a few years or centuries. Each species is moreover usually limited to one continuous area, over the whole of which it is more or less frequently to be met with; but there are many partial exceptions to this rule. Some animals are so adapted to certain kinds of country—as to forests or marshes, mountains or deserts—that they cannot live long elsewhere. These may be found scattered over a wide area in suitable spots only, but can hardly on that account be said to have several distinct areas of distribution. As an example, we may name the chamois, which lives only on high mountains, but is found in the Pyrenees, the Alps, the Carpathians, in some of the Greek mountains and the Caucasus. The variable hare is another and more remarkable case, being found all over Northern Europe and Asia, beyond lat. 55°, and also in Scotland and Ireland. In Central Europe it is unknown till we come to the Alps, the Pyrenees, and the Caucasus, where it again appears. This is one of the best cases known of the discontinuous distribution of a *species*, there being a gap of about a thousand miles between its southern limits in Russia and its reappearance in the Alps. There are, of course, numerous instances in which species occur in two or more islands, or in an island and continent, and are thus rendered discontinuous by the sea, but these involve questions of changes in sea and land which we shall have to consider further on. Other cases are believed to exist of still wider separation of a species, as with the marsh titmice and the reed buntings of Europe and Japan, where similar forms are found in the extreme localities, while a distinct variety, race, or sub-species inhabits the intervening district.

Extent and Limitations of Specific Areas.—Leaving for the present these cases of want of continuity in a species, we find the most wide difference between the extent of country occupied, varying, in fact, from a few square miles to almost the entire land surface of the globe. Among the mammalia, however, the same species seldom inhabits both the old and new worlds, unless they are strictly arctic animals, as the reindeer, elk, and arctic fox, the glutton, the ermine, and some others. The common wolf of Europe and Northern Asia is thought by many naturalists to be identical with the variously colored wolves of North America extending from the Arctic Ocean to Mexico, in which case this will have, perhaps, the widest range of any species of mammal. Little doubt exists as to the identity of the brown bears and the beavers of Europe and North America; but all these species range up to the Arctic circle, and there is no example of a mammal universally admitted to be identical yet confined to the temperate zones of the two hemispheres. Among the undisputed species of mammalia, the leopard has an enormous range, extending all over Africa and South Asia to Borneo and the east of China, and thus having probably the widest range of any known mammal. The winged mammalia have not usually very wide ranges, there being only one bat common to the Old and New Worlds. This is a British species, *Vesperugo serotinus*, which is found over the larger part of North America, Europe, and Asia, as far as Pekin, and even extends into tropical Africa, thus rivalling the leopard and the wolf in the extent of country it occupies.

Of very restricted ranges there are many examples, but some of these are subject to doubts as to the distinctness of the species or as to its geographical limits being really known. In Europe we have a distinct species of ibex (*Capra Pyrenaica*) confined to the Pyrenean mountains, while the true marmot is restricted to the Alpine range. More remarkable is the Pyrenean water-mole (*Mygale Pyrenaica*), a curious small insectivorous animal found only in a few places in the northern valleys of the Pyrenees. In islands there are many cases of undoubted restriction of species to a small area, but these involve a different question from the range of species on continents where there is no *apparent* obstacle to their wider extension.

Specific Range of Birds.—Among birds we find instances of much wider range of species, which is only what might be expected considering their powers of flight; but, what is very curious, we also find more striking (though perhaps not more frequent) examples of extreme limitation of range among birds than among mammals. Of the former phenomenon perhaps the most remarkable case is that afforded by the osprey, or fishing-hawk, which ranges over the greater portion of all the continents, as far as Brazil, South Africa, the Malay Islands, and Tasmania. The barn-owl (*Strix flammea*) has nearly as wide a range, but in this case there is more diversity of opinion as to the specific difference of many of the forms inhabiting remote countries, some of which seem undoubtedly to be distinct. Among passerine birds the raven has probably the widest range, extending from the Arctic regions to Texas and New Mexico in America, and to North India and Lake Baikal in Asia; while the little northern willow-wren (*Phylloscopus borealis*) ranges from Norway across Asia to Alaska, and southward to Ceylon, China, Borneo, and Timor.

Of very restricted continental ranges the best examples in Europe are the little blue magpie (*Cyanopica Cooki*) confined to the central portions of the Spanish peninsula; and the Italian sparrow found only in Italy and Corsica. In Asia, Palestine affords some examples of birds of very restricted range—a beautiful sun-bird (*Nectarinea osea*), a peculiar starling (*Amydrus Tristramii*), and some others, being almost or quite confined to the warmer portions of the valley of the Jordan. In the Himalayas there are numbers of birds which have very restricted ranges; but those of the Neilgherries are perhaps better known, several species of laughing thrushes and other birds being found only on the summits of these mountains. The most wonderfully restricted ranges are, however, to be found among the humming-birds of tropical America. The great volcanic peaks of Chimborazo and Pichincha have each a peculiar species of humming-bird confined to a belt just below the limits of perpetual snow, while the extinct volcano of Chiriqui, in Veragua, has a species confined to its wooded crater. One of the most strange and beautiful of the humming-birds (*Loddigesia mirabilis*) was obtained once only, more than forty years ago, near Chachapoyas, in the Andes of Northern Peru; and though Mr. Gould has sent many drawings of the bird to people visiting the district, and has for many years offered a high reward for a specimen, no other has ever been seen![1]

The above details will sufficiently explain what is meant by the "specific area" or range of a species. The very wide and very narrow ranges are exceptional, the great majority of species both of mammals and birds ranging over moderately wide areas, which present no striking contrasts in climate and physical conditions. Thus a large proportion of European birds range over the whole continent in an east and west direction, but considerable numbers are restricted either to the northern or the southern half. In Africa some species range over all the continent south of the desert, while large numbers are restricted to the equatorial forests, or to the upland plains. In North America, if we exclude the tropical and the arctic portions, a considerable number of species range over all the temperate parts of the continent, while still more are restricted to the east, the centre, or the west, respectively.

Generic Areas.—Having thus obtained a tolerably clear idea of the main facts as to the distribution of isolated species, let us now consider those collections of closely allied species termed genera. What a genus is will be sufficiently understood by a few illustrations. All the different kinds of dogs, jackals, and wolves belong to the dog genus, Canis; the tiger, lion, leopard, jaguar, and the wild-cats, to the cat genus, Felis; the blackbird, song-thrush, missel-thrush, fieldfare, and many others, to the thrush genus, Turdus; the crow, rook, raven, and jackdaw, to the crow genus, Corvus; but the magpie belongs to another,

---

[1] Since these lines were written, the report comes that fresh specimens have been found in the same locality.

though closely allied genus, Pica, distinguished by the different form and proportions of its wings and tail from all the species of the crow genus. The number of species in a genus varies greatly from one up to several hundreds. The giraffe, the glutton, the walrus, the bearded reedling, the secretary-bird, and many others, have no close allies, and each forms a genus by itself. The beaver genus, Castor, and the camel genus, Camelus, each consist of two species. On the other hand, the deer genus, Cervus, has forty species; the mouse and rat genus, Mus, more than a hundred species; and there is about the same number of the thrush genus; while among the lower classes of animals genera are often very extensive, the fine genus Papilio, or swallow-tailed butterflies, containing more than four hundred species; and Cicindela, which includes our native tiger beetles, has about the same number. Many genera of shells are very extensive, and one of them—the genus Helix, including the commonest snails, and ranging all over the world—is probably the most extensive in the animal kingdom, numbering about two thousand described species.

Separate and Overlapping Areas.—The species of a genus are distributed in two ways. Either they occupy distinct areas which do not touch each other and are sometimes widely separated, or they touch and occasionally overlap each other, each species occupying an area of its own which rarely coincides exactly with that of any other species of the same genus. In some cases, when a river, a mountain-chain, or a change of conditions, as from pasture to desert or forest, determines the range of species, the areas of two species of the same genus may just meet, one beginning where the other ends; but this is comparatively rare. It occurs, however, in the Amazon valley, where several species of monkeys, birds, and insects come up to the south bank of the river, but do not pass it, while allied species come to the north bank, which in like manner forms their boundary. As examples we may mention that one of the Saki monkeys (*Pithecia monachus ?*) comes up to the south bank of the Upper Amazon, while immediately we cross over to the north bank we find another species (*Pithecia rufibarbata ?*). Among birds we have the green jacamar (*Galbula viridis*) abundant on the north bank

of the Lower Amazon, while on the south bank we have two
allied species (*Galbula rufoviridis* and *G. cyaneicollis*); and
among insects we have at Santarem, on the south bank of the
Amazon, the beautiful blue butterfly *Callithea sapphira*, while
almost opposite to it, at Monte-alegre, an allied species, *Callithea
Leprieuri*, is alone found. Perhaps the most interesting and
best-known case of a series of allied species whose ranges are
separate but conterminous is that of the beautiful South American wading birds, called trumpeters, and forming the genus
Psophia. There are five species, all found in the Amazon valley,
but each limited to a well-marked district bounded by great
rivers. On the north bank of the Amazon there are two species,
one in its lower valley extending up to the Rio Negro, and the
other in the central part of the valley beyond that river; while
to the south of the Amazon there are three, one above the Madeira, one below it, and a third near Para, probably separated
from the last by the Tocantins River.

Overlapping areas among the species of a genus is a more
common phenomenon, and is almost universal where these species are numerous in the same continent. It is, however, exceedingly irregular, so that we often find one species extending
over a considerable portion of the area occupied by the genus,
and including the entire areas of some of the other species. So
little has been done to work out accurately the limits of species
that it is very difficult to give examples. One of the best is to
be found in the genus Dendrœca, a group of American wood-
warblers. These little birds all migrate in the winter into the
tropical regions, but in the summer they come north, each having its particular range. Thus, *D. Dominica* comes as far as
South Carolina, *D. cærulea* to Virginia, *D. discolor* to Southern
Maine and Canada; four other species go farther north in Canada, while five more extend to the borders of the Arctic zone.

The Species of Tits as Illustrating Areas of Distribution.—In
our own hemisphere the overlapping of allied species may be
well illustrated by the various kinds of titmice, several of which
are among our best-known English birds. The great titmouse
(*Parus major*) has the widest range of all, extending from the
Arctic circle to Algeria, Palestine, and Persia, and from Ireland

right across Siberia to the Ochotsk Sea, probably following the great northern forest belt. It does not extend into China and Japan, where distinct species are found. Next in extent of range is the coal tit (*Parus ater*), which inhabits all Europe, from the Mediterranean to about 64° N. latitude ; in Asia Minor to the Lebanon and Caucasus ; and across Siberia to Amoorland. The marsh tit (*Parus palustris*) inhabits temperate and south Europe from 61° N. latitude in Norway to Poland and South-west Russia, and in the south from Spain to Asia Minor. Close-ly allied to this—of which it is probably only a variety or sub-species—is the northern marsh tit (*Parus borealis*), which over-laps the last in Norway and Sweden, and also in South Russia and the Alps, but extends farther north into Lapland and North Russia, and thence probably in a southeasterly direction across Central Asia to North China. Yet another closely allied species (*Parus Camtschatkensis*) ranges from Northeastern Russia across Northern Siberia to Lake Baikal and to Hakodadi in Japan, thus overlapping *Parus borealis* in the western portion of its area. Our little favorite, the blue tit (*Parus cœruleus*), ranges over all Europe from the Arctic circle to the Mediterranean, and on to Asia Minor and Persia, but does not seem to pass beyond the Ural Mountains. Its lovely eastern ally, the azure tit (*Parus cyaneus*), overlaps the range of *P. cœruleus* in Western Europe as far as St. Petersburg and Austria, rarely straggling to Den-mark, while it stretches all across Central Asia between the lati-tudes 35° and 56° N. as far as the Amoor valley. Besides these wide-ranging species, there are several others which are more re-stricted. *Parus Teneriffæ*, a beautiful dark-blue form of our blue tit, inhabits Northwest Africa and the Canaries ; *Parus Le-douci*, closely allied to our coal tit, is found only in Algeria ; *Parus lugubris*, allied to the marsh tit, is confined to Southeast Europe and Asia Minor, from Hungary and South Russia to Pal-estine ; and *Parus cinctus*, another allied form, is confined to the extreme north in Lapland, Finland, and perhaps Northern Russia and Siberia. Another beautiful little bird, the crested titmouse (*Parus cristatus*), is sometimes placed in a separate ge-nus. It inhabits nearly all Central and South Europe, wherever there are pine forests, from 64° N. latitude to Austria and North

Italy, and in the west to Spain and Gibraltar, while in the east
it does not pass the Urals and the Caucasus range. Its nearest
allies are in the high Himalayas.

These are all the European tits, but there are many others inhabiting Asia, Africa, and North America; so that the genus
Parus has a very wide range, in Asia to Ceylon and the Malay
Islands, in Africa to the Cape, and in North America to the
highlands of Mexico.

The Distribution of the Species of Jays.—Owing to the very
wide range of several of the tits, the uncertainty of the specific
distinction of others, and the difficulty in many cases of ascertaining their actual distribution, it has not been found practica-
ble to illustrate this genus by means of a map. For this purpose we have chosen the genus Garrulus, or the jays, in which
the species are less numerous, the specific areas less extensive,
and the species generally better defined; while, being large and
handsome birds, they are sure to have been collected, or at least
noticed, wherever they occur. There are, so far as yet known,
twelve species of true jays, occupying an area extending from
Western Europe to Eastern Asia and Japan, and nowhere passing the Arctic circle to the north, or the Tropic of Cancer to the
south, so that they constitute one of the most typical of the Palæarctic[1] genera. The following are the species, beginning with
the most westerly and proceeding towards the east. The numbers prefixed to each species correspond to those on the colored
map which forms the frontispiece to this volume.

1. *Garrulus glandarius*, the common jay, inhabits the British Isles and all Europe except the extreme north, extending
also into North Africa, where it has been observed in many
parts of Algeria. It occurs near Constantinople, but apparently
not in Asia Minor, and in Russia up to, but not beyond, the
Urals. The jays, being woodland birds, are not found in open
plains or barren uplands, and their distribution is hence by no
means uniform within the area they actually occupy.

2. *Garrulus cervicalis*, the Algerian jay, is a very distinct

---

[1] The Palæarctic region includes temperate Asia and Europe, as will be explained
in the next chapter.

species inhabiting a limited area in North Africa, and found in some places along with the common species.

3. *Garrulus Krynicki*, the black-headed jay, is closely allied to the common species, but quite distinct, inhabiting a comparatively small area in Southeastern Europe and Western Asia.

4. *Garrulus atricapillus*, the Syrian jay, is very closely allied to the last, and inhabits an adjoining area in Syria, Palestine, and Southern Persia.

5. *Garrulus hyrcanus*, the Persian jay, is a small species allied to our jay, and only known from the Elburz Mountains in the north of Persia.

6. *Garrulus Brandti*, Brandt's jay, is a very distinct species, having an extensive range across Asia from the Ural Mountains to North China, Mandchuria, and the northern island of Japan, and also crossing the Urals into Russia, where it has been found as far west as Kazan in districts where the common jay also occurs.

7. *Garrulus lanceolatus*, the black-throated jay, is a very distinct form known only from the Northwestern Himalayas and Nepal, common about Simla, and extending into Cashmere beyond the range of the next species.

8. *Garrulus bispecularis*, the Himalayan jay, is also very distinct, having the head colored like the back, and not striped as in all the western species. It inhabits the Himalayas east of Cashmere, but is more abundant in the western than the eastern division, though, according to the Abbé David, it reaches Moupin in East Thibet.

9. *Garrulus Sinensis*, the Chinese jay, is very closely allied to the Himalayan, of which it is sometimes classed as a sub-species. It seems to be found in all the southern mountains of China, from Foochow on the east to Sze-chuen and East Thibet on the west, as it is recorded from Moupin by the Abbé David as well as the Himalayan bird—a tolerable proof that it is a distinct form.

10. *Garrulus taivanus*, the Formosan jay, is a very close ally of the preceding, confined to the island of Formosa.

11. *Garrulus Japonicus*, the Japanese jay, is very closely allied to our common British species, being somewhat smaller

and less brightly colored, and with black orbits; yet these are
the most widely separated species of the genus.

12. *Garrulus Lidthi.*—This is the handsomest of all the jays,
the head, neck, and wings being azure blue. Its locality was
long doubtful, but it has now been ascertained to inhabit Japan,
where it is evidently very rare, its exact habitat being still unknown.

In the accompanying map (see frontispiece) we have laid down
the distribution of each species so far as it can be ascertained
from the works of Sharpe and Dresser for Europe, Jerdon for
India, Swinhoe for China, and Messrs. Blakiston and Pryer for
Japan. There is, however, much uncertainty in many places, and
gaps have to be filled up conjecturally, while such a large part
of Asia is still very imperfectly explored that considerable modifications may have to be made when the country becomes more
accurately known. But though details may be modified, we can
hardly suppose that the great features of the several specific
areas, or their relations to each other, will be much affected; and
these are what we have chiefly to consider as bearing on the
questions here discussed.

The first thing that strikes us on looking at the map is the
small amount of overlapping of the several areas, and the isolation of many of the species; while the next most striking feature is the manner in which the Asiatic species almost surround
a vast area in which no jays are found. The only species with
large areas are the European *G. glandarius* and the Asiatic *G.
Brandti.* The former has three species overlapping it—in Algeria, in Southeastern and in Northeastern Europe respectively.
The Syrian jay (No. 4) is not known to occur anywhere with the
black-headed jay (No. 3), and perhaps the two areas do not meet.
The Persian jay (No. 5) is quite isolated. The Himalayan and
Chinese jays (Nos. 7, 8, and 9) form a group which are isolated
from the rest of the genus; while the Japanese jay (No. 11) is
also completely isolated as regards the European jays, to which
alone it is closely allied. These peculiarities of distribution are
no doubt in part dependent on the habits of the jays, which live
only in well-wooded districts, among deciduous trees, and are essentially non-migratory in their habits, though sometimes moving

southward in winter. This will explain their absence from the vast desert area of Central Asia, but it will not account for the gap between the North and South Chinese species, nor for the absence of jays from the wooded hills of Turkestan, where Mr. N. A. Severtzoff collected assiduously, obtaining 384 species of birds, but no jay. These peculiarities, and the fact that jays are never very abundant anywhere, seem to indicate that the genus is now a decaying one, and that it has at no very distant epoch occupied a larger and more continuous area, such as that of the genus Parus at the present day.

Discontinuous Generic Areas.—It is not very easy to find good examples of genera whose species occupy two or more quite disconnected areas, for though such cases may not be rare, we are seldom in a position to mark out the limits of the several species with sufficient accuracy. The best and most remarkable case among European birds is that of the blue magpies, forming the genus Cyanopica. One species (*C. Cooki*) is confined (as already stated) to the wooded and mountainous districts of Spain and Portugal, while the only other species of the genus (*C. cyanus*) is found far away in Northeastern Asia and Japan, so that the two species are separated by about 5000 miles of continuous land. Another case is that of the curious little water-moles forming the genus Mygale, one species, *M. Muscovitica*, being found only on the banks of the Volga and Don in Southeastern Russia, while the other, *M. Pyrenaica*, is confined to streams on the northern side of the Pyrenees. In tropical America there are four different kinds of bell-birds belonging to the genus Chasmorhynchus, each of which appears to inhabit a restricted area completely separated from the others. The most northerly is *C. tricarunculatus* of Costa Rica and Veragua, a brown bird with a white head and three long caruncles growing upwards at the base of the beak. Next comes *C. variegatus*, in Venezuela, a white bird with a brown head and numerous caruncles on the throat, perhaps conterminous with the last; in Guiana, extending to near the mouth of the Rio Negro, we have *C. niveus*, the bell-bird described by Waterton, which is pure white, with a single long fleshy caruncle at the base of the beak; the last species, *C. nudicollis*, inhabits Southeast Brazil, and is also white, but with

black stripes over the eyes, and with a naked throat. These birds
are about the size of thrushes, and are all remarkable for their
loud-ringing notes like a bell or a blow on an anvil, as well as
for their peculiar colors. They are therefore known to the native Indians wherever they exist, and we may be the more sure
that they do not spread over the intervening areas where they
have never been found, and where the natives know nothing of
them.

A good example of isolated species of a group nearer home is
afforded by the snow-partridges of the genus Tetraogallus. One
species inhabits the Caucasus range and nowhere else, keeping to
the higher slopes from 6000 to 11,000 feet above the sea, and
accompanying the ibex in its wanderings, as both feed on the
same plants. Another has a wider range in Asia Minor and
Persia from the Taurus Mountains to the southeast corner of
the Caspian Sea; a third species inhabits the Western Himalayas, between the forests and perpetual snow, extending eastward to Nepal, while a fourth is found on the north side of the
mountains in Thibet, and the ranges of these two perhaps overlap; the last species inhabits the Altai Mountains, and like the
two first appears to be completely separated from all its allies.

There are some few still more extraordinary cases in which
the species of one genus are separated in remote continents or
islands. The most striking of these is that of the tapirs, forming
the genus Tapirus, of which there are two or three species in
South America, and one very distinct species in Malacca and
Borneo, separated by nearly half the circumference of the globe.
Another example among quadrupeds is a peculiar genus of moles
named Urotrichus, of which one species inhabits Japan and the
other British Columbia. The cuckoo-like honey-guides, forming
the genus Indicator, are tolerably abundant in tropical Africa,
but there are two outlying species, one in the Eastern Himalaya
Mountains, the other in Borneo, both very rare, and quite recently an allied species has been found in the Malay peninsula.
The beautiful blue and green thrush-tits, forming the genus
Cochoa, have two species in the Eastern Himalayas, while the
third is confined to Java; the curious genus Eupetes, supposed
to be allied to the dippers, has two species in Sumatra, and the

other species two thousand miles distant in New Guinea; lastly, the lovely ground-thrushes of the genus Pitta range from Hindostan to Australia, while a single species, far removed from all its near allies, inhabits West Africa.

Peculiarities of Generic and Family Distribution.—The examples now given sufficiently illustrate the mode in which the several species of a genus are distributed. We have next to consider genera as the component parts of families, and families of orders, from the same point of view.

All the phenomena presented by the species of a genus are reproduced by the genera of a family, and often in a more marked degree. Owing, however, to the extreme restriction of genera by modern naturalists, there are not many among the higher animals that have a world-wide distribution. Among the mammalia there is no such thing as a truly cosmopolitan genus. This is owing to the absence of all the higher orders except the mice from Australia, while the genus Mus, which occurs there, is represented by a distinct group, Hesperomys, in America. If, however, we consider the Australian dingo as a native animal, we might class the genus Canis as cosmopolite, but the wild dogs of South America are now formed into separate genera by some naturalists. Many genera, however, range over three or more continents, as Felis (the cat genus), absent only from Australia; Ursus (the bear genus), absent from Australia and tropical Africa; Cervus (the deer genus), with nearly the same range; and Sciurus (the squirrel genus), found in all the continents but Australia. Among birds, Turdus, the thrush, and Hirundo, the swallow genus, are the only perching birds which are truly cosmopolites; but there are many genera of hawks, owls, wading and swimming birds which have a world-wide range.

As a great many genera consist of single species, there is no lack of cases of great restriction, such as the curious lemur called the " potto," which is found only at Sierra Leone, and forms the genus Perodicticus; the true chinchillas, found only in the Andes of Peru and Chili south of 9° S. lat. and between 8000 and 12,000 feet elevation; several genera of finches, each confined to limited portions of the higher Himalayas; the blood-pheasants (Ithaginis), found only above 10,000 feet from Nepal to East Thibet; the bald-headed starling of the Philippine Islands, the lyre-birds of East Australia, and a host of others.

It is among the different genera of the same family that we meet with the most striking examples of discontinuity, although these genera are often as unmistakably allied as are the species of a genus; and it is these cases that furnish the most interesting problems to the student of distribution. We must, therefore, consider them somewhat more fully.

Among mammalia the most remarkable of these divided families is that of the camels, of which one genus, Camelus, the true camels, comprising the camel and dromedary, is confined to Asia, while the other, Auchenia, comprising the llamas and alpacas, is found only in the high Andes and in the plains of temperate South America. Not only are these two genera separated by the Atlantic and by the greater part of the land of two continents, but one is confined to the Northern and the other to the Southern Hemisphere. The next case, though not so well known, is equally remarkable; it is that of the Centetidæ, a family of small insectivorous animals, which are wholly confined to Madagascar and the large West Indian islands Cuba and Hayti, the former containing five genera and the latter a single genus with a species in each island. Here again we have the whole continent of Africa as well as the Atlantic Ocean separating allied genera. Two families of rat-like animals, Octodontidæ and Echimyidæ, are also divided by the Atlantic. Both are mainly South American, but the former has two genera in North and East Africa, and the latter also two in South and West Africa. Two other families of mammalia, though confined to the Eastern Hemisphere, are yet markedly discontinuous. The Tragulidæ are small deer-like animals, known as chevrotains or mouse-deer, abundant in India and the larger Malay islands, and forming the genus Tragulus; while another genus, Hyomoschus, is confined to West Africa. The other family is the Simiidæ or anthropoid apes, in which we have the gorilla and chimpanzee confined to West and Central Africa, while the allied orangs are found only in the islands of Sumatra and Borneo, the two groups being separated by a greater space than the Echimyidæ and other rodents of Africa and South America.

Among birds and reptiles we have several families, which, from being found only within the tropics of Asia, Africa, and America, have been termed tropicopolitan groups. The Megalæmidæ, or barbets, are gayly colored fruit-eating birds, almost equally abundant in tropical Asia and Africa, but less plentiful in America, where they probably suffer from the competition of the larger sized toucans. The genera of each country are distinct, but all are closely allied, the family being a very natural one. The trogons form a family of very gorgeously colored and remarkable insect-eating birds very abundant in tropical America, less so in Asia, and with a single genus of two species in Africa.

Among reptiles we have two families of snakes — the Dendrophidæ, or tree-snakes, and the Dryiophidæ, or green whip-snakes — which are also found in the three tropical regions of Asia, Africa, and America, but in these cases even some of the genera are common to Asia and Africa, or to Africa and America. The lizards forming the small family Lepidosternidæ are divided between tropical Africa and South America, while even the peculiarly American family of the iguanas is represented by two genera in Madagascar. Passing on to the Amphibians, the worm-like Cæciliadæ are tropicopolitan, as are also the toads of the family Phryniscidæ. Insects also furnish some analogous cases, three genera of Cicindelidæ (Pogonostoma, Ctenostoma, and Peridexia), showing a decided connection between this family in South America and Madagascar; while the beautiful genus of diurnal moths, Urania, is confined to the same two countries. A somewhat similar but better-known illustration is afforded by the two genera of ostriches, one confined to Africa and Arabia, the other to the plains of temperate South America.

General Features of Overlapping and Discontinuous Areas.— These numerous examples of discontinuous genera and families form an important section of the facts of animal dispersal which any true theory must satisfactorily account for. In greater or less prominence they are to be found all over the world, and in

every group of animals, and they grade imperceptibly into
those cases of conterminous and overlapping areas which we
have seen to prevail in most extensive groups of species, and
which are perhaps even more common in those large families
which consist of many closely allied genera. A sufficient proof
of the overlapping of generic areas is the occurrence of a number of genera of the same family together. Thus in France or
Italy about twenty genera of warblers (Sylviadæ) are found,
and as each of the thirty-three genera of this family inhabiting
temperate Europe and Asia has a different area, a great number
must here overlap. So, in most parts of Africa at least, ten or
twelve genera of antelopes may be found, and in South America a large proportion of the genera of monkeys of the family
Cebidæ occur in many districts; and still more is this the case
with the larger bird families, such as the tanagers, the tyrant
shrikes, or the tree-creepers, so that there is in all these extensive families no genus whose area does not overlap that of many
others. Then among the moderately extensive families we find
a few instances of one or two genera isolated from the rest, as
the spectacled bear, Tremarctos, found only in Chili, while the
remainder of the family extends from Europe and Asia over
North America to the mountains of Mexico, but no farther
south; the Bovidæ, or hollow-horned ruminants, which have a
few isolated genera in the Rocky Mountains and the islands of
Sumatra and Celebes; and from these we pass on to the cases
of wide separation already given.

Restricted Areas of Families.—As families sometimes consist of single genera and even single species, they often present
examples of very restricted range; but what is perhaps more
interesting are those cases in which a family contains numerous
species and sometimes even several genera, and yet is confined
to a narrow area. Such are the golden moles (Chrysochloridæ),
consisting of two genera and three species, confined to extra-
tropical South Africa; the hill-tits (Liotrichidæ), a family of
eleven genera and thirty-five species almost wholly limited to
the Himalayas, but with a few straggling species in the Malay
countries; the Pteroptochidæ, large wren-like birds, consisting
of eight genera and nineteen species, almost entirely confined

to temperate South America and the Andes; and the birds-of-paradise, consisting of nineteen or twenty genera and about thirty-five species, almost all inhabitants of New Guinea and the immediately surrounding islands, while a few, doubtfully belonging to the family, extend to East Australia. Among reptiles the most striking case of restriction is that of the rough-tailed burrowing snakes (Uropeltidæ), the five genera and eighteen species being strictly confined to Ceylon and the southern parts of the Indian Peninsula.

The Distribution of Orders.—When we pass to the larger groups, termed orders, comprising several families, we find comparatively few cases of restriction and many of world-wide distribution; and the families of which they are composed are strictly comparable to the genera of which families are composed, inasmuch as they present examples of overlapping, or conterminous, or isolated areas, though the latter are comparatively rare. Among mammalia the Insectivora offer the best example of an order several of whose families inhabit areas more or less isolated from the rest; while the Marsupialia have six families in Australia, and one, the opossums, far off in America.

Perhaps, more important is the limitation of some entire orders to certain well-defined portions of the globe. Thus the Proboscidea, comprising the single family and genus of the elephants, and the Hyracoidea, that of the Hyrax or Syrian cony, are confined to parts of Africa and Asia; the Marsupials to Australia and America; and the Monotremata, the lowest of all mammals—comprising the duck-billed Platypus and the spiny Echidna—to Australia. Among birds the Struthiones, or ostrich tribe, are almost confined to the three southern continents, South America, Africa, and Australia; and among Amphibia the tailed Batrachia—the newts and salamanders—are similarly restricted to the Northern Hemisphere.

These various facts will receive their explanation in a future chapter.

CHAPTER III.

CLASSIFICATION OF THE FACTS OF DISTRIBUTION.—ZOOLOGICAL REGIONS.

The Geographical Divisions of the Globe do not Correspond to Zoological Divisions. —The Range of British Mammals as Indicating a Zoological Region.—Range of East Asian and North African Mammals.—The Range of British Birds.—Range of East Asian Birds.—The Limits of the Palæarctic Region.—Characteristic Features of the Palæarctic Region.—Definition and Characteristic Groups of the Ethiopian Region.—Of the Oriental Region.—Of the Australian Region.—Of the Nearctic Region.—Of the Neotropical Region.—Comparison of Zoological Regions with the Geographical Divisions of the Globe.

HAVING now obtained some notion of how animals are dispersed over the earth's surface, whether as single species or as collected in those groups termed genera, families, and orders, it will be well, before proceeding further, to understand something of the classification of the facts we have been considering, and some of the simpler conclusions these facts lead to.

We have hitherto described the distribution of species and groups of animals by means of the great geographical divisions of the globe in common use; but it will have been observed that in hardly any case do these define the limits of anything beyond species, and very seldom, or perhaps never, even those accurately. Thus the term "Europe" will not give, with any approach to accuracy, the range of any one genus of mammals or birds, and perhaps not that of half-a-dozen species. Either they range into Siberia, or Asia Minor, or Palestine, or North Africa; and this seems to be always the case when their area of distribution occupies a large portion of Europe. There are, indeed, a few species limited to Central or Western or Southern Europe, and these are almost the only cases in which we can use the word for zoological purposes without having to add to it some portion of another continent. Still less useful is the

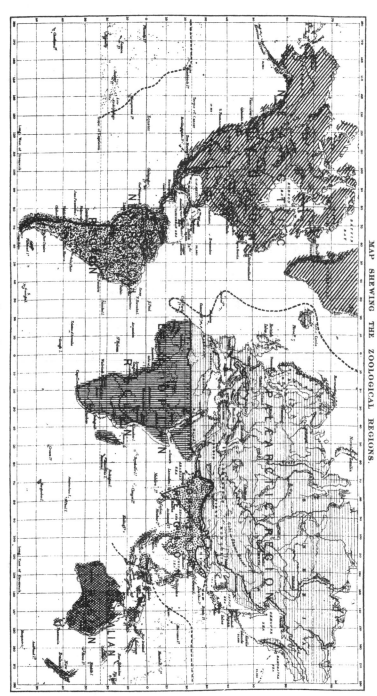

MAP SHEWING THE ZOOLOGICAL REGIONS.

Harper & Brothers New York.

term Asia for this purpose, since there is probably no single animal or group confined to Asia which is not also more or less nearly confined to the tropical or the temperate portion of it. The only exception is perhaps the tiger, which may really be called an Asiatic animal, as it occupies nearly two thirds of the continent; but this is a unique example, while the cases in which Asiatic animals and groups are strictly limited to a portion of Asia, or extend also into Europe or into Africa or to the Malay Islands, are exceedingly numerous. So, in Africa, very few groups of animals range over the whole of it without going beyond, either into Europe or Asia Minor or Arabia, while those which are purely African are generally confined to the portion south of the tropic of Cancer. Australia and America are terms which better serve the purpose of the zoologist. The former defines the limit of many important groups of animals; and the same may be said of the latter, but the division into North and South America introduces difficulties, for almost all the groups especially characteristic of South America are found also beyond the isthmus of Panama, in what is geographically part of the northern continent.

It being thus clear that the old and popular divisions of the globe are very inconvenient when used to describe the range of animals, we are naturally led to ask whether any other division can be made which will be more useful, and will serve to group together a considerable number of the facts we have to deal with. Such a division was made by Mr. P. L. Sclater more than twenty years ago, and it has, with some slight modification, come into pretty general use in this country, and to some extent also on the Continent; we shall therefore proceed to explain its nature and the principles on which it is established, as it will have to be often referred to in future chapters of this work, and will take the place of the old geographical divisions, whose extreme inconvenience has already been pointed out. The primary zoological divisions of the globe are called " regions," and we will begin by ascertaining the limits of the region of which our own country forms a part.

The Range of British Mammals as Indicating a Zoological Region.—We will first take our commonest wild mammalia and

see how far they extend, and especially whether they are confined to Europe or range over parts of other continents:

1. Wild-cat...............	Europe	North Africa	Siberia, Afghanistan.
2. Fox...................	"	" "	Central Asia to Amoor.
3. Weasel...............	"	" "	" " "
4. Otter.................	"	" "	Siberia.
5. Badger...............	"	" "	Central Asia to Amoor.
6. Stag.................	"	" "	" " "
7. Hedgehog............	"		" " "
8. Mole................	"		Central Asia.
9. Squirrel.............	"		Central Asia to Amoor.
10. Dormouse............	"		
11. Water-rat............	"		Central Asia to Amoor.
12. Hare...............	"		West Siberia, Persia.
13. Rabbit...............	"	North Africa	

We thus see that out of thirteen of our commonest quadrupeds only one is confined to Europe, while seven are found also
in Northern Africa and eleven range into Siberia, most of them
stretching quite across Asia to the valley of the Amoor on the
extreme eastern side of that continent. Two of the above-named
British species, the fox and weasel, are also inhabitants of the
New World, being as common in the northern parts of North
America as they are with us; but with these exceptions the entire range of our commoner species is given, and they clearly
show that all Northern Asia and Northern Africa must be added
to Europe in order to form the region which they collectively
inhabit. If now we go into Central Europe and take, for example, the quadrupeds of Germany, we shall find that these too, although much more numerous, are confined to the same limits,
except that some of the more arctic kinds, as already stated, extend into the colder regions of North America.

Range of East Asian and North African Mammals.—Let us
now pass to the other side of the great northern continent, and
examine the list of the quadrupeds of Amoorland, in the same
latitude as Germany. We find that there are forty-four terrestrial species (omitting the bats, the seals, and other marine animals), and of these no less than twenty-six are identical with
European species, and twelve or thirteen more are closely allied
representatives, leaving only five or six which are peculiarly
Asiatic. We can hardly have a more convincing proof of the essential oneness of the mammalia of Europe and Northern Asia.

In Northern Africa we do not find so many European species (though even here they are very numerous), because a considerable number of West Asiatic and Desert forms occur. Having, however, shown that Europe and Western Asia have almost identical animals, we may treat all these as really European, and we shall then be able to compare the quadrupeds of North Africa with those of Europe and West Asia. Taking those of Algeria as the best known, we find that there are thirty-three species identical with those of Europe and West Asia, while twenty-four more, though distinct, are closely allied, belonging to the same genera; thus making a total of fifty-seven of European type. On the other hand, we have seven species which are either identical with species of tropical Africa or allied to them, and six more which are especially characteristic of the African and Asiatic deserts, which form a kind of neutral zone between the temperate and tropical regions. If now we consider that Algeria and the adjacent countries bordering the Mediterranean form part of Africa, while they are separated from Europe by a wide sea, and are only connected with Asia by a narrow isthmus, we cannot but feel surprised at the wonderful preponderance of the European and West Asiatic elements in the mammalia which inhabit the district.

The Range of British Birds.—As it is very important that no doubt should exist as to the limits of the zoological region of which Europe forms a part, we will now examine the birds, in order to see how far they agree in their distribution with the mammalia. Of late years great attention has been paid to the distribution of European and Asiatic birds, many ornithologists having travelled in North Africa, in Palestine, in Asia Minor, in Persia, in Siberia, in Mongolia, and in China; so that we are now able to determine the exact ranges of many species in a manner that would have been impossible a few years ago. These ranges are given for all British species in the new edition of Yarrell's "History of British Birds," now in course of publication under the editorship of Professor Newton, while those of all European birds are given in still more detail in Mr. Dresser's beautiful work on the birds of Europe just completed. In order to confine our examination within reasonable limits, and at the same time give it the interest attaching to familiar objects, we will take the whole series of British Passeres, or perching birds, given in Professor Newton's work (118 in number), and arrange them in series according to the extent of their range. These include not only the permanent residents and regular migrants to our country, but also those which occasionally straggle here, so that it really comprises a large proportion of all European birds.

1. BRITISH BIRDS WHICH EXTEND TO NORTH AFRICA AND CENTRAL OR NORTH-EAST ASIA.

1. *Lanius collurio*..................Red-backed Shrike (also all Africa)
2. *Oriolus galbula*.....................Golden Oriole (also all Africa).
3. *Turdus musicus*...................Song-thrush.
4.   "   *iliacus*...................Redwing.
5.   "   *pilaris*...................Fieldfare.
6. *Monticola saxatilis*...............Blue-rock Thrush.
7. *Ruticilla Suecica*................Blue-throat (also India in winter).
8. *Saxicola rubicola*...............Stone-chat (also India in winter).
9.   "   *œnanthe*................Wheat-ear (also North America).
10. *Acrocephalus arundinaceus*........Great Reed-warbler.
11. *Sylvia curruca*...................Lesser White-throat.
12. *Parus major*......................Great Titmouse.
13. *Motacilla sulphurea*..............Gray Wagtail (also China and Malaya).
14.   "   *Raii* ....Yellow Wagtail.
15. *Anthus trivialis*..................Tree-pipit.
16.   "   *spiloletta*.................Water-pipit.
17.   "   *campestris*................Tawny Pipit.
18. *Alauda arvensis*..................Skylark.
19.   "   *cristata*..................Crested Lark.
20. *Emberiza schœniclus*.............Reed-bunting.
21.   "   *citrinella*...............Yellow-hammer.
22. *Fringilla montifringilla*..........Brambling.
23. *Passer montanus*.................Tree-sparrow (also South Asia).
24.   "   *domesticus*...............House-sparrow.
25. *Coccothraustes vulgaris*...........Hawfinch.
26. *Carduelis spinus*.................Siskin (also China).
27. *Loxia curvirostra*...............Crossbill.
28. *Sturnus vulgaris*.................Starling.
29. *Pyrrhocorax graculus*............Chough.
30. *Corvus corone*...................Crow.
31. *Hirundo rustica*.................Swallow (all Africa and Asia).
32. *Cotyle riparia*....................Sand-martin (also India and North America).

2. BRITISH BIRDS WHICH RANGE TO CENTRAL OR NORTHEAST ASIA.

1. *Lanius excubitor*.................Great Gray Shrike.
2. *Turdus varius*....................White's Thrush (also to Japan).

3. *Turdus atrigularis.* .... Black-throated Thrush.
4. *Acrocephalus nævius.* .... Grasshopper-warbler.
5. *Phylloscopus superciliosus.* .... Yellow-browed Warbler.
6. *Certhia familiaris.* .... Tree-creeper.
7. *Parus cæruleus.* .... Blue Titmouse.
8. "   *ater.* .... Coal Titmouse.
9. "   *palustris.* .... Marsh Titmouse.
10. *Acredula caudata* .... Long-tailed Titmouse.
11. *Ampelis garrulus.* .... Waxwing.
12. *Anthus Richardi.* .... Richard's Pipit.
13. *Alauda alpestris.* .... Shore-lark (also North America).
14. *Plectrophanes nivalis.* .... Snow-bunting (also North America).
15. "   *Lapponicus.* .... Lapland Bunting.
16. *Emberiza rustica.* .... Rustic Bunting (also China).
17. "   *pusilla.* .... Little Bunting.
18. *Linota linaria.* .... Mealy Redpoll (also North America).
19. *Pyrrhula Erythrina.* .... Scarlet Grossbeak (also North India, China).
20. "   *enucleator* .... Pine Grossbeak (also North America).
21. *Loxia bifasciata.* .... Two-barred Crossbill.
22. *Pastor roseus.* .... Rose-colored Starling (also India).
23. *Corvus corax.* .... Raven (also North America).
24. *Pica rustica.* .... Magpie.
25. *Nucifraga caryocatactes.* .... Nut-cracker.

3. British Birds Ranging into North Africa and West Asia.

1. *Lanius minor.* .... Lesser Gray Shrike.
2. "   *auriculatus.* .... Wood-chat (also tropical Africa).
3. *Muscicapa grisola.* .... Spotted Flycatcher (also E. and S. Africa).
4. "   *atricapilla.* .... Pied Flycatcher (also Central Africa).
5. *Turdus viscivorus.* .... Mistletoe Thrush (North India in winter).
6. "   *merula.* .... Blackbird.
7. "   *torquatus.* .... Ring-ouzel.
8. *Accentor modularis.* .... Hedge-sparrow.
9. *Erithacus rubecula.* .... Redbreast.
10. *Daulias luscinia.* .... Nightingale.
11. *Ruticilla phœnicurus.* .... Redstart.
12. "   *Tithys.* .... Black Redstart.
13. *Saxicola rubetra.* .... Whinchat.
14. *Aëdon galactodes.* .... Rufous Warbler.
15. *Acrocephalus streperus.* .... Reed-warbler.
16. "   *schœnobenus* .... Sedge-warbler.
17. *Melizophilus undatus.* .... Dartford Warbler.
18. *Sylvia rufa.* .... Greater White-throat.
19. "   *salicaria.* .... Garden-warbler.
20. "   *atricapilla.* .... Blackcap.
21. "   *orphea* .... Orphean Warbler.
22. *Phylloscopus sibilatrix.* .... Wood-wren.

23. *Phylloscopus trochilus* .... Willow-wren.
24.   "   *collybita* .... Chiffchaff
25. *Regulus cristatus* .... Golden-crested Wren.
26.   "   *ignicapillus* .... Fire-crested Wren.
27. *Troglodytes parvulus* .... Wren.
28. *Sitta cæsia* .... Nuthatch.
29. *Motacilla alba* .... White Wagtail (also West Africa).
30.   "   *flava* .... Blue-headed Wagtail.
31. *Anthus pratensis* .... Meadow-pipit.
32. *Alauda arborea* .... Woodlark.
33. *Calandrella brachydactyla* .... Short-toed Lark.
34. *Emberiza milaria* .... Common Bunting.
35.   "   *cirlus* .... Cirl Bunting.
36.   "   *hortulana* .... Ortolan.
37. *Fringilla cœlebs* .... Chaffinch.
38. *Coccothraustes chloris* .... Greenfinch.
39. *Serinus hortulanus* .... Serin.
40. *Carduelis elegans* .... Goldfinch.
41. *Linota cannabina* .... Linnet.
42. *Corvus monedula* .... Jackdaw.
43. *Chelidon urbica* .... House-martin.

4. British Birds Ranging to North Africa.

1. *Hypolais icterina* .... Icterine Warbler.
2. *Acrocephalus aquaticus* .... Aquatic Warbler.
3.   "   *luscinioides* .... Savi's Warbler.
4. *Motacilla lugubris* .... Pied Wagtail.
5. *Pyrrhula Europœa* .... Bullfinch.
6. *Garrulus glandarius* .... Jay.

5. British Birds Ranging to West Asia only.

1. *Muscicapa parva* .... Red-breasted Flycatcher (to Northwest India).
2. *Panurus biarmicus* .... Bearded Titmouse.
3. *Melanocorypha Sibirica* .... White-winged Lark.
4. *Euspiza melanocephala* .... Black-headed Bunting.
5. *Linota flavirostris* .... Twite.
6. *Corvus frugilegus* .... Rook.

6. British Birds Confined to Europe.

1. *Cinclus aquaticus* .... Dipper.
2. *Accentor collaris* .... Alpine Accentor.
3. *Parus cristatus* .... Crested Titmouse.
4. *Anthus obscurus* .... Rock-pipit.
5. *Linota rufescens* .... Lesser Redpoll.
6. *Loxia pityopsittacus* .... Parrot Crossbill.

We find that out of a total of 118 British Passeres there are:

32 species which range to North Africa and Central or East Asia.
25 " " " " Central or East Asia, but not to North Africa.
43 " " " " North Africa and Western Asia.
6 " " " " North Africa, but not at all into Asia.
6 " " " " West Asia, but not to North Africa.
6 " " do not range out of Europe.

These figures agree essentially with those furnished by the mammalia, and complete the demonstration that all the temperate portions of Asia and North Africa must be added to Europe to form a natural zoological division of the earth. We must also note how comparatively few of these overpass the limits thus indicated; only seven species extending their range occasionally into tropical or South Africa, eight into some parts of tropical Asia, and six into arctic or temperate North America.

Range of East Asian Birds.—To complete the evidence, we only require to know that the East Asiatic birds are as much like those of Europe as we have already shown to be the case when we take the point of departure from our end of the continent. This does not follow necessarily, because it is possible that a totally distinct North Asiatic fauna might there prevail; and, although our birds go eastward to the remotest parts of Asia, their birds might not come westward to Europe. The birds of Eastern Siberia have been carefully studied by Russian naturalists, and afford us the means of making the required comparison. There are 151 species belonging to the orders Passeres and Picariæ (the perching and climbing birds), and of these no less than 77, or more than half, are absolutely identical with European species; 63 are peculiar to North Asia, but all except five or six of these are allied to European forms; the remaining 11 species are migrants from Southeastern Asia. The resemblance is therefore equally close whichever extremity of the Euro-Asiatic continent we take as our starting-point, and is equally remarkable in birds as in mammalia. We have now only to determine the limits of this our first zoological region, which has been termed the "Palæarctic" by Mr. Sclater, meaning the "northern old-world" region—a name now well known to naturalists.

The Limits of the Palæarctic Region.—The boundaries of

this region, as nearly as they can be ascertained, are shown on our general map at the beginning of this chapter, but it will be evident on consideration that, except in a few places, its limits can only be approximately defined. On the north, east, and west it extends to the ocean, and includes a number of islands whose peculiarities will be pointed out in a subsequent chapter; so that the southern boundary alone remains; but as this runs across the entire continent from the Atlantic to the Pacific Ocean, often traversing little-known regions, we may perhaps never be able to determine it accurately, even if it admits of such determination. In drawing the boundary-line across Africa we meet with our first difficulty. The Euro-Asiatic animals undoubtedly extend to the northern borders of the Sahara, while those of tropical Africa come up to its southern margin, the desert itself forming a kind of dry sea between them. Some of the species on either side penetrate and even cross the desert, but it is impossible to balance these with any accuracy, and it has therefore been thought best, as a mere matter of convenience, to consider the geographical line of the tropic of Cancer to form the boundary. We are thus enabled to define the Palæarctic region as including all north temperate Africa; and a similar intermingling of animal types occurring in Arabia, the same boundary-line is continued to the southern shore of the Persian Gulf. Persia and Afghanistan undoubtedly belong to the Palæarctic region, and Beloochistan should probably go with these. The boundary in the northwestern part of India is again difficult to determine, but it cannot be far one way or the other from the river Indus as far up as Attock, opposite the mouth of the Cabool River. Here it will bend to the southeast, passing a little south of Cashmere, and along the southern slopes of the Himalayas into East Thibet and China, at heights varying from 9000 to 11,000 feet, according to soil, aspect, and shelter. It may, perhaps, be defined as extending to the upper belt of forests as far as coniferous trees prevail; but the temperate and tropical faunas are here so intermingled that to draw any exact parting-line is impossible. The two faunas are, however, very distinct. In and above the pine woods there are abundance of warblers of northern genera, with wrens, numerous titmice, and

a great variety of buntings, grossbeaks, bullfinches, and rosefinch-
es, all more or less nearly allied to the birds of Europe and
Northern Asia; while a little lower down we meet with a host
of peculiar birds allied to those of tropical Asia and the Malay
Islands, but often of distinct genera. There can be no doubt,
therefore, of the existence here of a pretty sharp line of demar-
cation between the temperate and tropical faunas, though this
line will be so irregular, owing to the complex system of valleys
and ridges, that in our present ignorance of much of the country it cannot be marked in detail on any map.

Farther east in China it is still more difficult to determine
the limits of the region, owing to the great intermixture of migrating birds; tropical forms passing northward in summer as
far as the Amoor River, while the northern forms visit every
part of China in winter. From what we know, however, of the
distribution of some of the more typical northern and southern
species, we are able to fix the limits of the Palæarctic region a
little south of Shanghai on the coast. Several tropical genera
come as far as Ningpo or even Shanghai, but rarely beyond;
while in Formosa and Amoy tropical forms predominate. Such
decidedly northern forms as bullfinches and hawfinches are
found at Shanghai; hence we may commence the boundary-
line on the coast between Shanghai and Ningpo, but inland it
probably bends a little southward, and then northward to the
mountains and valleys of West China and East Thibet in about
32° north latitude; where, at Moupin, a French missionary, Père
David, made extensive collections showing this district to be at
the junction of the tropical and temperate faunas. Japan, as a
whole, is decidedly Palæarctic, although its extreme southern
portion, owing to its mild insular climate and evergreen vegetation, gives shelter to a number of tropical forms.

Characteristic Features of the Palæarctic Region.—Having
thus demonstrated the unity of the Palæarctic region by tracing
out the distribution of a large proportion of its mammalia and
birds, it only remains to show how far it is characterized by peculiar groups such as genera and families, and to say a few words
on the lower forms of life which prevail in it.

Taking first the mammalia, we find this region is distinguished by its possession of the entire family of Talpidæ, or moles, consisting of eight genera and sixteen species, all of which are confined to it except one which is found in Northwest America, and two which extend to Assam and Formosa. Among carnivorous animals the lynxes (nine species) and the badgers (two species) are peculiar to it in the Old World, while in the New the lynxes are found only in the colder regions of North America. It has six peculiar genera (with seven species) of deer; seven peculiar genera of Bovidæ, chiefly antelopes; while the entire group of goats and sheep, comprising twenty-two species, is almost confined to it, one species only occurring in the Rocky Mountains of North America and another in the Neilgherries of Southern India. Among the rodents there are nine genera with twenty-seven species wholly confined to it, while several others, as the voles, the dormice, and the pikas, have only a few species elsewhere.

In birds there are a large number of peculiar genera, of which we need only mention a few of the more important, as the grasshopper-warblers (Locustella) with seven species, the Accentors with twelve species, and about a dozen other genera of warblers, including the robins; the bearded titmouse and several allied genera; the long-tailed titmice forming the genus Acredula; the magpies, choughs, and nut-crackers; a host of finches, among which the bullfinches (Pyrrhula) and the buntings (Emberiza) are the most important. The true pheasants (Phasianus) are wholly Palæarctic, except one species in Formosa, as are several genera of wading birds. Though the reptiles of cold countries are few as compared with those of the tropics, the Palæarctic region in its warmer portions has a considerable number, and among these are many which are peculiar to it. Such are two genera of snakes, seven of lizards, eight of frogs and toads, and eight of newts and salamanders; while of fresh-water fishes there are about twenty peculiar genera. Among insects we may mention the elegant Apollo butterflies of the Alps as forming a peculiar genus (Parnassius), only found elsewhere in the Rocky Mountains of North America; while the beautiful genus Thais of the South of Europe and Sericinus of North China are equally remarkable. Among other insects we can now only refer to the great family of Carabidæ, or predaceous ground beetles, which are immensely numerous in this region, there being about fifty peculiar genera; while the large and handsome genus Carabus, with its allies Procerus and Procrustes, containing nearly three hundred species, is almost wholly confined to this region, and would alone serve to distinguish it zoologically from all other parts of the globe.

Having given so full an exposition of the facts which determine the extent and boundaries of the Palæarctic region, there is less need of entering into much detail as regards the other regions of the Eastern Hemisphere; their boundaries being easily defined, while their forms of animal life are well marked and strongly contrasted.

Definition and Characteristic Groups of the Ethiopian Region.—The Ethiopian region consists of all tropical and South Africa, to which is appended the large island of Madagascar, and the Mascarene Islands to the east and north of it, though these differ materially from the continent, and will have to be discussed in a separate chapter. For the present, then, we will take Africa south of the tropic of Cancer, and consider how far its animals are distinct from those of the Palæarctic region.

Taking first the mammalia, we find the following remarkable animals at once separating it from the Palæarctic and every other region. The gorilla and chimpanzee, the baboons, numerous lemurs, the lion, the spotted hyena, the aard-wolf and hyena-dog, zebras, the hippopotamus, giraffe, and more than seventy peculiar antelopes. Here we have a wonderful collection of large and peculiar quadrupeds; but the Ethiopian region is also characterized by the absence of others which are not only abundant in the Palæarctic region, but in many tropical regions as well. The most remarkable of these deficiencies are the bears, the deer, and wild oxen, all of which abound in the tropical parts of Asia, while bears and deer extend into both North and South America. Besides the large and conspicuous animals mentioned above, Africa possesses a number of completely isolated groups; such are the potamogale, a curious otter-like wa-

ter-shrew, discovered by Du Chaillu in West Africa, so distinct
as to constitute a new family, Potamogalidæ; the golden moles,
also forming a peculiar family, Chrysochloridæ; as do the ele-
phant-shrews, Macroscelididæ; the singular aard-varks, or earth-
pigs, forming a peculiar family of Edentata, called Orycteropo-
didæ; while there are numerous peculiar genera of monkeys,
swine, civets, and rodents.

Among birds the most conspicuous and remarkable are the
great-billed vulture-crows (Corvultur), the long-tailed whydah
finches (Vidua), the curious ox-peckers (Buphaga), the splendid
metallic starlings (Lamprocolius), the handsome plantain-eaters
(Musophaga), the ground-hornbills (Bucorvus), the numerous
guinea-fowls belonging to four distinct genera, the serpent-eat-
ing secretary-bird (Serpentarius), the huge boat-billed heron
(Balæniceps), and the true ostriches. Besides these there are
three quite peculiar African families, the Musophagidæ, or plantain-eaters, including the elegant crested touracos; the curious
little finch-like colies (Coliidæ), and the Irrisoridæ, insect-eating
birds allied to the hoopoes, but with glossy metallic plumage
and arboreal habits.

In reptiles, fishes, insects, and land shells, Africa is very rich,
and possesses an immense number of peculiar forms. These are
not sufficiently known to require notice in a work of this character, but we may mention a few as mere illustrations; the puff-
adders, the most hideous of poisonous snakes; the chameleons,
the most remarkable of lizards; the goliath-beetles, the largest
and handsomest of the Cetoniidæ; and some of the Achatinæ,
which are the largest of all known land shells.

Definition and Characteristic Groups of the Oriental Region.
—The Oriental region comprises all Asia south of the Palæarctic limits, and along with this the Malay Islands as far as the
Philippines, Borneo, and Java. It was called the Indian region
by Mr. Sclater; but this term has been objected to because the
Indo-Chinese and Malayan districts are the richest and most
characteristic, while the peninsula of India is the poorest portion of it. The name "Oriental" has therefore been adopted in
my work on "The Geographical Distribution of Animals" as
preferable to either Malayan or Indo-Australian, both of which

have been proposed, but are objectionable, as being already in use in a different sense.

The great features of the Oriental region are the long-armed apes, the orang-outangs, the tiger, the sun-bears and honey-bears, the tapir, the chevrotains or mouse-deer, and the Indian elephant. Its most conspicuous birds are the immense number and variety of babbling-thrushes (Timaliidæ), its beautiful little hill-tits (Liotrichidæ), its green bulbuls (Phyllornithidæ), its many varieties of the crow family, its beautiful gapers and pittas adorned with the most delicate colors, its great variety of hornbills, and its magnificent Phasianidæ, comprising the peacocks, argus-pheasants, fire-backed pheasants, and jungle-fowl. Many of these are, it is true, absent from the peninsula of Hindostan, but sufficient remain there to ally it with the other parts of the region.

Among the remarkable but less conspicuous forms of mammalia which are peculiar to this region are, monkeys of the genus Presbyter, extending to every part of it; lemurs of three peculiar genera—Nycticebus and Loris (slow lemurs) and Tarsius (spectre lemurs); the flying lemur (Galeopithecus), now classed as a peculiar family of Insectivora and found only in the Malay Islands; the family of the Tupaias, or squirrel-shrews, curious little arboreal Insectivora somewhat resembling squirrels; no less than twelve peculiar genera of the civet family, three peculiar antelopes, five species of rhinoceros, and the round-tailed flying squirrels forming the genus Pteromys.

Of the peculiar groups of birds we can only mention a few. The curious little tailor-birds, of the genus Orthotomus, are found over the whole region, and almost alone serve to characterize it, as do the fine laughing-thrushes, forming the genus Garrulax; while the beautiful grass-green fruit-thrushes (Phyllornis), and the brilliant little minivets (Pericrocotus), are almost equally universal. Woodpeckers are abundant, belonging to a dozen peculiar genera; while gaudy barbets and strange forms of cuckoos and hornbills are also to be met with everywhere. Among game birds, the only genus that is universally distributed, and which may be said to characterize the region, is Gallus, comprising the true jungle-fowl, one of which, *Gallus Ban-*

kiva, is found from the Himalayas and Central India to Malacca, Java, and even eastward to Timor, and is the undoubted origin of almost all our domestic poultry. Southern India and Ceylon each possess distinct species of jungle-fowl, and a third very handsome green bird (*Gallus æneus*) inhabits Java.

Reptiles are as abundant as in Africa, but they present no well-known groups which can be considered as specially characteristic. Among insects we may notice the magnificent golden and green Papilionidæ of various genera as being unequalled in the world; while the great Atlas moth is probably the most gigantic of Lepidoptera, being sometimes ten inches across the wings, which are also very broad. Among the beetles the strange flat-bodied Malayan mormolyce is the largest of all the Carabidæ, while the catoxantha is equally a giant among the Buprestidæ. On the whole, the insects of this region probably surpass those of any other part of the world, except South America, in size, variety, and beauty.

Definition and Characteristic Groups of the Australian Region.—The Australian region is so well marked off from the Oriental, as well as from all other parts of the world, by zoological peculiarities that we need not take up much time in describing it, especially as some of its component islands will come under review at a subsequent stage of our work. Its most important portions are Australia and New Guinea, but it also includes all the Malayan and Pacific Islands to the east of Borneo, Java, and Bali, the Oriental region terminating with the submarine bank on which those islands are situated. The island of Celebes is included in this region from a balance of considerations, but it almost equally well belongs to the Oriental, and must be left out of the account in our general sketch of the zoological features of the Australian region.

The great feature of the Australian region is the almost total absence of all the forms of mammalia which abound in the rest of the world, their place being supplied by a great variety of marsupials. In Australia and New Guinea there are no Insectivora, Carnivora, nor Ungulata, while even the rodents are only represented by a few small rats and mice. In the Pacific Islands mammals are altogether absent (except perhaps in New Zealand), but in the Moluccas and other islands bordering on the Oriental region the higher mammals are represented by a few deer, civets, and pigs, though it is doubtful whether the two former may not have been introduced by man, as was almost certainly the case with the semi-domesticated dingo of Australia. These peculiarities in the mammalia are so great that every naturalist agrees that Australia must be made a separate region, the only difference of opinion being as to its extent, some thinking that New Zealand should form another separate region; but this question need not now delay us.

In birds Australia is by no means so isolated from the rest of the world, as it contains great numbers of warblers, thrushes, flycatchers, shrikes, crows, and other familiar types of the Eastern Hemisphere; yet a considerable number of the most characteristic Oriental families are absent. Thus there are no vultures, woodpeckers, pheasants, bulbuls, or barbets in the Australian region; and the absence of these is almost as marked a feature as that of cats, deer, or monkeys among mammalia. The most conspicuous and characteristic birds of the Australian region are, the piping-crows; the honey-suckers (Meliphagidæ), a family quite peculiar to the region; the lyre-birds; the great terrestrial kingfishers (Dacelo); the great goat-suckers, called more-porks in Australia, and forming the genus Podargus; the wonderful abundance of parrots, including such remarkable forms as the white and the black cockatoos, and the gorgeously colored brush-tongued lories; the almost equal abundance of fine pigeons more gayly colored than any others on the globe; the strange brush-turkeys and mound-builders, the only birds that never sit upon their eggs, but allow them to be hatched, reptile-like, by the heat of the sand or of fermenting vegetable matter; and, lastly, the emus and cassowaries, in which the wings are far more rudimentary than in the ostriches of Africa and South America. New Guinea and the surrounding islands are remarkable for their tree-kangaroos, their birds-of-paradise, their raquet-tailed kingfishers, their great crown-pigeons, their crimson lories, and many other remarkable birds. This brief outline being sufficient to show the distinctness and isolation of the Australian region, we will now pass to the consideration of the Western Hemisphere.

Definition and Characteristic Groups of the Nearctic Region.
—The Nearctic region comprises all temperate and arctic North
America, including Greenland, the only doubt being as to its
southern boundary, many northern types penetrating into the
tropical zone by means of the highlands and volcanic peaks of
Mexico and Guatemala, while a few which are characteristic of
the tropics extend northward into Texas and California. There
is, however, considerable evidence showing that on the east
coast the Rio Grande del Norte, and on the west a point nearly
opposite Cape St. Lucas, form the most natural boundary; but
instead of being drawn straight across the line bends to the
southeast as soon as it rises on the flanks of the table-land, forming a deep loop which extends some distance beyond the city
of Mexico, and perhaps ought to be continued along the higher
ridges of Guatemala.

The Nearctic region is so similar to the Palæarctic in position and climate, and the two so closely approach each other at
Behring Strait, that we cannot wonder at there being a certain
amount of similarity between them—a similarity which some
naturalists have so far overestimated as to think that the two
regions ought to be united. Let us therefore carefully examine
the special zoological features of this region, and see how far it
resembles, and how far differs from, the Palæarctic.

At first sight the mammalia of North America do not seem
to differ much from those of Europe or Northern Asia. There
are cats, lynxes, wolves and foxes, weasels, bears, elk and deer,
voles, beavers, squirrels, marmots, and hares, all very similar to
those of the Eastern Hemisphere, and several hardly distinguishable. Even the bison or "buffalo" of the prairies, once so
abundant and characteristic, is a close ally of the now almost
extinct "aurochs" of Lithuania. Here, then, we undoubtedly
find a very close resemblance between the two regions; and if
this were all, we should have great difficulty in separating them.
But along with these we find another set of mammals, not quite
so conspicuous, but nevertheless very important. We have, first,
three peculiar genera of moles, one of which, the star-nosed
mole, is a most extraordinary creature, quite unlike anything
else. Then there are three genera of the weasel family, includ-

ing the well-known skunk (Mephitis), all quite different from Eastern forms. Then we come to a peculiar family of carnivora, the raccoons, very distinct from anything in Europe or Asia; and in the Rocky Mountains we find the prong-horn antelope (Antilocapra) and the mountain-goat of the trappers (Aplocerus), both peculiar genera. Coming to the rodents, we find that the mice of America differ in some dental peculiarities from those of the rest of the world, and thus form several distinct genera; the jumping-mouse (Xapus) is a peculiar form of the jerboa family, and then we come to the pouched rats, Geomyidæ, a very curious family, consisting of four genera and nineteen species, peculiar to North America, though not confined to the Nearctic region. The prairie-dogs (Cynomys), the tree-porcupine (Erethizon), the curious sewellel (Haploodon), and the opossum (Didelphys) complete the list of peculiar mammalia which distinguish the northern region of the New World from that of the Old. We must add to these peculiarities some remarkable deficiencies. The Nearctic region has no hedgehogs, nor wild-pigs, nor dormice, and only one wild-sheep in the Rocky Mountains, as against twenty species of sheep and goats in the Palæarctic region.

In birds also the similarities to our own familiar songsters first strike us, though the differences are perhaps really greater than in the quadrupeds. We see thrushes and wrens, tits and finches, and what seem to be warblers and flycatchers and starlings in abundance; but a closer examination shows the ornithologist that what he took for the latter are really quite distinct, and that there is not a single true flycatcher of the family Muscicapidæ, or a single starling of the family Sturnidæ, in the whole continent; while there are very few true warblers (Sylviidæ), their place being taken by the very distinct families Mniotiltidæ, or wood-warblers, and Vireonidæ, or greenlets. In like manner the flycatchers of America belong to the totally distinct family of tyrant-birds, Tyrannidæ, and those that look like starlings to the hang-nests, Icteridæ; and these four peculiar families comprise more than a hundred species, and give a special character to the ornithology of the country. Add to these such peculiar birds as the mocking-thrushes (Mimus), the

blue jays (Cyanocitta), the tanagers, the peculiar genera of cuckoos (Coccygus and Crotophaga), the humming - birds, the wild-
turkeys (Meleagris), and the turkey-buzzards (Cathartes), and we
see that if there is any doubt as to the mammals of North America being sufficiently distinct to justify the creation of a separate region, the evidence of the birds would alone settle the
question.

The reptiles, and some others of the lower animals, add still
more to this weight of evidence. The true rattlesnakes are
highly characteristic, and among the lizards are several genera
of the peculiar American family the Iguanidæ. Nowhere in
the world are the tailed batrachians so largely developed as in
this region, the Sirens and the Amphiumidæ forming two peculiar families; while there are nine peculiar genera of salamanders, and two others allied respectively to the Proteus of Europe
and the Sieboldia, or giant salamander, of Japan. There are
about twenty-nine peculiar genera of fresh-water fishes; while
the fresh-water mollusks are more numerous than in any other
region, more than thirteen hundred species and varieties having
been described.

Combining the evidence derived from all these classes of animals, we find the Nearctic region to be exceedingly well characterized, and to be amply distinct from the Palæarctic. The
few species that are common to the two are almost all arctic, or
at least northern, types, and may be compared with those desert
forms which occupy the debatable ground between the Palæarctic, Ethiopian, and Oriental regions. If, however, we compare
the number of species which are common to the Nearctic and
Palæarctic regions with the number common to the western and
eastern extremities of the latter region, we shall find a wonderful difference between the two cases; and if we further call to
mind the number of important groups characteristic of the one
region but absent from the other, we shall be obliged to admit
that the relation that undoubtedly exists between the faunas of
North America and Europe is of a very distinct nature from
that which connects together Western Europe and Northeastern
Asia in the bonds of zoological unity.

Definition and Characteristic Groups of the Neotropical Re-

gion. — The Neotropical region requires very little definition, since it comprises the whole of America south of the Nearctic region, with the addition of the Antilles or West Indian Islands. Its zoological peculiarities are almost as marked as those of Australia, which, however, it far exceeds in the extreme richness and variety of all its forms of life. To show how distinct it is from all the other regions of the globe, we need only enumerate some of the best known and more conspicuous of the animal forms which are peculiar to it. Such are, among mammalia, the prehensile-tailed monkeys and the marmosets, the blood-sucking bats, the coati-mundis, the peccaries, the llamas and alpacas, the chinchillas, the agoutis, the sloths, the armadillos, and the ant-eaters—a series of types more varied, and more distinct from those of the rest of the world, than any other continent can boast of. Among birds we have the charming sugar-birds, forming the family Cœrebidæ, the immense and wonderfully varied group of tanagers, the exquisite little manakins, and the gorgeously colored chatterers; the host of tree-creepers of the family Dendrocolaptidæ, the wonderful toucans, the puff-birds, jacamars, todies, and motmots; the marvellous assemblage of four hundred distinct kinds of humming-birds, the gorgeous macaws, the curassows, the trumpeters, and the sun-bitterns. Here again there is no other continent or region that can produce such an assemblage of remarkable and perfectly distinct groups of birds; and no less wonderful is its richness in species, since these fully equal, if they do not surpass, those of the two great tropical regions of the Eastern Hemisphere (the Ethiopian and the Oriental) combined.

As an additional indication of the distinctness and isolation of the Neotropical region from all others, and especially from the whole Eastern Hemisphere, we must say something of the otherwise widely distributed groups which are absent. Among mammalia we have first the order Insectivora, entirely absent from South America, though a few species are found in Central America and the West Indies; the Viverridæ, or civet family, are wholly wanting, as is every form of sheep, oxen, or antelopes; while the swine, the elephants, and the rhinoceroses of the Old World are represented by the diminutive peccaries and tapirs.

Among birds we have to notice the absence of tits, true fly-catchers, shrikes, sun-birds, starlings, larks (except a solitary species in the Andes), rollers, bee-eaters, and pheasants; while warblers are very scarce, and the almost cosmopolitan wagtails are represented by a single species of pipit.

We must also notice the preponderance of low or archaic types among the animals of South America. Edentates, marsupials, and rodents form the majority of the terrestrial mammalia; while such higher groups as the carnivora and hoofed animals are exceedingly deficient. Among birds a low type of Passeres, characterized by the absence of the singing-muscles, is excessively prevalent, the enormous groups of the ant-thrushes, tyrants, tree-creepers, manakins, and chatterers belonging to it. The Picariæ (a lower group) also prevail to a far greater extent than in any other regions, both in variety of forms and number of species; and the chief representatives of the gallinaceous birds —the curassows and tinamous—are believed to be allied, the former to the brush-turkeys of Australia, the latter (very remotely) to the ostriches, two of the least-developed types of birds.

Whether, therefore, we consider its richness in peculiar forms of animal life, its enormous variety of species, its numerous deficiencies as compared with other parts of the world, or the prevalence of a low type of organization among its higher animals, the Neotropical region stands out as undoubtedly the most remarkable of the great zoological divisions of the earth.

In reptiles, amphibia, fresh-water fishes, and insects, this region is equally peculiar, but we need not refer to these here, our only object now being to establish by a sufficient number of well-known and easily remembered examples the distinctness of each region from all others, and its unity as a whole. The former has now been sufficiently demonstrated, but it may be well to say a few words as to the latter point.

The only outlying portions of the region about which there can be any doubt are, Central America, or that part of the region north of the Isthmus of Panama, the Antilles, or West Indian Islands, and the temperate portion of South America, including Chili and Patagonia.

In Central America, and especially in Mexico, we have an intermixture of South American and North American animals, but the former undoubtedly predominate, and a large proportion of the peculiar Neotropical groups extend as far as Costa Rica. Even in Guatemala and Mexico we have howling and spider monkeys, coati-mundis, tapirs, and armadillos; while chatterers, manakins, ant-thrushes, and other peculiarly Neotropical groups of birds are abundant. There is therefore no doubt as to Mexico forming part of this region, although it is comparatively poor, and exhibits the intermingling of temperate and tropical forms.

The West Indies are less clearly Neotropical, their poverty in mammals as well as in most other groups being extreme, while great numbers of North American birds migrate there in winter. The resident birds, however, comprise trogons, sugar-birds, chatterers, with many humming-birds and parrots, representing eighteen peculiar Neotropical genera—a fact which decides the region to which the islands belong.

South temperate America is also very poor as compared with the tropical parts of the region, and its insects contain a considerable proportion of north temperate forms. But it contains armadillos, cavies, and opossums; and its birds are all of American groups, though, owing to the inferior climate and deficiency of forests, a number of the families of birds peculiar to tropical America are wanting. Thus there are no manakins, chatterers, toucans, trogons, or motmots; but there are abundance of hangnests, tyrant-birds, ant-thrushes, tree-creepers, and a fair proportion of humming-birds, tanagers, and parrots. The zoology is therefore thoroughly Neotropical, although somewhat poor; and it has a number of peculiar forms, as the chinchillas, alpacas, etc., which are not found in the tropical regions except in the high Andes.

Comparison of Zoological Regions with the Geographical Divisions of the Globe.—Having now completed our survey of the great zoological regions of the globe, we find that they do not differ so much from the old geographical divisions as our first example might have led us to suppose. Europe, Asia, Africa, Australia, North America, and South America really correspond, each to a zoological region, but their boundaries require to be modified more or less considerably; and if we remember this, and keep their extensions or limitations always in our mind, we may use the terms "South American" or "North American" as being equivalent to Neotropical and Nearctic, without much inconvenience; while "African" and "Australian" equally well serve to express the zoological types of the Ethiopian and Australian regions. Europe and Asia require more important modifications. The European fauna does indeed well represent the Palæarctic in all its main features; and if instead of Asia we say tropical Asia, we have the Oriental region very fairly defined; so that the relation of the geographical and the zoological primary divisions of the earth is sufficiently clear. In order to make these relations visible to the eye and more easily remembered, we will put them in a tabular form:

Regions.   Geographical Equivalent.
Palæarctic........Europe, with north temperate Africa and Asia.
Ethiopian.........Africa (south of the Sahara), with Madagascar.
Oriental..........Tropical Asia, to Philippines and Java.
Australian........Australia, with Pacific islands, Moluccas, etc.
Nearctic..........North America, to North Mexico.
Neotropical.......South America, with tropical North America and West Indies.

The following arrangement of the regions will indicate their geographical position, and to a considerable extent their relation to each other:

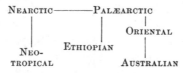

CHAPTER IV.

EVOLUTION THE KEY TO DISTRIBUTION.

Importance of the Doctrine of Evolution.—The Origin of New Species.—Variation in Animals.—The Amount of Variation in North American Birds.—How New Species Arise from a Variable Species.—Definition and Origin of Genera.—Cause of the Extinction of Species.—The Rise and Decay of Species and Genera.—Discontinuous Specific Areas, why Rare.—Discontinuity of the Area of *Parus palustris.* —Discontinuity of *Emberiza schœniclus.*—The European and Japanese Jays.— Supposed Examples of Discontinuity among North American Birds.—Distribution and Antiquity of Families. — Discontinuity a Proof of Antiquity. — Concluding Remarks.

In the preceding chapters we have explained the general nature of the phenomena presented by the distribution of animals, and have illustrated and defined the new geographical division of the earth which is found best to agree with them. Before we go further into the details of our subject, and especially before we attempt to trace the causes which have brought about the existing biological relations of the islands of the globe, it is absolutely necessary to have a clear comprehension of the collateral facts and general principles to which we shall most frequently have occasion to refer. These may be briefly defined as the powers of dispersal of animals and plants under different conditions—geological and climatal changes—and the origin and development of species and groups by natural selection. This last is of the most fundamental importance, and its bearing on the dispersal of animals has been much neglected. We therefore devote the present chapter to its consideration.

As we have already shown in our first chapter that the distribution of species, of genera, and of families presents almost exactly the same general phenomena in varying degrees of complexity, and that almost all the interesting problems we have to deal with depend upon the mode of dispersal of one or other of these; and as, further, our knowledge of most of these groups,

in the higher animals at least, is confined to the Tertiary period
of geology, it is therefore unnecessary for us to enter into any
questions involving the origin of more comprehensive groups,
such as classes or orders.   This enables us to avoid most of the
disputed questions as to the development of animals, and to confine ourselves to those general principles regulating the origin
and development of species and genera which were first laid
down by Mr. Darwin twenty years ago, and have now come to
be adopted by naturalists as established propositions in the theory of evolution.

   The Origin of New Species.—How, then, do new species arise,
supposing the world to have been, physically, much as we now
see it ? and what becomes of them after they have arisen ?   In
the first place, we must remember that new species can only be
formed when and where there is room for them.   If a continent
is fully stocked with animals, each species being so well adapted to its mode of life that it can overcome all the dangers to
which it is exposed, and maintain on the average a tolerably
uniform population, then, so long as no change takes place, no
new species will arise.   For every place or station is supposed
to be filled by creatures perfectly adapted to all surrounding
conditions, able to defend themselves from all enemies, and to
obtain food notwithstanding the rivalry of many competitors.
But such a perfect balance of organisms nowhere exists upon
the earth, and probably never has existed.   The well-known
fact that some species are very common, while others are very
rare, is an almost certain proof that the one is better adapted to
its position than the other ; and this belief is strengthened when
we find the individuals of one species ranging into different
climates, subsisting on different food, and competing with different sets of animals, while the individuals of another species will
be limited to a small area beyond which they seem unable to extend.   When a change occurs, either of climate or geography,
some of the small and ill-adapted species will probably die out
altogether, and thus leave room for others to increase, or for new
forms to occupy their places.

   But the change will most likely affect even flourishing species
in different ways, some beneficially, others injuriously.   Or,

again, it may affect a great many injuriously, to such an extent as to require some change in their structure or habits to enable them to get on as well as before. Now "variation" and the "struggle for existence" come into play. All the weaker and less perfectly organized individuals die out, while those which vary in such a way as to bring them into more harmony with the new conditions constantly survive. If the change of conditions has been considerable, then, after a few centuries, or perhaps even a few generations, one or more *new species* will be almost sure to be formed.

Variation in Animals.—To make this more intelligible to those who have not considered the subject, and to obviate the difficulty many feel about "favorable variations occurring at the right time," it will be well to discuss this matter a little more fully. Few persons consider how largely and universally all animals are varying. We know, however, that in every generation, if we would examine all the individuals of any common species, we should find considerable differences, not only in size and color, but in the form and proportions of all the parts and organs of the body. In our domesticated animals we know this to be the case, and it is by means of the continual selection of such slight varieties to breed from that all our extremely different domestic breeds have been produced. Think of the difference in every limb, and every bone and muscle, and probably in every part, internal and external, of the whole body, between a greyhound and a bull-dog! Yet if we had the whole series of ancestors of these two breeds before us, we should probably find that in no one generation was there a greater difference than now occurs in the same breed, or sometimes even the same litter. It is often thought, however, that wild species do not vary sufficiently to bring about any such change as this in the same time; and though naturalists are well aware that this is a mistake, it is only recently that they are able to adduce positive proof of their opinion.

The Amount of Variation in North American Birds.—An American naturalist, Mr. J. A. Allen, has made elaborate observations and measurements of the birds of the United States, and he finds a wonderful and altogether unsuspected amount of variation between individuals of the same species. They differ in the general tint, and in the markings and distribution of the colors; in size and proportions; in the length of the wings, tail, bill, and feet; in the length of particular feathers, altering the shape of the wing or tail; in the length of the tarsi and of the separate toes; and in the length, width, thickness, and curvature of the bill. These variations are very considerable, often reaching to one sixth or one seventh of the average dimensions, and sometimes more. Thus *Turdus fuscescens* (Wilson's thrush) varied in length of wing from 3.58 to 4.16 inches, and in the tail from 3.55 to 4 inches; and in twelve specimens, all taken in the same locality, the wing varied in length from 14.5 to 21 per cent., and the tail from 14 to 22.5 per cent. In *Sialia sialis* (the blue-bird) the middle toe varied from .77 to .91 inch, and the hind toe from .58 to .72 inch, or more than 21.5 per cent. on the mean; while the bill varied from .45 to .56 inch in length, and from .30 to .38 inch in width, or about 20 per cent. in both cases. In *Dendrœca coronata* (the yellow-crowned warbler) the quills vary in proportionate length, so that the first, the second, the third, or the fourth is sometimes longest; and a similar variation of the wing involving a change of proportion between two or more of the feathers is recorded in eleven species of birds. Color and marking vary to an equal extent; the dark streaks on the under-surface of *Melospiza melodia* (the American song-sparrow) being sometimes reduced to narrow lines, while in other specimens they are so enlarged as to cover the greater part of the breast and sides of the body, sometimes uniting on the middle of the breast into a nearly continuous patch. In one of the small spotted wood-thrushes, *Turdus fuscescens*, the colors are sometimes very pale, and the markings on the breast reduced to indistinct narrow lines; while in other specimens the general color is much darker, and the breast-markings dark, broad, and triangular. All the variations here mentioned occur between adult males, so that there is no question of differences of age or sex, and the pair last referred to were taken at the same place and on the same day.[1]

---

[1] These facts are taken from a memoir on "The Mammals and Winter Birds of Florida," by J. A. Allen, forming Vol. II., No. 3, of the "Bulletin of the Museum of Comparative Zoology at Harvard College," Cambridge, Massachusetts.

These interesting facts entirely support the belief in the variability of all animals in all their parts and organs, to an extent amply sufficient for natural selection to work with. We may, indeed, admit that these are extreme cases, and that the majority of species do not vary half or a quarter so much as shown in the examples quoted, and we shall still have ample variation for all purposes of specific modification. Instead of an extreme variation in the dimensions and proportions of the various organs of from 10 to 25 per cent., as is here proved to occur, we may assume from 3 to 6 per cent. as generally occurring in the majority of species ; and if we further remember that the above excessive variations were found by comparing a number of specimens of each species varying from 50 to 150 only, we may be sure that the smaller variations we require must occur in considerable numbers among the thousands or millions of individuals of which all but the very rare species consist. If, therefore, we were to divide the population of any species into three groups of equal extent, with regard to any particular character—as length of wing or of toes, or thickness or curvature of bill, or strength of markings—we should have one group in which the mean or average character prevailed with little variation, one in which the character was greatly, and one in which it was little, developed. If we formed our groups, not by equal numbers, but by equal amount of variation, we should probably find, in accordance with the law of averages, that the central group, in which the mean characteristics prevailed, was much more numerous than the extremes ; perhaps twice, or even three times, as great as either of them, and forming such a series as the following : 10 maximum, 30 mean, 10 minimum development. In ordinary cases we have no reason to believe that the mean characters or the amount of variation of a species changes materially from year to year or from century to century, and we may therefore look upon the central group as the type of the species which is best adapted to the conditions in which it has actually to exist. This type will therefore always form the majority, because the struggle for existence will lead to the continual suppression of the less perfectly adapted extremes. But sometimes a species has a wide range into countries which differ in physical conditions, and then it often happens that one or other of the extremes will predominate in a portion of its range. These form local varieties; but as they occur mixed with the other forms, they are not considered to be distinct species, although they may differ from the other extreme form quite as much as species often do from each other.

How New Species Arise from a Variable Species.—It is now very easy to understand how, from such a variable species, one or more new species may arise. The peculiar physical or organic conditions that render one part of the area better adapted to an extreme form may become intensified, and the most extreme variations thus having the advantage, they will multiply at the expense of the rest. If the change of conditions spreads over the whole area occupied by the species, this one extreme form will replace the others; while if the area should be cut in two by subsidence or elevation, the conditions of the two parts may be modified in opposite directions, so as to be each adapted to one extreme form; in which case the original type will become extinct, being replaced by two species, each formed by a combination of certain extreme characters which had before existed in some of its varieties.

The changes of conditions which lead to such selection of varieties are very diverse in nature, and new species may thus be formed, diverging in many ways from the original stock. The climate may change from moist to dry, or the reverse, or the temperature may increase or diminish for long periods, in either case requiring a corresponding change of constitution, of covering, of vegetable or of insect food, to be met by the selection of variations of color or of swiftness, of length of bill or of strength of claws. Again, competitors or enemies may arrive from other regions, giving the advantage to such varieties as can change their food, or by swifter flight or greater wariness can escape their new foes. We may thus easily understand how a series of changes may occur at distant intervals, each leading to the selection and preservation of a special set of variations, and thus what was a single species may become transformed into a group of allied species differing from each other in a variety of ways, just as we find them in nature.

Among these species, however, there will be some which will have become adapted to very local or special conditions, and will therefore be comparatively few in number and confined to a limited area; while others, retaining the more general characters of the parent form, but with some important change of structure, will be better adapted to succeed in the struggle for existence with other animals, will spread over a wider area, and increase so as to become common species. Sometimes these will acquire such a perfection of organization by successive favorable modifications that they will be able to spread greatly beyond the range of the parent form. They then become what are termed dominant species, maintaining themselves in vigor and abundance over very wide areas, displacing other species with which they come into competition, and, under still further changes of conditions, becoming the parents of a new set of diverging species.

Definition and Origin of Genera.—As some of the most important and interesting phenomena of distribution relate to genera rather than to single species, it will be well here to explain what is meant by a genus, and how genera are supposed to arise.

A genus is a group of allied species which differs from all other groups in some well-marked characters, usually of a structural rather than a superficial nature. Species of one genus usually differ from each other in size, in color or marking, in the proportions of the limbs or other organs, and in the form and size of such superficial appendages as horns, crests, manes, etc.; but they generally agree in the form and structure of important organs, as the teeth, the bill, the feet, and the wings. When two groups of species differ from each other constantly in one or more of these latter particulars, they are said to belong to different genera. We have already seen that species vary in these more important as well as in the more superficial characters. If, then, in any part of the area occupied by a species some change of habits becomes useful to it, all such structural variations as facilitate the change will be accumulated by natural selection; and when they have become fixed in the proportions most beneficial to the animal, we shall have the first species of a new genus.

A creature which has been thus modified in important characters will be a new type, specially adapted to fill its place in the economy of nature. It will almost certainly have arisen from an extensive or dominant group, because only such are sufficiently rich in individuals to afford an ample supply of the necessary variations, and it will inherit the vigor of constitution and adaptability to a wide range of conditions which gave success to its ancestors. It will therefore have every chance in its favor in the struggle for existence; it may spread widely and displace many of its nearest allies, and in doing so will itself become modified superficially and become the parent of a number of subordinate species. It will now have become a dominant *genus*, occupying an entire continent, or perhaps even two or more continents, spreading in every direction till it comes in contact with competing forms better adapted to the different environments. Such a genus may continue to exist during long geological epochs; but the time will generally come when either physical changes, or competing forms, or new enemies are too much for it, and it begins to lose its supremacy. First one, then another, of its component species will dwindle away and become extinct, till at last only a few species remain. Sometimes these soon follow the others, and the whole genus dies out, as thousands of genera have died out during the long course of the earth's life-history; but it will also sometimes happen that a few species will continue to maintain themselves in areas where they are removed from the influences that exterminated their fellows.

Cause of the Extinction of Species.—There is good reason to believe that the most effective agent in the extinction of species is the pressure of other species, whether as enemies or merely as competitors. If therefore any portion of the earth is cut off from the influx of new or more highly organized animals, we may there expect to find the remains of groups which have elsewhere become extinct. In islands which have been long separated from their parent continents these conditions are exactly fulfilled, and it is in such that we find the most striking examples of the preservation of fragments of primeval groups of animals, often widely separated from each other, owing to their having been preserved at remote portions of the area of the once wide-spread parental group. There are many other ways in which portions of dying-out groups may be saved. Nocturnal or sub-terranean modes of life may save a species from enemies or competitors, and many of the ancient types still existing have such habits. The dense gloom of equatorial forests also affords means of concealment and protection, and we sometimes find in such localities a few remnants of low types in the midst of a general assemblage of higher forms. Some of the most ancient types now living inhabit caves, like the Proteus; or bury themselves in mud, like the Lepidosiren; or in sand, like the Amphioxus, the last being the most ancient of all vertebrates; while the Galeopithecus and Tarsius of the Malay Islands, and the potto of West Africa, survive amidst the higher mammalia of the Asiatic and African continents, owing to their nocturnal habits and concealment in the densest forests.

The Rise and Decay of Species and Genera.—The preceding sketch of the mode in which species and genera have arisen, have come to maturity, and then decay, leads us to some very important conclusions as to the mode of distribution of animals. When a species or a genus is increasing and spreading, it necessarily occupies a continuous area which gets larger and larger till it reaches a maximum; and we accordingly find that almost all extensive groups are thus continuous. When decay commences, and the group, ceasing to be in harmony with its environment, is encroached upon by other forms, the continuity may frequently be broken. Sometimes the outlying species may be the first to become extinct, and the group may simply diminish in area while keeping a compact central mass; but more often the process of extinction will be very irregular, and may even divide the group into two or more disconnected portions. This is the more likely to be the case because the most recently formed species, probably adapted to local conditions, and therefore most removed from the general type of the group, will have the best chance of surviving, and these may exist at several isolated points of the area once occupied by the whole group. We may thus understand how the phenomenon of discontinuous areas has come about, and we may be sure that when allied species or varieties of the same species are found widely separated

from each other, they were once connected by intervening forms
or by each extending till it overlapped the other's area.

Discontinuous Specific Areas, why Rare.—But although discontinuous generic areas, or the separation from each other of
species whose ancestors must once have occupied conterminous
or overlapping areas, are of frequent occurrence, yet undoubted
cases of discontinuous specific areas are very rare, except, as already stated, when one portion of a species inhabits an island.
A few examples among mammalia have been referred to in our
first chapter, but it may be said that these are examples of the
very common phenomenon of a species being only found in the
station for which its organization adapts it; so that forest or
marsh or mountain animals are of course only found where there
are forests, marshes, or mountains. This may be true; and when
the separate forests or mountains inhabited by the same species
are not far apart, there is little that needs explanation: but in
one of the cases referred to there was a gap of a thousand miles
between two of the areas occupied by the species; and this being
too far for the animal to traverse through an uncongenial territory, we are forced to the conclusion that it must at some former
period, and under different conditions, have occupied a considerable portion of the intervening area.

Among birds such cases of specific discontinuity are very rare,
and hardly ever quite satisfactory. This may be owing to birds
being more rapidly influenced by changed conditions, so that
when a species is divided the two portions almost always become modified into varieties or distinct species; while another
reason may be that their powers of flight cause them to occupy,
on the average, wider and less precisely defined areas than do
the species of mammalia. It will be interesting, therefore, to
examine the few cases on record, as we shall thereby obtain additional knowledge of the steps and processes by which the distribution of varieties and species has been brought about.

Discontinuity of the Area of Parus palustris.—Mr. Seebohm,
who has travelled and collected in Europe, Siberia, and India,
and possesses extensive and accurate knowledge of Palæarctic
birds, has recently called attention to the varieties and sub-species of the marsh tit (*Parus palustris*), of which he has exam-

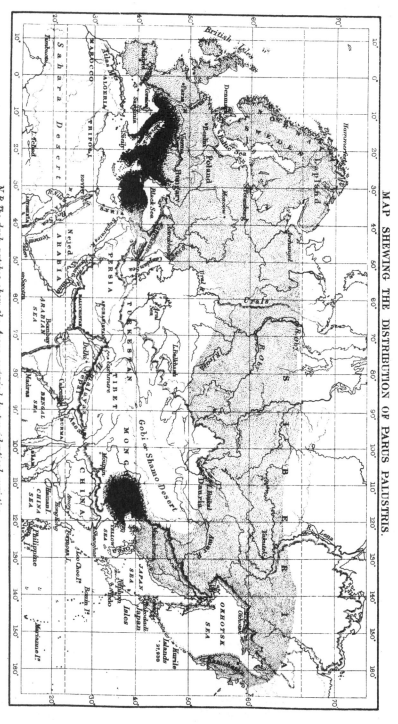

MAP SHEWING THE DISTRIBUTION OF PARUS PALUSTRIS.

N.B. The dark patches show the Areas occupied by two identical varieties.

Harper & Brothers New York.

ined numerous specimens ranging from England to Japan.[1] The curious point is that those of Southern Europe and of China are exactly alike, while all over Siberia a very distinct form occurs, the sub-species *P. borealis*. In Japan and Kamtschatka other varieties are found, which have been named respectively *P. Japonicus* and *P. Camtschatkensis*. Now it all depends upon these forms being classed as sub-species or as true species whether this is or is not a case of discontinuous specific distribution. If *Parus borealis* is a distinct species from *Parus palustris*, as it is reckoned in Gray's " Hand List of Birds," and also in Sharpe and Dresser's "Birds of Europe," then *Parus palustris* has a most remarkable discontinuous distribution as shown in the accompanying map, one portion of its area comprising Central and South Europe and Asia Minor, the other an undefined tract in Northern China, the two portions being thus situated in about the same latitude and having a very similar climate, but with a distance of about four thousand miles between them. If, however, these two forms are reckoned as sub-species only, then the area of the species becomes continuous, while only one of its varieties or sub-species has a discontinuous area. It is a curious fact that *P. palustris* and *P. borealis* are found together in Southern Scandinavia and in some parts of Central Europe, and are said to differ somewhat in their note and their habits, as well as in coloration.

Discontinuity of Emberiza schœniclus.—The other case is that of our reed-bunting (*Emberiza schœniclus*), which ranges over almost all Europe and Western Asia as far as the Yenisei valley and Northwest India. It is then replaced by another smaller species, *E. passerina*, which ranges eastward to the Lena River, and in winter as far south as Amoy in China; but in Japan the original species appears again, receiving a new name (*E. pyrrhulina*), but Mr. Seebohm assures us that it is quite indistinguishable from the European bird.[2] Although the distance between these two portions of the species is not so great as in the last example, being about two thousand miles, in other respects the case is a most satisfactory one, because the forms which occupy the intervening space are recognized by Mr. Seebohm himself as undoubted species.

---

[1] *Ibis*, 1879, p. 32. [2] *Ibis*, 1879, p. 40.

The European and Japanese Jays.—Another case somewhat resembling that of the marsh tit is afforded by the European and Japanese jays (*Garrulus glandarius* and *G. Japonicus*). Our common jay inhabits the whole of Europe except the extreme north, but is not known to extend anywhere into Asia, where it is represented by several quite distinct species. (See Map, frontispiece.) But the great central island of Japan is inhabited by a jay (*G. Japonicus*) which is very like ours, and was formerly classed as a sub-species only, in which case our jay would be considered to have a discontinuous distribution. But the specific distinctness of the Japanese bird is now universally admitted, and it is certainly a very remarkable fact that among the twelve species of jays which together range over all temperate Europe and Asia, one which is so closely allied to our English bird should be found at the remotest possible point from it. Looking at the map exhibiting the distribution of the several species, we can hardly avoid the conclusion that a bird very like our jay once occupied the whole area of the genus, that in various parts of Asia it became gradually modified into a variety of distinct species in the manner already explained, a remnant of the original type being preserved almost unchanged in Japan, owing probably to favorable conditions of climate and protection from competing forms.

Supposed Examples of Discontinuity among North American Birds.—In North America the eastern and western provinces are so different in climate and vegetation, and are besides separated by such remarkable physical barriers — the arid central plains and the vast ranges of the Rocky Mountains and Sierra Nevada—that we can hardly expect to find species whose areas may be divided maintaining their identity. Towards the north, however, the above-named barriers disappear, the forests being almost continuous from east to west, while the mountain-range is broken up by passes and valleys. It thus happens that most species of birds which inhabit both the eastern and western coasts of the North American continent have maintained their continuity towards the north, while even when differentiated into two or more allied species their areas are often conterminous or overlapping.

Almost the only bird that seems to have a really discontinuous range is the species of wren *Thryothorus Bewickii*, of which the type-form ranges from the east coast to Kansas and Minnesota, while a longer-billed variety is found in the wooded parts of California and as far north as Puget Sound. If this really represents the range of the species, there remains a gap of about one thousand miles between its two disconnected areas. Other cases are those of the greenlet, *Vireosylvia gilvus*, of the Eastern States, and its variety *V. Swainsonii*, of the Western; and of the purple redfinch, *Carpodacus purpureus*, with its variety *C. Californicus*. But, unfortunately, the exact limits of these varieties are in neither case known; and though each one is characteristic of its own province, it is possible they may somewhere become conterminous, though in the case of the redfinches this does not seem likely to be the fact.

In a later chapter we shall have to point out some remarkable cases of this kind where one portion of the species inhabits an island; but the facts now given are sufficient to prove that the discontinuity of the area occupied by a single homogeneous species, by two varieties of a species, by two well-marked sub-species, and by two closely allied but distinct species, are all different phases of one phenomenon—the decay of ill-adapted and their replacement by better-adapted forms, under the pressure of a change of conditions either physical or organic. We may now proceed with our sketch of the mode of distribution of higher groups.

Distribution and Antiquity of Families.—Just as *genera* are groups of allied species distinguished from all other groups by some well-marked structural characters, so *families* are groups of allied genera distinguished by more marked and more important characters, which are generally accompanied by a peculiar outward form and style of coloration, and by distinctive habits and mode of life. As a genus is usually more ancient than any of the species of which it is composed, because during its growth and development the original rudimentary species becomes supplanted by more and more perfectly adapted forms, so a family is usually older than its component genera, and during the long period of its life-history may have survived many and great terrestrial and organic changes. Many families of the higher animals have now an almost world-wide extension, or at least range over several continents; and it seems probable that all families which have survived long enough to develop a considerable variety of generic and specific forms have also at one time or other occupied an extensive area.

Discontinuity a Proof of Antiquity. — Discontinuity will therefore be an indication of antiquity; and the more widely the fragments are scattered, the more ancient we may usually presume the parent group to be. A striking example is furnished by the strange reptilian fishes forming the order or suborder Dipnoi, which includes the Lepidosiren and its allies. Only three or four living species are known, and these inhabit tropical rivers situated in the remotest continents. The *Lepidosiren paradoxa* is only known from the Amazon and some other South American rivers. An allied species, *Lepidosiren annectens*, sometimes placed in a distinct genus, inhabits the Gambia in West Africa; while the recent discovery in Eastern Australia of the Ceratodus, or mud-fish, of Queensland adds another form to the same isolated group. Numerous fossil teeth long known from the Triassic beds of this country, and also found in Germany and India in beds of the same age, agree so closely with those of the living Ceratodus that both are referred to the same genus. No more recent traces of any such animal have been discovered, but the Carboniferous Ctenodus and the Devonian Dipterus evidently belong to the same group, while in North America the Devonian rocks have yielded a gigantic allied form which has been named Heliodus by Professor Newberry. Thus an enormous range in time is accompanied by a very wide and scattered distribution of the existing species.

Whenever, therefore, we find two or more living genera belonging to the same family or order, but not very closely allied to each other, we may be sure that they are the remnants of a once extensive group of genera; and if we find them now isolated in remote parts of the globe, the natural inference is that the family of which they are fragments once had an area embracing the countries in which they are found. Yet this simple and very obvious explanation has rarely been adopted by naturalists, who have instead imagined changes of land and sea to afford a direct passage from the one fragment to the other. If there were no cosmopolitan or very wide-spread families still existing, or even if such cases were rare, there would be some justification for such a proceeding; but as about one fourth of the existing families of land mammalia have a range extending to at least three or four continents, while many which are now represented by disconnected genera are known to have occupied intervening lands or to have had an almost continuous distribution in Tertiary times, all the presumptions are in favor of the former continuity of the group. We have also in many cases direct evidence that this former continuity was effected by means of existing continents, while in no single case has it been shown that such a continuity was impossible, and that it either was or must have been effected by means of continents now sunk beneath the ocean.

Concluding Remarks.—When writing on the subject of distribution, it usually seems to have been forgotten that the theory of evolution absolutely necessitates the former existence of a whole series of extinct genera filling up the gap between the isolated genera which in many cases now alone exist; while it is almost an axiom of " natural selection" that such numerous forms of one type could only have been developed in a wide area and under varied conditions, implying a great lapse of time. In our succeeding chapters we shall show that the known and probable changes of sea and land, the known changes of climate, and the actual powers of dispersal of the different groups of animals were such as would have enabled all the now disconnected groups to have once formed parts of a continuous series. Proofs of such former continuity are continually being obtained by the discovery of allied extinct forms in intervening lands; but the extreme imperfection of the geological record as regards land animals renders it unlikely that this proof will be forthcoming in the majority of cases. The notion that if such animals ever existed their remains would certainly be found is a superstition which, notwithstanding the efforts of Lyell and Darwin, still largely prevails among naturalists; but until it is got rid of, no true notions of the former distribution of life upon the earth can be attained.

CHAPTER V.

THE POWERS OF DISPERSAL OF ANIMALS AND PLANTS.

Statement of the General Question of Dispersal.—The Ocean as a Barrier to the
Dispersal of Mammals.—The Dispersal of Birds.—The Dispersal of Reptiles.—
The Dispersal of Insects.—The Dispersal of Land Mollusca.—Great Antiquity of
Land Shells.—Causes Favoring the Abundance of Land Shells.—The Dispersal
of Plants.—Special Adaptability of Seeds for Dispersal.—Birds as Agents in the
Dispersal of Seeds.—Ocean Currents as Agents in Plant-dispersal.—Dispersal
along Mountain-chains.—Antiquity of Plants as Affecting their Distribution.

In order to understand the many curious anomalies we meet
with in studying the distribution of animals and plants, and to
be able to explain how it is that some species and genera have
been able to spread widely over the globe, while others are confined to one hemisphere, to one continent, or even to a single
mountain or a single island, we must make some inquiry into
the different powers of dispersal of animals and plants, into the
nature of the barriers that limit their migrations, and into the
character of the geological or climatal changes which have favored or checked such migrations.

The first portion of the subject—that which relates to the
vario·s modes by which organisms can pass over wide areas of
sea and land—has been fully treated by Sir Charles Lyell, by
Mr. Darwin, and many other writers, and it will be only necessary here to give a very brief notice of the best-known facts on
the subject, which will be further referred to when we come to
discuss the particular cases that arise in regard to the faunas
and floras of remote islands. But the other side of the question
of dispersal—that which depends on geological and climatal
changes — is in a far less satisfactory condition ; for, though
much has been written upon it, the most contradictory opinions
still prevail, and at almost every step we find ourselves on the
battle-field of opposing schools in geological or physical science.

As, however, these questions lie at the very root of any general solution of the problems of distribution, I have given much time to a careful examination of the various theories that have been advanced, and the discussions to which they have given rise; and have arrived at some definite conclusions which I venture to hope may serve as the foundation for a better comprehension of these intricate problems. The four chapters which follow this are devoted to a full examination of these profoundly interesting and important questions, after which we shall enter upon our special inquiry—the nature and origin of insular faunas and floras.

The Ocean as a Barrier to the Dispersal of Mammals.—A wide extent of ocean forms an almost absolute barrier to the dispersal of all land animals, and of most of those which are aerial, since even birds cannot fly for thousands of miles without rest and without food, unless they are aquatic birds which can find both rest and food on the surface of the ocean. We may be sure, therefore, that without artificial help neither mammalia nor land birds can pass over very wide oceans. The exact width they can pass over is not determined, but we have a few facts to guide us. Contrary to the common notion, pigs can swim very well, and have been known to swim over five or six miles of sea, and the wide distribution of pigs in the islands of the Eastern Hemisphere may be due to this power. It is almost certain, however, that they would never voluntarily swim away from their native land, and if carried out to sea by a flood they would certainly endeavor to return to the shore. We cannot, therefore, believe that they would ever swim over fifty or a hundred miles of sea, and the same may be said of all the larger mammalia. Deer also swim well, but there is no reason to believe that they would venture out of sight of land. With the smaller, and especially with the arboreal, mammalia there is a much more effectual way of passing over the sea, by means of floating trees, or those floating islands which are often formed at the mouths of great rivers. Sir Charles Lyell describes such floating islands which were encountered among the Moluccas, on which trees and shrubs were growing on a stratum of soil, which even formed a white beach round the margin of each raft. Among the Philippine Islands similar rafts with trees growing on them have been seen after hurricanes; and it is easy to understand how, if the sea were tolerably calm, such a raft might be carried along by a current, aided by the wind acting on the trees, till, after a passage of several weeks, it might arrive safely on the shores of some land hundreds of miles away from its starting-point. Such small animals as squirrels and mice might have been carried away on the trees which formed part of such a raft, and might thus colonize a new island; though, as it would require a pair of the same species to be carried away together, such accidents would no doubt be rare. Insects, however, and land shells would almost certainly be abundant on such a raft or island, and in this way we may account for the wide dispersal of many species of both these groups.

Notwithstanding the occasional action of such causes, we cannot suppose that they have been effective in the dispersal of mammalia as a whole; and whenever we find that a considerable number of the mammals of two countries exhibit distinct marks of relationship, we may be sure that an actual land connection, or, at all events, an approach to within a very few miles of each other, has at one time existed. But a considerable number of identical mammalian families, and even genera, are actually found in all the great continents, and the present distribution of land upon the globe renders it easy to see how they have been able to disperse themselves so widely. All the great land masses radiate from the arctic regions as a common centre, the only break being at Behring Strait, which is so shallow that a rise of less than a thousand feet would form a broad isthmus connecting Asia and America as far south as the parallel of 60° N. Continuity of land, therefore, may be said to exist already for all parts of the world (except Australia and a number of large islands, which will be considered separately), and we have thus no difficulty in the way of that former wide diffusion of many groups which we maintain to be the only explanation of most anomalies of distribution other than such as may be connected with unsuitability of climate.

The Dispersal of Birds.—Wherever mammals can migrate, other vertebrates can generally follow with even greater facility. Birds, having the power of flight, can pass over wide arms of the sea, or even over extensive oceans, when these are, as in the Pacific, studded with islands to serve as resting-places. Even the smaller land birds are often carried by violent gales of wind from Europe to the Azores, a distance of nearly a thousand miles, so that it becomes comparatively easy to explain the exceptional distribution of certain species of birds. Yet on the whole it is remarkable how closely the majority of birds follow the same laws of distribution as mammals, showing that they generally require either continuous land or an island-strewn sea as a means of dispersal to new homes.

The Dispersal of Reptiles.—Reptiles appear at first sight to be as much dependent on land for their dispersal as mammalia; but they possess two peculiarities which favor their occasional transmission across the sea—the one being their greater tenacity of life, the other their oviparous mode of reproduction. A large boa-constrictor was once floated to the island of St. Vincent twisted round the trunk of a cedar-tree, and was so little injured by its voyage that it captured some sheep before it was killed. The island is nearly two hundred miles from Trinidad and the coast of South America, whence it almost certainly came.[1] Snakes are, however, comparatively scarce on islands far from continents, but lizards are often abundant; and though these might also travel on floating trees, it seems more probable that there is some as yet unknown mode by which their eggs are safely, though perhaps very rarely, conveyed from island to island. Examples of their peculiar distribution will be given when we treat of the fauna of some islands in which they abound.

The Dispersal of Amphibia and Fresh-water Fishes.—The two lower groups of vertebrates, amphibia and fresh-water fishes, possess special facilities for dispersal, in the fact of their eggs being deposited in water, and in their aquatic or semi-aquatic habits. They have another advantage over reptiles in being capable of flourishing in arctic regions, and in the power possessed by their eggs of being frozen without injury. They have thus, no doubt, been assisted in their dispersal by floating ice, and by that approximation of all the continents in high northern lati-

---

[1] Lyell's "Principles of Geology," II., p. 369.

tudes which has been the chief agent in producing the general uniformity in the animal productions of the globe. Some genera of Batrachia have almost a world-wide distribution; while the Tailed Batrachia, such as the newts and salamanders, are almost entirely confined to the Northern Hemisphere, some of the genera spreading over the whole of the north temperate zone. Freshwater fishes have often a very wide range, the same species being sometimes found in all the rivers of a continent. This is no doubt chiefly due to the want of permanence in river basins, especially in their lower portions, where streams belonging to distinct systems often approach each other and may be made to change their course from one to the other basin by very slight elevations or depressions of the land. Hurricanes and waterspouts also often carry considerable quantities of water from ponds and rivers, and thus disperse eggs and even small fishes. As a rule, however, the same species are not often found in countries separated by a considerable extent of sea, and in the tropics rarely the same genera. The exceptions are in the colder regions of the earth, where the transporting power of ice may have come into play. High ranges of mountains, if continuous for long distances, rarely have the same species of fish in the rivers on their two sides. Where exceptions occur, it is often due to the great antiquity of the group, which has survived so many changes in physical geography that it has been able, step by step, to reach countries which are separated by barriers impassable to more recent types. Yet another and more efficient explanation of the distribution of this group of animals is the fact that many families and genera inhabit both fresh and salt water; and there is reason to believe that many of the fishes now inhabiting the tropical rivers of both hemispheres have arisen from allied marine forms becoming gradually modified for a life in fresh water. By some of these various causes, or a combination of them, most of the facts in the distribution of fishes can be explained without much difficulty.

The Dispersal of Insects.—In the enormous group of insects the means of dispersal among land animals reach their maximum. Many of them have great powers of flight, and from their extreme lightness they can be carried immense distances by gales of wind. Others can survive exposure to salt water for many days, and may thus be floated long distances by marine currents. The eggs and larvæ often inhabit solid timber, or lurk under bark or in crevices of logs, and may thus reach any countries to which such logs are floated. Another important factor in the problem is the immense antiquity of insects, and the long persistence of many of the best-marked types. The rich insect fauna of the Miocene period in Switzerland consisted largely of genera still inhabiting Europe, and even of a considerable number identical, or almost so, with living species. Out of 156 genera of Swiss fossil beetles, no less than 114 are still living; and the general character of the species is exactly like that of the existing fauna of the Northern Hemisphere in a somewhat more southern latitude. There is, therefore, evidently no difficulty in accounting for any amount of dispersal among insects; and it is all the more surprising that with such powers of migration they should yet be often as restricted in their range as the reptiles or even the mammalia. The cause of this wonderful restriction to limited areas is undoubtedly the extreme specialization of most insects. They have become so exactly adapted to one set of conditions that when carried into a new country they cannot live. Many can only feed in the larva state on one species of plant; others are bound up with certain groups of animals on which they are more or less parasitic. Climatal influences have a great effect on their delicate bodies; while, however well a species may be adapted to cope with its enemies in one locality, it may be quite unable to guard itself against those which elsewhere attack it. From this peculiar combination of characters it happens that among insects are to be found examples of the widest and most erratic dispersal and also of the extremest restriction to limited areas; and it is only by bearing these considerations in mind that we can find a satisfactory explanation of the many anomalies we meet with in studying their distribution.

The Dispersal of Land Mollusca.—The only other group of animals we need now refer to is that of the air-breathing mollusca, commonly called land shells. These are almost as ubiquitous as insects, though far less numerous; and their wide distribution is by no means so easy to explain. The genera have usually a very wide, and often a cosmopolitan, range, while the species are rather restricted, and sometimes wonderfully so. Not only do single islands, however small, often possess peculiar species of land shells, but sometimes single mountains or valleys, or even a particular mountain-side, possess species or varieties found nowhere else upon the globe. It is pretty certain that they have no means of passing over the sea but such as are very rare and exceptional. Some which possess an operculum, or which close the mouth of the shell with a diaphragm of secreted mucus, may float across narrow arms of the sea, especially when protected in the crevices of logs of timber; while in the young state when attached to leaves or twigs they may be carried long distances by hurricanes.[1] Owing to their exceedingly slow motion, their powers of voluntary dispersal, even on land, are very limited, and this will explain the extreme restriction of their range in many cases.

Great Antiquity of Land Shells.—The clew to the almost universal distribution of the several families and of many genera is to be found, however, in their immense antiquity. In the Pliocene and Miocene formations most of the land shells are either identical with living species or closely allied to them; while even in the Eocene almost all are of living genera, and one British Eocene fossil still lives in Texas. Strange to say, no true land shells have been discovered in the Secondary formations; but they must certainly have abounded, for in the far more ancient Palæozoic coal measures of Nova Scotia two species belonging to the living genera Pupa and Zonites have been found in considerable abundance.

Land shells have therefore survived all the revolutions the earth has undergone since Palæozoic times. They have been

---

[1] Mr. Darwin found that the large *Helix pomatia* lived after immersion in seawater for twenty days. It is hardly likely that this is the extreme limit of their powers of endurance, but even this would allow of their being floated many hundred miles at a stretch; and if we suppose the shell to be partially protected in the crevice of a log of wood, and to be thus out of water in calm weather, the distance might extend to a thousand miles or more. The eggs of fresh-water mollusca are known to attach themselves to the feet of aquatic birds, and this is supposed to account for their very wide diffusion.

able to spread slowly but surely into every land that has ever been connected with a continent, while the rare chances of transfer across the ocean, to which we have referred as possible, have again and again occurred during the almost unimaginable ages of their existence. The remotest and most solitary of the islands of the mid-ocean have thus become stocked with them, though the variety of species and genera bears a direct relation to the facilities of transfer, and the shell fauna is never very rich and varied, except in countries which have at one time or other been united to some continental land.

Causes Favoring the Abundance of Land Shells.—The abundance and variety of land shells are also, more than those of any other class of animals, dependent on the nature of the surface and the absence of enemies; and where these conditions are favorable, their forms are wonderfully luxuriant. The first condition is the presence of lime in the soil, and a broken surface of country with much rugged rock offering crevices for concealment and hybernation. The second is a limited bird and mammalian fauna, in which such species as are especially shell-eaters shall be rare or absent. Both these conditions are found in certain large islands, and pre-eminently in the Antilles, which possess more species of land shells than any single continent. If we take the whole globe, more species of land shells are found on the islands than on the continents—a state of things to which no approach is made in any other group of animals whatever, but which is perhaps explained by the considerations now suggested.

The Dispersal of Plants.—The ways in which plants are dispersed over the earth, and the special facilities they often possess for migration, have been pointed out by eminent botanists, and a considerable space might be occupied in giving a summary of what has been written on the subject. In the present work, however, it is only in two or three chapters that I discuss the origin of insular floras in any detail; and it will therefore be advisable to adduce any special facts when they are required to support the argument in particular cases. A few general remarks only will therefore be made here.

Special Adaptability of Seeds for Dispersal.—Plants possess many great advantages over animals as regards the power of dispersal, since they are all propagated by seeds or spores, which are hardier than the eggs of even insects, and retain their vitality for a much longer time. Seeds may lie dormant for many years and then vegetate, while they endure extremes of heat, of cold, of drought, or of moisture which would almost always be fatal to animal germs. Among the causes of the dispersal of seeds De Candolle enumerates the wind, rivers, ocean currents, icebergs, birds and other animals, and human agency. Great numbers of seeds are specially adapted for transport by one or other of these agencies. Many are very light and have winged appendages, pappus, or down, which enables them to be carried enormous distances. It is true, as De Candolle remarks, that we have no actual proofs of their being so carried; but this is not surprising when we consider how small and inconspicuous most seeds are. Supposing every year a million seeds were brought by the wind to the British Isles from the Continent, this would be only ten to a square mile, and the observation of a lifetime might never detect one; yet a hundredth part of this number would serve in a few centuries to stock an island like Britain with a great variety of Continental plants.

When, however, we consider the enormous quantity of seeds produced by plants; that great numbers of these are more or less adapted to be carried by the wind; and that winds of great violence and long duration occur in most parts of the world, we are as sure that seeds must be carried to great distances as if we had seen them so carried. Such storms carry leaves, hay, dust, and many small objects to a great height in the air, while many insects have been conveyed by them for hundreds of miles out to sea and far beyond what their unaided powers of flight could have effected.

Birds as Agents in the Dispersal of Plants.—Birds are undoubtedly important agents in the dispersal of plants over wide spaces of ocean, either by swallowing fruits and rejecting the seeds in a state fit for germination, or by the seeds becoming attached to the plumage of ground-nesting birds, or to the feet of aquatic birds embedded in small quantities of mud or earth. Illustrations of these various modes of transport will be found in Chapter XII. when discussing the origin of the flora of the Azores and Bermuda.

Ocean Currents as Agents in Plant-dispersal.—Ocean currents are undoubtedly more important agents in conveying seeds of plants than they are in the case of any other organisms, and a considerable body of facts and experiments have been collected proving that seeds may sometimes be carried in this way many thousand miles and afterwards germinate. Mr. Darwin made a series of interesting experiments on this subject, some of which will be given in the chapter above referred to.

Dispersal along Mountain-chains.—These various modes of transport are, as will be shown when discussing special cases, amply sufficient to account for the vegetation found on oceanic islands, which almost always bears a close relation to that of the nearest continent; but there are other phenomena presented by the dispersal of species and genera of plants over very wide areas, especially when they occur in widely separated portions of the Northern and Southern hemispheres, that are not easily explained by such causes alone. It is here that transmission along mountain-chains has probably been effective; and the exact mode in which this has occurred is discussed in Chapter XXIII., where a considerable body of facts is given showing that extensive migrations may be effected by a succession of moderate steps, owing to the frequent exposure of fresh surfaces of soil or *débris* on mountain sides and summits, offering stations on which foreign plants can temporarily establish themselves.

Antiquity of Plants as Affecting their Distribution.—We have already referred to the importance of great antiquity in enabling us to account for the wide dispersal of some genera and species of insects and land shells, and recent discoveries in fossil botany show that this cause has also had great influence in the case of plants. Rich floras have been discovered in the Miocene, the Eocene, and the Upper Cretaceous formation, and these consist almost wholly of living genera, and many of them of species very closely allied to existing forms. We have therefore every reason to believe that a large number of our plant species have survived great geological, geographical, and climatal changes; and this fact, combined with the varied and wonderful powers of dispersal many of them possess, renders it far less difficult to understand the examples of wide distribution of the genera and species of plants than in the case of similar instances among animals.   This subject will be further alluded to when discussing the origin of the New Zealand flora in Chapter XXII.

CHAPTER VI.

GEOGRAPHICAL AND GEOLOGICAL CHANGES: THE PERMANENCE OF CONTINENTS.

Changes of Land and Sea, their Nature and Extent.—Shore-deposits and Stratified Rocks.—The Movements of Continents.—Supposed Oceanic Formations; the Origin of Chalk.—Fresh-water and Shore Deposits as Proving the Permanence of Continents.—Oceanic Islands as Indications of the Permanence of Continents and Oceans.—General Stability of Continents with Constant Change of Form.—Effect of Continental Changes on the Distribution of Animals.—Changed Distribution Proved by the Extinct Animals of Different Epochs.—Summary of Evidence for the General Permanence of Continents and Oceans.

THE changes of land and sea which have occurred in particular cases will be described when we discuss the origin and relations of the faunas of the different classes of islands. We have here only to consider the general character and extent of such changes, and to correct some erroneous ideas which are prevalent on the subject.

Changes of Land and Sea, their Nature and Extent.—It is a very common belief that geological evidence proves a complete change of land and sea to have taken place over and over again. Every foot of dry land has undoubtedly, at one time or other, formed part of a sea-bottom; and we can hardly exclude the surfaces occupied by volcanic and fresh-water deposits, since in many cases, if not in all, these rest upon a substratum of marine formations. At first sight, therefore, it seems a necessary inference that when the present continents were under water there must have been other continents situated where we now find the oceans, from which the sediments came to form the various deposits we now see. This view was held by so acute and learned a geologist as Sir Charles Lyell, who says, "Continents, therefore, although permanent for whole geological epochs, shift their positions entirely in the course of ages."[1] Mr. T. Mellard

---

[1] "Principles of Geology," 11th ed., Vol. I., p. 258.

Reade, late President of the Geological Society of Liverpool, so recently as 1878, says, "While believing that the ocean depths are of enormous age, it is impossible to resist other evidences that they have once been land. The very continuity of animal and vegetable life on the globe points to it. The molluscous fauna of the eastern coast of North America is very similar to that of Europe, and this could not have happened without littoral continuity; yet there are depths of 1500 fathoms between these continents."[1] It is certainly strange that a geologist should not remember the recent and long-continued warm climates of the arctic regions, and see that a connection of Northern Europe by Iceland with Greenland and Labrador over a sea far less than a thousand fathoms deep would furnish the "littoral continuity" required. Again, in the same pamphlet Mr. Reade says, "It can be mathematically demonstrated that the whole, or nearly the whole, of the sea-bottom has been at one time or other dry land. If it were not so, and the oscillations of the level of the land with respect to the sea were confined within limits near the present continents, the results would have been a gradual diminution instead of development of the calcareous rocks. To state the case in common language, the calcareous portion of the rocks would have been washed out during the mutations, the destruction and re-deposit of the continental rocks, and eventually deposited in the depths of the immutable sea far from land. Immense beds of limestone would now exist at the bottom of the ocean, while the land would be composed of sandstones and argillaceous shales. The evidence of chemistry thus confirms the inductions drawn from the distribution of animal life upon the globe."

So far from this being a "mathematical demonstration," it appears to me to be a complete misinterpretation of the facts. Animals did not create the lime which they secrete from the sea-water, and therefore we have every reason to believe that the inorganic sources which originally supplied it still keep up that supply, though perhaps in diminished quantity. Again, the great lime-secreters—corals—work in water of moderate depth

---

[1] "On Limestone as an Index of Geological Time."

(that is, near land), while there is no proof whatever that there is any considerable accumulation of limestone at the bottom of the deep ocean. On the contrary, the fact ascertained by the *Chal-lenger*, that beyond a certain depth the "calcareous" ooze ceases, and is replaced by red and gray clays, although the calcareous organisms still abound in the surface waters of the ocean, shows that the lime is dissolved again by the excess of carbonic acid usually found at great depths, and its accumulation thus prevented. As to the increase of limestones in recent as compared with older formations, it may be readily explained by two considerations: in the first place, the growth and development of the land in longer and more complex shore-lines, and the increase of sedimentary over volcanic formations, may have offered more stations favorable to the growth of coral, while the solubility of limestone in rain-water renders the destruction of such rocks more rapid than that of sandstones and shales, and would thus lead to their comparative abundance in later as compared with earlier formations.

However weak we may consider the above-quoted arguments against the permanence of oceans, the fact that these arguments are so confidently and authoritatively put forward renders it advisable to show how many and what weighty considerations can be adduced to justify the opposite belief, which is now rapidly gaining ground among students of earth-history.

Shore - deposits and Stratified Rocks. — If we go round the shores of any of our continents, we shall always find a considerable belt of shallow water, meaning thereby water from a hundred to a hundred and fifty fathoms deep. The distance from the coast-line at which such depths are reached is seldom less than twenty miles, and is very frequently more than a hundred, while in some cases such shallow seas extend several hundred miles from existing continents. The great depth of a thousand fathoms is often reached at thirty miles from shore, but more frequently at about sixty or a hundred miles. Round the entire African coast, for example, this depth is reached at distances varying from forty to a hundred and fifty miles (except in the Red Sea and the Strait of Mozambique), the average being about eighty miles.

Now the numerous specimens of sea-bottoms collected during the voyage of the *Challenger* show that true shore-deposits—that is, materials denuded from the land and carried down as sediment by rivers—are almost always confined within a distance of fifty or a hundred miles of the coast, the finest mud only being sometimes carried a hundred and fifty or, rarely, two hundred miles. As the sediment varies in coarseness and density, it is evident that it will sink to the bottom at unequal distances, the bulk of it sinking comparatively near shore, while only the very finest and almost impalpable mud will be carried out to the farthest limits. Beyond these limits the only deposits (with few exceptions) are organic, consisting of the shells of minute calcareous or siliceous organisms with some decomposed pumice and volcanic dust which floats out to mid-ocean. It follows, therefore, that by far the larger part of all stratified deposits, especially those which consist of sand or pebbles or any visible fragments of rock, must have been formed within fifty or a hundred miles of then existing continents; or if at a greater distance, in shallow inland seas receiving deposits from more sides than one, or in certain exceptional areas where deep ocean currents carry the débris of land to greater distances.[1]

If we now examine the stratified rocks found in the very centre of all our great continents, we find them to consist of sandstones, limestones, conglomerates, or shales, which must, as we have seen, have been deposited within a comparatively short distance of a sea-shore. Professor Archibald Geikie says, "Among the thickest masses of sedimentary rock—those of the ancient Palæozoic systems—no features recur more continually than the alternations of different sediments, and the recurrence of surfaces covered with well-preserved ripple-marks, trails and

---

[1] In his "Preliminary Report on Oceanic Deposit," Mr. Murray says, "It has been found that the deposits taking place near continents and islands received their chief characteristics from the presence of the débris of adjacent lands. In some cases these deposits extend to a distance of over a hundred and fifty miles from the coast."—*Proceedings of the Royal Society*, Vol. XXIV., p. 519.

"The materials in suspension appear to be almost entirely deposited within two hundred miles of the land."—*Proceedings of the Royal Society of Edinburgh*, 1876–77, p. 253.

burrows of annelides, polygonal and irregular desiccation-marks, like the cracks at the bottom of a sun-dried muddy pool. These phenomena unequivocally point to shallow and even littoral waters. They occur from bottom to top of formations which reach a thickness of several thousand feet. They can be interpreted only in one way—viz., that the formations in question began to be laid down in shallow water; that during their formation the area of deposit gradually subsided for thousands of feet; yet that the rate of accumulation of sediment kept pace, on the whole, with this depression; and hence that the original shallow-water character of the deposits remained, even after the original sea-bottom had been buried under a vast mass of sedimentary matter." He goes on to say that this general statement applies to the more recent as well as to the more ancient formations, and concludes, "In short, the more attentively the stratified rocks of the earth are studied, the more striking becomes the absence of any formations among them which can legitimately be considered those of a deep sea. They have all been deposited in comparatively shallow water." [1]

The arrangement and succession of the stratified rocks also indicate the mode and place of their formation. We find them stretching across the country in one general direction, in belts of no great width, though often of immense length, just as we should expect in shore-deposits; and they often thin out and change from coarse to fine in a definite manner, indicating the position of the adjacent land from the débris of which they were originally formed. Again quoting Professor Geikie, "The materials carried down to the sea would arrange themselves then as they do still, the coarser portions nearest the shore, the finer silt and mud farthest from it. From the earliest geological times the great area of deposit has been, as it still is, the marginal belt of sea-floor skirting the land. It is there that nature has always strewn the dust of continents to be."

The Movements of Continents.—As we find these stratified rocks of different periods spread over almost the whole surface of existing continents where not occupied by igneous or meta-

---

[1] "Geographical Evolution," *Proceedings of the Royal Geographical Society*, 1879, p. 426.

morphic rocks, it follows that at one period or another each part of the continent has been under the sea, but at the same time not far from the shore. Geologists now recognize two kinds of movements by which the deposits so formed have been elevated into dry land—in the one case the strata remain almost level and undisturbed, in the other they are contorted and crumpled, often to an enormous extent. The former often prevails in plains and plateaus, while the latter is almost always found in the great mountain-ranges. We are thus led to picture the land of the globe as a flexible area in a state of slow but incessant change; the changes consisting of low undulations which creep over the surface so as to elevate and depress limited portions in succession without perceptibly affecting their nearly horizontal position, and also of intense lateral compression, supposed to be produced by partial subsidence along certain lines of weakness in the earth's crust, the effect of which is to crumple the strata and force up certain areas in great contorted masses, which, when carved out by subaerial denudation into peaks and valleys, constitute our great mountain systems.[1] In this way every part of a continent may again and again have sunk beneath the sea, and yet as a whole may never have ceased to exist as a continent or a vast continental archipelago. And as subsidence will always be accompanied by deposition, piles of marine strata many thousand feet thick may have been formed in a sea which was never very deep, by means of a slow depression either continuous or intermittent, or through alternate subsidences and elevations, each of moderate amount.

Supposed Oceanic Formations; the Origin of Chalk.—There

---

[1] Professor Dana points out that the regions which, after long undergoing subsidence, and accumulating vast piles of sedimentary deposits, have been elevated into mountain-ranges, have thereby become stiff and unyielding, and that the next depression and subsequent upheaval will be situated on one or the other side of it; and he shows that in North America this is the case with all the mountains of the successive geological formations. Thus, depressions and elevations of extreme slowness, but often of vast amount, have occurred successively in restricted adjacent areas; and the effect has been to bring each portion in succession beneath the ocean, but always bordered on one or both sides by the remainder of the continent, from the denudation of which the deposits are formed which, on the subsequent upheaval, become mountain-ranges ("Manual of Geology," 2d ed., p. 751).

seems very good reason to believe that few, if any, of the rocks known to geologists correspond exactly to the deposits now forming at the bottom of our great oceans. The white oceanic mud, or Globigerina ooze, found in all the great oceans at depths varying from 250 to nearly 3000 fathoms, and almost constantly in depths under 2000 fathoms, has, however, been supposed to be an exception, and to correspond exactly to our white and gray chalk. Hence some naturalists have maintained that there has probably been one continuous formation of chalk in the Atlantic from the Cretaceous epoch to the present day. This view has been adopted chiefly on account of the similarity of the minute organisms found to compose a considerable portion of both deposits, more especially the pelagic foraminifera, of which several species of Globigerina appear to be identical in the chalk and the modern Atlantic mud. Other extremely minute organisms, whose nature is doubtful, called coccoliths and discoliths, are also found in both formations, while there is a considerable general resemblance between the higher forms of life. Sir Wyville Thomson tells us that "sponges are abundant in both, and the recent chalk-mud has yielded a large number of examples of the group *Porifera vitrea*, which find their nearest representatives among the Ventriculites of the white chalk. The echinoderm fauna of the deeper parts of the Atlantic basin is very characteristic, and yields an assemblage of forms which represent in a remarkable degree the corresponding group in the white chalk. Species of the genus Cidaris are numerous; some remarkable flexible forms of the Diademidæ seem to approach Echinothuria." [1] Now, as some explanation of the origin of chalk had long been desired by geologists, it is not surprising that the amount of resemblance shown to exist between it and some kinds of oceanic mud should have been at once seized upon, and the conclusion arrived at that chalk is a deep-sea oceanic formation exactly analogous to that which has been shown to cover large areas of the Atlantic, Pacific, and Southern oceans.

But there are several objections to this view which seem fatal

---

[1] *Nature*, Vol. II., p. 297.

to its acceptance. In the first place, no specimens of Globigerina ooze from the deep ocean-bed yet examined agree even approximately with chalk in chemical composition, only containing from 44 to 79 per cent. of carbonate of lime, with from 5 to 11 per cent. of silica, and from 8 to 33 per cent. of alumina and oxide of iron.[1] Chalk, on the other hand, contains usually from 94 to 99 per cent. of carbonate of lime, and a very minute quantity of alumina and silica. This large proportion of carbonate of lime implies some other source of this mineral, and it is probably to be found in the excessively fine mud produced by the decomposition and denudation of coral reefs. Mr. Dana, the geologist of the United States Exploring Expedition, found in the elevated coral reef of Oahu, one of the Sandwich Islands, a deposit closely resembling chalk in color, texture, etc. ; while in several growing reefs a similar formation of modern chalk, undistinguishable from the ancient, was observed.[2] Sir Charles

---

[1] Sir W. Thomson, " Voyage of the Challenger," Vol. II., p. 374.

[2] The following is the analysis of the chalk at Oahu :

Carbonate of lime	92.800 per cent.	
Carbonate of magnesia	2.385	"
Alumina	0.250	"
Oxide of iron	0.543	"
Silica	0.750	"
Phosphoric acid and fluorine	2.113	"
Water and loss	1.148	"

This chalk consists simply of comminuted corals and shells of the reef. It has been examined microscopically and found to be destitute of the minute organisms abounding in the chalk of England (" Geology of the United States Exploring Expedition," p. 150).

This absence of Globigerinæ is a local phenomenon. They are quite absent in the Arafura Sea, and no Globigerina ooze was found in any of the enclosed seas of the Pacific ; but with these exceptions the Globigerinæ " are really found all over the bottom of the ocean " (Murray on " Oceanic Deposits," Proceedings of the Royal Society, Vol. XXIV., p. 523).

The above analysis shows a far closer resemblance to chalk than that of the Globigerina ooze of the Atlantic, four specimens of which, given by Sir W. Thomson ("Voyage of the Challenger," Vol. II., Appendix, pp. 374–376, Nos. 9, 10, 11, and 12), from the mid-Atlantic, show the following proportions :

Carbonate of lime	43.93	to 79.17 per cent.	
Carbonate of magnesia	1.40	to 2.58	"
Alumina and oxide of iron	6.00?	to 32.98	"
Silica	4.60	to 11.23	"

Lyell well remarks that the pure calcareous mud produced by the decomposition of the shelly coverings of mollusca and zoophytes would be much lighter than argillaceous or arenaceous mud, and being thus transported to greater distances would be completely separated from all impurities.

Now the Globigerinæ have been shown by the *Challenger* explorations to abound in all moderately warm seas; living both at the surface, at various depths in the water, and at the bottom. It was long thought that they were surface-dwellers only, and that their dead tests sank to the bottom, producing the Globigerina ooze in those areas where other deposits were absent or scanty. But the examination of the whole of the dredgings and surface-gatherings of the *Challenger* by Mr. H. B. Brady has led him to a different conclusion; for he finds numerous forms at the bottom quite distinct from those which inhabit the surface, while, when the same species live both at surface and bottom, the latter are always larger and have thicker and stronger cell-walls. This view is also supported by the fact that in many stations not far from our own shores Globigerinæ are abundant in bottom dredgings, but are never found on the surface in the tow-

---

In addition to the above, there is a quantity of insoluble residue consisting of small particles of sanidine, augite, hornblende, and magnetite, supposed to be the product of volcanic dust or ashes carried either in the air or by ocean currents. This volcanic matter amounts to from 4.60 to 8.33 per cent. of the Globigerina ooze of the mid-Atlantic, where it seems to be always present; and the small proportion of similar matter in true chalk is another proof that its origin is different, and that it was deposited far more rapidly than the oceanic ooze.

The following analysis of chalk by Mr. D. Forbes will show the difference between the two formations:

	Gray Chalk. *Folkestone.*	White Chalk. *Shoreham.*
Carbonate of lime	94.09	98.40
Carbonate of magnesia	0.31	0.08
Alumina and phosphoric acid	a trace	0.42
Chloride of sodium	1.29	——
Insoluble débris	3.61	1.10

(From *Quarterly Journal of the Geological Society*, Vol. XXVII.)

The large proportion of carbonate of lime, and the very small quantity of silica, alumina, and insoluble débris, at once distinguish true chalk from the Globigerina ooze of the deep ocean-bed.

ing-nets.[1] These organisms then exist almost universally where the waters are pure and are not too cold, and they would naturally abound most where the diffusion of carbonate of lime both in suspension and solution afforded them an abundant supply of material for their shelly coverings. Dr. Wallich believes that they flourish best where the warm waters of the Gulf Stream bring organic matter from which they derive nutriment, since they are wholly wanting in the course of the arctic current between Greenland and Labrador. Dr. Carpenter also assures us that they are rigorously limited to warm areas.

Now, with regard to the depth at which our chalk was formed, we have evidence of several distinct kinds to show that it was not profoundly oceanic. Mr. J. Murray, in the Report already referred to, says, "The Globigerina oozes which we get in shallow water resemble the chalk much more than those in deeper water, say over 1000 fathoms."[2] This is important and weighty evidence, and it is supported in a striking manner by the nature of the molluscan fauna of the chalk. Mr. Gwyn Jeffries, one of our greatest authorities on shells, who has himself dredged largely both in deep and shallow water, and who has no theory to support, has carefully examined this question. Taking the whole series of genera which are found in the Chalk formation, seventy-one in number, he declares that they are all comparatively shallow-water forms, many living at depths not exceeding forty to fifty fathoms, while some are confined to still shallower waters. Even more important is the fact that the genera especially characteristic of the deep Atlantic ooze—Leda, Verticordia, Neæra, and the Bulla family—are either very rare or entirely wanting in the ancient Cretaceous deposits.[3]

Let us now see how the various facts already adduced will enable us to explain the peculiar characteristics of the Chalk formation. Sir Charles Lyell tells us that "pure Chalk, of nearly uniform aspect and composition, is met with in a north-

[1] " Notes on Reticularian Rhizopoda," in *Microscopical Journal*, Vol. XIX., New Series, p. 84.

[2] *Proceedings of the Royal Society*, Vol. XVIV., p. 532.

[3] See Presidential Address in Sect. D of British Association at Plymouth, 1877.

west and southeast direction, from the north of Ireland to the Crimea, a distance of about 1140 geographical miles; and in an opposite direction it extends from the south of Sweden to the south of Bordeaux, a distance of about 840 geographical miles." This marks the extreme limits within which true chalk is found, though it is by no means continuous. It probably implies, however, the existence across Central Europe of a sea somewhat larger than the Mediterranean. It may have been much larger, because this pure chalk formation would only be formed at a considerable distance from land, or in areas where there was no other shore-deposit. This sea was probably bounded on the north by the old Scandinavian highlands, extending to Northern Germany and Northwestern Russia, where Palæozoic and ancient Secondary rocks have a wide extension, though now partially concealed by late Tertiary deposits; while on the south it appears to have been limited by land extending through Austria, South Germany, and the South of France, as shown in the map of Central Europe during the Cretaceous period in Professor Heer's "Primeval World of Switzerland," p. 175. To the north the sea may have had an outlet to the Arctic Ocean between the Ural range and Finland. South of the Alps there was probably another sea, which may have communicated with the northern one just described, and there was also a narrow strait across Switzerland, north of the Alps, but, as might be expected, in this only marls, clays, sandstones, and limestones were deposited instead of true chalk. It is also a suggestive fact that both above and below the true chalk, in almost all the countries where it occurs, are extensive deposits of marls, clays, and even pure sands and sandstones, characterized by the same general types of fossil remains as the chalk itself. These beds imply the vicinity of land, and this is even more clearly proved by the occurrence, both in the Upper and Lower Cretaceous, of deposits containing the remains of land plants in abundance, indicating a rich and varied flora.

Now all these facts are totally opposed to the idea of anything like oceanic conditions having prevailed in Europe during the Cretaceous period; but it is quite consistent with the existence of a great Mediterranean sea of considerable depth in its

central portions, and occupying, either at one or successive periods, the whole area of the Cretaceous formation. We may also
note that the Maestricht beds in Belgium and the Faxöe chalk
in Denmark are both highly coralline, the latter being, in fact,
as completely composed of corals as a modern coral reef; so that
we have here a clear indication of the source whence the white
calcareous mud was derived which forms the basis of chalk. If
we suppose that during this period the comparatively shallow
sea-bottom between Scandinavia and Greenland was elevated,
forming a land connection between these countries, the result
would be that a large portion of the Gulf Stream would be diverted into the inland European sea, and would bring with it
that abundance of Globigerinæ and other foraminifera which
form such an important constituent of chalk. This sea was
probably bordered with islands and coral reefs; and if no very
large rivers flowed into it, we should have all the conditions for
the production of the true chalk, as well as the other members
of the Cretaceous formation. The products of the denudation
of its shores and islands would form the various sandstones,
marls, and clays which would be deposited almost wholly within
a few miles of its coasts; while the great central sea, perhaps at
no time more than a few thousand feet deep, would receive only
the impalpable mud of the coral reefs and the constantly falling
tests of foraminifera. These would embed and preserve for us
the numerous echinoderms, sponges, and mollusca which lived
upon the bottom, the fishes and turtles which swam in its
waters, and sometimes the winged reptiles that flew overhead.
The abundance of ammonites and other cephalopods in the
Chalk is another indication that the water in which they lived
was not very deep, since Dr. S. P. Woodward thinks that these
organisms were limited to a depth of about thirty fathoms.

The best example of the modern formation of chalk is perhaps to be found on the coasts of subtropical North America,
as described in the following passage:

"The observations of Pourtales show that the steep banks of
Bahama are covered with soft white lime-mud. The lime-bot-
tom, which consists almost entirely of Polythalamia, covers in
greater depths the entire channel of Florida. This formation

extends without interruption over the whole bed of the Gulf Stream in the Gulf of Mexico, and is continued along the Atlantic coast of America. The commonest genera met with in this deposit are Globigerina, Rotalia cultrata, in large numbers, several Textilariæ, Marginulinæ, etc. Besides these, small free corals, Alcyonidæ, Ophiuræ, Mollusca, Crustacea, small fishes, etc., are found living in these depths. The whole sea-bottom appears to be covered with a vast deposit of white chalk still in formation." [1]

There is yet another consideration which seems to have been altogether overlooked by those who suppose that a deep and open island-studded ocean occupied the place of Europe in Cretaceous times. No fact is more certain than the considerable break, indicative of a great lapse of time, intervening between the Cretaceous and Tertiary formations. A few deposits of intermediate age have indeed been found, but these have been generally allocated either with the Chalk or the Eocene, leaving the gap almost as pronounced as before. Now, what does this gap mean? It implies that when the deposition of the various Cretaceous beds of Europe came to an end, they were raised above the sea-level and subject to extensive denudation, and that for a long but unknown period no extensive portion of what is now European land was below the sea-level. It was only when this period terminated that large areas in several parts of Europe became submerged and received the earliest Tertiary deposits known as Eocene. If, therefore, Europe at the close of the Cretaceous period was generally identical with what it is now, and perhaps even more extensive, it is absurd to suppose that it was all, or nearly all, under water during that period; or, in fact, that any part of it was submerged except those areas on which we actually find Cretaceous deposits, or where we have good reason to believe they have existed.

The several considerations now adduced are, I think, sufficient to show that the view put forth by some naturalists (and which has met with a somewhat hasty acceptance by geologists) that our white chalk is an oceanic formation strictly comparable

---

[1] *Geological Magazine*, 1871, p. 426.

with that now forming at depths of a thousand fathoms and upwards in the centre of the Atlantic, gives a totally erroneous
idea of the actual condition of Europe during that period. Instead of being a wide ocean, with a few scattered islands, comparable to some parts of the Pacific, it formed as truly a portion
of the great northern continent as it does now, although the inland seas of that epoch may have been more extensive and more
numerous than they are at the present day.[1]

*Fresh-water and Shore Deposits as Proving the Permanence
of Continents.*—The view here maintained, that all known marine deposits have been formed near the coasts of continents and
islands, and that our actual continents have been in continuous
existence under variously modified forms during the whole
period of known geological history, is further supported by another and totally distinct series of facts. In almost every period
of geology, and in all the continents which have been well examined, there are found lacustrine, estuarine, or shore deposits,
containing the remains of land animals or plants, thus demonstrating the continuous existence of extensive land areas on or
adjoining the sites of our present continents. Beginning with
the Miocene, or Middle Tertiary, period, we have such deposits
with remains of land animals or plants in Devonshire and Scotland, in France, Switzerland, Germany, Croatia, Vienna, Greece,
North India, Central India, Burmah, North America (both east
and west of the Rocky Mountains), Greenland, and other parts
of the arctic regions. In the older Eocene period similar for-

---

[1] In his lecture on " Geographical Evolution " (which was published after the greater
part of this chapter had been written), Professor Geikie expresses views in complete
accordance with those here advocated. He says, " The next long era, the Cretaceous, was more remarkable for slow accumulation of rock under the sea than for the
formation of new land. During that time the Atlantic sent its waters across the
whole of Europe and into Asia. But they were probably nowhere more than a few
hundred feet deep over the site of our continent, even at their deepest part. Upon
their bottom there gathered a vast mass of calcareous mud, composed in great part
of foraminifera, corals, echinoderms, and mollusks. Our English chalk, which ranges
across the North of France, Belgium, Denmark, and the North of Germany, represents a portion of the deposits of that sea-floor." The weighty authority of the Di-
rector of the Geological Survey of Scotland may perhaps cause some geologists to
modify their views as to the deep-sea origin of chalk, who would have treated any
arguments advanced by myself as not worthy of consideration.

mations are widely spread in the South of England, in France, and to an enormous extent on the central plateau of North America; while in the Eastern States, from Maryland to Alabama, there are extensive marine deposits of the same age, which, from the abundance of fossil remains of a large cetacean (Zeuglodon), must have been formed in shallow gulfs or estuaries where these huge animals were stranded. Going back to the Cretaceous formation, we have the same indications of persisting lands in the rich plant-beds of Aix-la-Chapelle and a few other localities on the Continent, as well as in coniferous fruits from the Gault of Folkestone; while in North America Cretaceous plant-beds occur in New Jersey, Alabama, Kansas, the sources of the Missouri, the Rocky Mountains from New Mexico to the Arctic Ocean, Alaska, British Columbia, California, and in Greenland and Spitzbergen; while birds and land reptiles are found in the Cretaceous deposits of Colorado and other Western districts. Fresh-water deposits of this age are also found on the coast of Brazil. In the lower part of this formation we have the fresh-water Wealden deposits of England, extending into France, Hanover, and Westphalia. In the older Oolite or Jurassic formation we have abundant proofs of continental conditions in the fresh-water and " dirt "-beds of the Purbecks, in the south of England, with plants, insects, and mammals; the Bavarian lithographic stone, with fossil birds and insects; the earlier " forest marble " of Wiltshire, with ripple-marks, wood, and broken shells, indicative of an extensive beach; the Stonesfield slate, with plants, insects, and marsupials; and the Oolitic coal of Yorkshire and Sutherlandshire. Beds of the same age occur in the Rocky Mountains of North America, containing abundance of Dinosaurians and other reptiles, among which is the Atlantosaurus, the largest land-animal ever known to have existed. Professor O. C. Marsh describes it as having been between fifty and sixty feet long, and when standing erect at least thirty feet high![1] Such monsters could hardly have been developed except in an extensive land area. A small

---

[1] " Introduction and Succession of Vertebrate Life in America," by Professor O. C. Marsh. Reprinted from the *Popular Science Monthly*, March, April, 1878.

mammal, Dryolestes, has been discovered in the same deposits.
A rich Jurassic flora has also been found in East Siberia and
the Amoor valley. The older Triassic deposits are very extensively developed in America, and both in the Connecticut valley and the Rocky Mountains show tracks or remains of land
reptiles, amphibians, and mammalia; while coal-fields of the same
age in Virginia and Carolina produce abundance of plants.
Here, too, is found the ancient mammal Microlestes, of Würl-
temberg, with the ferns, conifers, and Labyrinthodonts of the
Bunter Sandstone in Germany; while the beds of rock-salt in
this formation, both in England and in many parts of the Continent, could only have been formed in inland seas or lakes, and
thus equally demonstrate continental conditions.

We now pass into the oldest or Palæozoic formations, but
find no diminution in the proofs of continental conditions. The
Permian formation has a rich flora often producing coal in England, France, Saxony, Thuringia, Silesia, and Eastern Russia.
Coal-fields of the same age occur in Ohio, in North America. In
the still more ancient Carboniferous formation we find the most
remarkable proofs of the existence of our present continents at
that remote epoch, in the wonderful extension of coal-beds in
all the known continents. We find them in Ireland, England,
and Scotland; in France, Spain, Belgium, Saxony, Prussia, Bo-
hemia, Hungary, Sweden, Spitzbergen, Siberia, Russia, Greece,
Turkey, and Persia; in many parts of continental India; extensively in China; and in Australia, Tasmania, and New Zealand.
In North America there are immense coal-fields in Nova Scotia
and New Brunswick, from Pennsylvania southward to Alabama,
in Indiana and Illinois, and in Missouri; and there is also a true
coal formation in South Brazil. This wonderfully wide distribution of coal, implying as it does a rich vegetation and extensive land areas, carries back the proof of the persistence and
general identity of our continents to a period so remote that
none of the higher animal types had probably been developed.
But we can go even further back than this, to the preceding
Devonian formation, which was almost certainly an inland deposit often containing remains of fresh-water shells, plants, and
even insects; while Professor Ramsay believes that he has

found "sun-cracks and rain-pittings" in the Longmynd beds of the still earlier Cambrian formation.[1] If now, in addition to the body of evidence here adduced, we take into consideration the fresh-water deposits that still remain to be discovered, and those extensive areas where they have been destroyed by denudation, or remain deeply covered up by later marine or volcanic formations, we cannot but be struck by the abounding proofs of the permanence of the great features of land and sea as they now exist; and we shall see how utterly gratuitous, and how entirely opposed to all the evidence at our command, are the hypothetical continents bridging over the deep oceans, by the help of which it is so often attempted to cut the Gordian knot presented by some anomalous fact in geographical distribution.

Oceanic Islands as Indications of the Permanence of Continents and Oceans.—Coming to the question from the other side, Mr. Darwin has adduced an argument of considerable weight in favor of the permanence of the great oceans. He says (" Origin of Species," 6th ed., p. 288), " Looking to existing oceans, which are thrice as extensive as the land, we see them studded with many islands; but hardly one truly oceanic island (with the exception of New Zealand, if this can be called a truly oceanic island) is as yet known to afford even a fragment of any Palæozoic or Secondary formation. Hence we may perhaps infer that during the Palæozoic and Secondary periods neither continents nor continental islands existed where our oceans now extend; for had they existed, Palæozoic and Secondary formations would in all probability have been accumulated from sediment derived from their wear and tear; and these would have been at least partially upheaved by the oscillations of level which must have intervened during these enormously long periods. If, then, we may infer anything from these facts, we may infer that where our oceans now extend, oceans have extended from the remotest period of which we have any record; and, on the other hand, that where continents now exist large tracts of land have existed, subjected, no doubt, to great oscillations of level, since the Cambrian period." This argument standing by itself has

---

[1] "Physical Geography and Geology of Great Britain," 5th ed., p. 61.

not received the attention it deserves, but coming in support of the long series of facts of an altogether distinct nature, going to show the permanence of continents, the cumulative effect of the whole must, I think, be admitted to be irresistible.[1]

General Stability of Continents with Constant Change of Form. —It will be observed that the very same evidence which has been adduced to prove the *general* stability and permanence of our continental areas also goes to prove that they have been subjected to wonderful and repeated changes in *detail*. Every square mile of their surface has been again and again under water, sometimes a few hundred feet deep, sometimes perhaps several thousands. Lakes and inland seas have been formed, have been filled up with sediment, and been subsequently raised into hills or even mountains. Arms of the sea have existed crossing the continents in various directions, and thus completely isolating the divided portions for varying intervals. Seas have been changed into deserts, and deserts into seas. Volcanoes have grown into mountains, have been degraded and sunk beneath the ocean, have been covered with sedimentary deposits,

---

[1] Of late it has been the custom to quote the so-called "ridge" down the centre of the Atlantic as indicating an extensive ancient land. Even Professor Judd adopts this view, for he speaks of the great belt of Tertiary volcanoes "which extended through Greenland, Iceland, the Faroe Islands, the Hebrides, Ireland, Central France, the Iberian Peninsula, the Azores, Madeira, Canaries, Cape de Verd Islands, Ascension, St. Helena, and Tristan d'Acunha, and which constituted, as shown by the recent soundings of H. M. S. *Challenger*, a mountain-range comparable in its extent, elevation, and volcanic character with the Andes of South America" (*Geological Mag.*, 1874, p. 71). On examining the diagram of the Atlantic Ocean in the *Challenger* Reports, No. 7. a considerable part of this ridge is found to be more than 1900 fathoms deep, while the portion called the "Connecting Ridge" seems to be due in part to the deposits carried out by the river Amazon. In the neighborhood of the Azores, St. Paul's Rocks, Ascension, and Tristan d'Acunha are considerable areas varying from 1200 to 1500 fathoms deep, while the rest of the ridge is usually 1800 or 1900 fathoms. The shallower water is no doubt due to volcanic upheaval and the accumulation of volcanic ejections, and there may be many other deeply submerged old volcanoes on the ridge; but that it ever formed a chain of mountains "comparable in elevation with the Andes" there seems not a particle of evidence to prove. It is, however, probable that this ridge indicates the former existence of some considerable Atlantic islands, which will serve to explain the presence of a few identical genera, and even species, of plants and insects in Africa and South America, while the main body of the fauna and flora of these two continents remains radically distinct.

and again raised up into mountain-ranges; while other mountains have been formed by the upraised coral reefs of inland seas. The mountains of one period have disappeared by denudation or subsidence, while the mountains of the succeeding period have been rising from beneath the waves. The valleys, the ravines, and the mountain-peaks have been carved out and filled up again; and all the vegetable forms which clothe the earth and furnish food for the various classes of animals have been completely changed again and again.

Effect of Continental Changes on the Distribution of Animals. —It is impossible to exaggerate, or even adequately to conceive, the effect of these endless mutations on the animal world. Slowly but surely the whole population of living things must have been driven backward and forward from east to west, or from north to south, from one side of a continent or a hemisphere to the other. Owing to the remarkable continuity of all the land masses, animals and plants must have often been compelled to migrate into other continents, where in the struggle for existence under new conditions many would succumb; while such as were able to survive would constitute those wide-spread groups whose distribution often puzzles us. Owing to the repeated isolation of portions of continents for long periods, special forms of life would have time to be developed, which, when again brought into competition with the fauna from which they had been separated, would cause fresh struggles of ever-increasing complexity, and thus lead to the development and preservation of every weapon, every habit, and every instinct, which could in any way conduce to the safety and preservation of the several species.

Changed Distribution Proved by the Extinct Animals of Different Epochs.—We thus find that while the inorganic world has been in a state of continual though very gradual change, the species of the organic world have also been slowly changing in form and in the localities they inhabit; and the records of these changes and these migrations are everywhere to be found, in the actual distribution of the species no less than in the fossil remains which are preserved in the rocks. Everywhere the animals which have most recently become extinct resemble more or less closely those which now live in the same country; and where there are exceptions to the rule we can generally trace them to some changed conditions which have led to the extinction of certain types. But when we go a little further back, to the late or middle Tertiary deposits, we almost always find, along with forms which might have been the ancestors of some now living, others which are now found only in remote regions, and often in distinct continents—clear indications of those extensive migrations which have ever been going on. Every large island contains in its animal inhabitants a record of the period when it was last separated from the adjacent continent; while some portions of existing continents still show by the comparative poverty and specialty of their animals that at no distant epoch they were cut off by arms of the sea, and formed islands. If the geological record were more perfect, or even if we had as good a knowledge of that record in all parts of the world as we have in Europe and North America, we could arrive at much more accurate results than we are able to do with our present very imperfect knowledge of extinct forms of life; but even with our present scanty information we are able to throw much light upon the past history of our globe and its inhabitants, and can sketch out with confidence many of the changes they must have undergone.

Summary of Evidence for the General Permanence of Continents and Oceans.—As this question of the permanence of our continents lies at the root of all our inquiries into the past changes of the earth and its inhabitants, and as it is at present completely ignored by many writers, and even by naturalists of eminence, it will be well to summarize the various kinds of evidence which go to establish it.[1] We know as a fact that all sedi-

---

[1] In a review of Mr. Reade's "Chemical Denudation and Geological Time," in *Nature* (October 2, 1879), the writer remarks as follows: "One of the funny notions of some scientific thinkers meets with no favor from Mr. Reade, whose geological knowledge is practical as well as theoretical. They consider that because the older rocks contain nothing like the present red clays, etc., of the ocean-floor, the oceans have always been in their present positions. Mr. Reade points out that the first proposition is not yet proved; and the distribution of animals and plants, and the fact that the bulk of the strata on land are of marine origin, are opposed to the hy-

mentary deposits have been formed under water, but we also know that they were largely formed in lakes or inland seas, or near the coasts of continents or great islands, and that deposits uniform in character and more than a hundred and fifty or two hundred miles wide were rarely, if ever, formed at the same time. The farther we go from the land, the less rapidly deposition takes place; hence the great bulk of all the strata must have been formed near land. Some deposits are, it is true, continually forming in the midst of the great oceans; but these are chiefly organic, and increase very slowly, and there is no proof that any part of the series of known geological formations exactly resembles them. Chalk, which is still believed to be such a deposit by many naturalists, has been shown, by its contained fossils, to be a comparatively shallow-water formation—that is, one formed at a depth measured by hundreds rather than by thousands of fathoms. The nature of the formations composing all our continents also proves the continuity of those continents. Everywhere we find clearly marked shore and estuarine deposits, showing that every part of the existing land has in turn been on the sea-shore; and we also find, in all periods, lacustrine formations of considerable extent with remains of plants and land animals, proving the existence of continents or extensive lands in which such lakes or estuaries could be formed. These lacustrine deposits can be traced back through every period, from the newer Tertiary to the Devonian and Cambrian, and in every continent which has been geologically explored; and thus complete the proof that our continents have been in existence under ever-changing forms throughout the whole of that enormous lapse of time.

---

pothesis." We must leave it to our readers to decide whether the "notion" developed in this chapter is "funny," or whether such hasty and superficial arguments as those here quoted from a "practical geologist" have any value as against the different classes of facts, all pointing to an opposite conclusion, which have now been briefly laid before them, supported as they are by the expressed opinion of so weighty an authority as Professor Archibald Geikie, who, in the lecture already quoted, says, "From all this evidence, we may legitimately conclude that the present land of the globe, though formed in great measure of marine formations, has never lain under the deep sea; but that its site must always have been near land. Even its thick marine limestones are the deposits of comparatively shallow water."

On the side of the oceans we have also a great weight of evidence in favor of their permanence and stability. In addition to their enormous depths and great extent, and the circumstance that the deposits now forming in them are distinct from anything found upon the land surface, we have the extraordinary fact that the countless islands scattered over their whole area (with one or two exceptions only) never contain any Palæozoic or Secondary rocks—that is, have not preserved any fragments of the supposed ancient continents, nor of the deposits which must have resulted from their denudation during the whole period of their existence! The exceptions are New Zealand and the Seychelles Islands, both situated near to continents, leaving almost the whole of the vast areas of the Atlantic, Pacific, Indian, and Southern oceans without a solitary relic of the great islands or continents supposed to have sunk beneath their waves.

CHAPTER VII.

CHANGES OF CLIMATE WHICH HAVE INFLUENCED THE DIS-
PERSAL OF ORGANISMS: THE GLACIAL EPOCH.

Proofs of the Recent Occurrence of a Glacial Epoch.—Moraines.—Travelled Blocks.
—Glacial Deposits of Scotland: the " Till."—Inferences from the Glacial Phenomena of Scotland.—Glacial Phenomena of North America.—Effects of the Glacial Epoch on Animal Life.—Warm and Cold Periods.—Palæontological Evidence
of Alternate Cold and Warm Periods.—Evidence of Interglacial Warm Periods
on the Continent and in North America.—Migrations and Extinctions of Organisms Caused by the Glacial Epoch.

We have now to consider another set of physical revolutions
which have profoundly affected the whole organic world. Besides the wonderful geological changes to which, as we have seen,
all continents have been exposed, and which must, with extreme
slowness, have brought about the greater features of the dispersal
of animals and plants throughout the world, there have been also a
long succession of climatal changes, which, though very slow and
gradual when measured by centuries, may have sometimes been
rapid as compared with the slow march of geological mutations.
These climatal changes may be divided into two classes, which
have been thought to be the opposite phases of the same great
phenomenon—cold or even glacial epochs in the temperate zones,
on the one hand ; and mild or even warm periods extending into
the arctic regions, on the other. The evidence for both these
changes having occurred is conclusive ; and as they must be
taken account of whenever we endeavor to explain the past migrations and actual distribution of the animal world, a brief outline of the more important facts and of the conclusions they lead
to must be here given.

Proofs of the Recent Occurrence of a Glacial Epoch.—The
phenomena that prove the recent occurrence of glacial epochs
in the temperate regions are exceedingly varied, and extend over

very wide areas. It will be well, therefore, to state, first, what those facts are as exhibited in our own country, referring afterwards to similar phenomena in other parts of the world.

Perhaps the most striking of all the evidences of glaciation are the grooved, scratched, or striated rocks. These occur abundantly in Scotland, Cumberland, and North Wales, and no rational explanation of them has ever been given except that they were formed by glaciers. In many valleys—as, for instance, that of Llanberris, in North Wales—hundreds of examples may be seen, consisting of deep grooves several inches wide, smaller furrows, and striæ of extreme fineness wherever the rock is of sufficiently close and hard texture to receive such marks. These grooves or scratches are often many yards long; they are found in the bed of the valley as well as high up on its sides, and they are almost all without exception in one general direction—that of the valley itself, even though the particular surface they are upon slopes in another direction. When the native covering of turf is cleared away from the rock, the grooves and striæ are often found in great perfection, and there is reason to believe that such markings cover, or have once covered, a large part of the surface. Accompanying these markings we find another hardly less curious phenomenon, the rounding-off or planing-down of the hardest rocks to a smooth undulating surface. Hard crystalline schists with their strata nearly vertical, and which one would expect to find exposing jagged edges, are found ground off to a perfectly smooth but never to a flat surface. These rounded surfaces are found not only on single rocks, but over whole valleys and mountain-sides, and form what are termed *roches moutonnées*, from their often having the appearance at a distance of sheep lying down.

Now these two phenomena are actually produced by existing glaciers, while there is no other known or even conceivable cause that could have produced them. Whenever the Swiss glaciers retreat a little, as they sometimes do, the rocks in the bed of the valley they have passed over are found to be rounded, grooved, and striated just as are those of Wales and Scotland. The two sets of phenomena are so exactly identical that no one who has ever compared them can doubt that they are due to the same causes. But we have further and even more convincing evidence. Glaciers produce many other effects besides these two; and whatever effects they produce in Switzerland, in Norway, or in Greenland, we find examples of similar effects having been produced in our own country. The most striking of these are moraines and travelled blocks.

A GLACIER WITH MORAINES.

Moraines.—Almost every existing glacier carries down with it great masses of rock, stones, and earth, which fall on its surface from the precipices and mountain-slopes which hem it in, or the rocky peaks which rise above it. As the glacier slowly moves downward, this débris forms long lines on each side, or on the centre whenever two glacier-streams unite, and is deposited at its termination in a huge mound called the terminal moraine. The decrease of a glacier may often be traced by successive old moraines across the valley up which it has retreated. When once seen and examined, these moraines can always be distinguished almost at a glance. Their position is most remarkable, having no apparent natural relation to the form of the valley or the surrounding slopes, so that they look like huge earthworks formed by man for purposes of defence. Their composition is equally peculiar, consisting of a mixture of earth and rocks of all sizes, usually without any arrangement, the rocks often being huge angular masses just as they had fallen from the surrounding precipices. Some of these rock masses often rest on the very top of the moraine in positions where no other natural force but that of ice could have placed them. Exactly similar mounds are found in the valleys of North Wales and Scotland, and always where the other evidences of ice-action occur abundantly.

Travelled Blocks.—The phenomenon of travelled or perched blocks is also a common one in all glacier countries, marking out very clearly the former extent of the ice. When a glacier fills a lateral valley, its foot will sometimes cross over the main valley and abut against its opposite slope, and it will deposit there some portion of its terminal moraine. But in these circumstances the end of the glacier, not being confined laterally, will spread out, and the moraine matter will be distributed over a large surface, so that the only well-marked token of its presence will be the larger masses of rock that may have been brought down. Such blocks are found abundantly in many of the districts of our own country where other marks of glaciation exist, and they often rest on ridges or hillocks over which the ice has passed, these elevations consisting sometimes of loose material and sometimes of rock *different from that of which the blocks are composed*. These are called travelled blocks, and can almost always be traced to their source in one of the higher valleys from which the glacier descended. Some of the most remarkable examples of such travelled blocks are to be found on the southern slopes of the Jura. These consist of enormous angular blocks of granite, gneiss, and other crystalline rocks quite foreign to the Jura Mountains, but exactly agreeing with those of the Alpine range fifty miles away across the great central valley of

Switzerland. One of the largest of these blocks is forty feet in diameter, and is situated 900 feet above the level of the Lake of Neufchâtel. These blocks have been proved by Swiss geologists to have been brought by the ancient glacier of the Rhone, which was fed by the whole Alpine range, from Mont Blanc to the Furka Pass. This glacier must have been many thousand feet

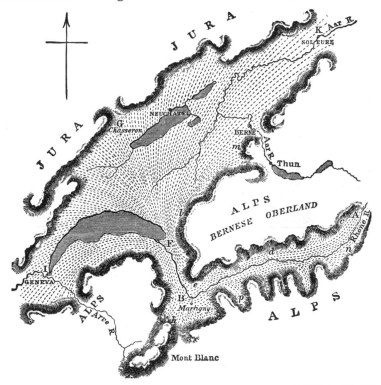

MAP SHOWING THE COURSE OF THE ANCIENT GLACIER OF THE RHONE, AND THE DISTRIBUTION OF ERRATIC BLOCKS ON THE JURA.

thick at the mouth of the Rhone valley near the head of the Lake of Geneva, since it spread over the whole of the great valley of Switzerland, extending from Geneva to Neufchâtel, Berne, and Soleure, and even on the flanks of the Jura reached a maximum height of 2015 feet above the valley. The numerous blocks scattered over the Jura for a distance of about a hundred miles vary considerably in the material of which they are composed; but they are found to be each traceable to a part of the Alps corresponding to their position, on the theory that they have been brought by a glacier spreading out from the Rhone valley. Thus, all the blocks situated to the east of a central point G (see map) can be traced to the eastern side of the Rhone valley ($l\ e\ d$), while those found towards Geneva have all come from the west side ($p\ h$). It is also very suggestive that the highest blocks on the Jura at G have come from the eastern shoulder of Mont Blanc in the direct line h B F G. Here the glacier would naturally preserve its greatest thickness, while, as it spread out eastward and westward, it would become thinner. We accordingly find that the travelled blocks on either side of the central point become lower and lower, till near Soleure and Geneva they are not more than about 500 feet above the valley. The evidence is altogether so conclusive that, after personal examination of the district in company with eminent Swiss geologists, Sir Charles Lyell gave up the view he had first adopted—that the blocks had been carried by ice during a period of submergence—as altogether untenable.[1]

The phenomena now described demonstrate a change of climate sufficient to cover all our higher mountains with perpetual snow, and fill the adjacent valleys with huge glaciers at least as extensive as those now found in Switzerland. But there are other phenomena, best developed in the northern part of our islands, which show that even this state of things was but the concluding phase of the glacial period, which, during its maximum development, must have reduced the northern half of our island to a condition only to be paralleled now in Greenland and the antarctic regions. As few persons besides professed geologists are acquainted with the weight of evidence for this statement, and as it is most important for our purpose to understand the amount of the climatal changes the Northern Hemisphere has undergone, I will endeavor to make the evidence intelligible, referring my readers for full details to Dr. James Geikie's descriptions and illustrations.[2]

---

[1] "Antiquity of Man," 4th ed., pp. 340-348.

[2] "The Great Ice Age and its Relation to the Antiquity of Man;" by James Geikie, F.R.S. (Ibister & Co., 1874.)

Glacial Deposits of Scotland: the " Till."—Over almost all
the lowlands and in most of the highland valleys of Scotland
there are immense superficial deposits of clay, sand, gravel, or
drift, which can be traced more or less directly to glacial action.
Some of these are moraine matter, others are lacustrine deposits,
while others again have been formed or modified by the sea
during periods of submergence. But below them all, and often
resting directly on the rock surface, there are extensive layers of
a very tough clayey deposit known as " till." The " till " is very
fine in texture, very tenacious, and often of a rock-like hardness.
It is always full of stones, all of which are of rude form, but
with the angles rubbed off, and almost always covered with
scratches and striæ often crossing each other in various directions. Sometimes the stones are so numerous that there seems
to be only just enough clay to unite them into a solid mass; and
they are of all sizes, from mere grit up to rocks many feet in
diameter. The " till " is found chiefly in the low-lying districts,
where it covers extensive areas sometimes to a depth of a hundred feet; while in the highlands it occurs in much smaller
patches, but in some of the broader valleys forms terraces which
have been cut through by the streams. Occasionally it is found
as high as 2000 feet above the sea, in hollows or hill-sides, where
it seems to have been protected from denudation.

The " till " is totally unstratified, and the rock surfaces on
which it almost always rests are invariably worn smooth, and
much grooved and striated when the rock is hard; but when it
is soft or jointed it frequently shows a greatly broken surface.
Its color and texture, and the nature of the stones it contains, all
correspond to the character of the rock of the district where it
occurs, so that it is clearly a local formation. It is often found
underneath moraines, drift, and other late glacial deposits, but
never overlies them (except in special cases to be hereafter referred to), so that it is certainly an earlier deposit.

Throughout Scotland, where " till " is found, the glacial striæ,
perched blocks, *roches moutonnées*, and other marks of glacial
action occur very high up the mountains to at least 3000, and
often even to 3500, feet above the sea, while all lower hills and
mountains are rounded and grooved on their very summits; and

these grooves always radiate outward from the highest peaks
and ridges towards the valleys or the sea.

Inferences from the Glacial Phenomena of Scotland.—Now all
these phenomena, taken together, render it certain that the whole
of Scotland was once buried in a vast sea of ice out of which
only the highest mountains raised their summits.   There is absolutely no escape from this conclusion ; for the facts which lead
to it are not local—found only in one spot or one valley—but
general throughout the entire length and breadth of Scotland ;
and are, besides, supported by such a mass of detailed corroborative evidence as to amount to absolute demonstration.   The
weight of this vast ice -sheet, at least 3000 feet in maximum
thickness, and continually moving seaward with a slow grinding
motion like that of all existing glaciers, must have ground down
the whole surface of the country, especially all the prominences,
leaving the rounded rocks as well as the grooves and striæ we
still see marking the direction of its motion.   All the loose
stones and rock masses which lay on the surface would be
pressed into the ice ; the harder blocks would serve as scratch-
ing and grinding tools, and would thus themselves become
rounded, scratched, and striated as we see them, while all the
softer masses would be ground up into impalpable mud along
with the material planed off the rocky projections of the country, leaving them in the condition of *roches moutonnées*.

The peculiar characters of the " till," its fineness and tenacity,
correspond closely with the fine matter which now issues from
under all glaciers, making the streams milky-white, yellow, or
brown, according to the nature of the rock.   The sediment from
such water is a fine unctuous sticky deposit, only needing pressure to form it into a tenacious clay ; and when " till " is exposed
to the action of water it dissolves into a similar soft sticky unctuous mud.   The present glaciers of the Alps, being confined to
valleys which carry off a large quantity of drainage-water, lose
this mud perhaps as rapidly as it is formed ; but when the ice
covered the whole country there was comparatively little drainage-water, and thus the mud and stones collected in vast compact masses in all the hollows, and especially in the lower flat
valleys, so that when the ice retreated the whole country was

more or less covered with it. It was then, no doubt, rapidly denuded by rain and rivers; but, as we have seen, great quantities
remain to the present day to tell the tale of its wonderful formation.[1] There is good evidence that when the ice was at its
maximum it extended not only over the land, but far out to sea,
covering all the Scottish islands, and stretching in one connected

[1] This view of the formation of " till " is that adopted by Dr. Geikie, and upheld
by almost all the Scotch, Swiss, and Scandinavian geologists. The objection, however, is made by many eminent English geologists, including Mr. Searles V. Wood,
Jr., that mud ground off the rocks cannot remain beneath the ice, forming sheets of
great thickness, because the glacier cannot at the same time grind down solid rock
and yet pass over the surface of soft mud and loose stones. But this difficulty will disappear if we consider the numerous fluctuations in the glacier with increasing size,
and the additions it must have been constantly receiving as the ice from one valley
after another joined together, and at last produced an ice-sheet covering the whole
country. The grinding power is the motion and pressure of the ice, and the pressure
will depend on its thickness. Now the points of maximum thickness must have often
changed their positions, and the result would be that the matter ground out in one
place would be forced into another place where the pressure was less. If there were
no lateral escape for the mud, it would necessarily support the ice over it, just as a
water-bed supports the person lying on it ; and when there was little drainage-water,
and the ice extended say twenty miles in every direction from a given part of a valley
where the ice was of less than the average thickness, the mud would necessarily accumulate at this part simply because there was no escape for it. Whenever the
pressure all round any area was greater than the pressure on that area, the débris of
the surrounding parts would be forced into it, and would even raise up the ice to give
it room. This is a necessary result of hydrostatic pressure. During this process the
superfluous water would, no doubt, escape through fissures or pores of the ice, and
would leave the mud and stones in that excessively compressed and tenacious condition in which the " till " is found. The unequal thickness and pressure of the ice
above referred to would be a necessary consequence of the inequalities in the valleys,
now narrowing into gorges, now opening out into wide plains, and again narrowed
lower down ; and it is just in these openings in the valleys that the " till " is said to
be found, and also in the lowlands, where an ice-sheet must have extended for many
miles in every direction. In these lowland valleys the " till " is both thickest and
most wide-spread, and this is what we might expect. At first, when the glaciers from
the mountains pushed out into these valleys, they would grind out the surface beneath them into hollows, and the drainage-water would carry away the débris. But
when they spread all over the surface from sea to sea, and there was little or no
drainage-water compared to the enormous area covered with ice, the great bulk of
the débris must gather under the ice wherever the pressure was least, and the ice
would necessarily rise as it accumulated. Some of the mud would, no doubt, be forced
out along lines of least resistance to the sea, but the friction of the stone-charged
" till " would be so enormous that it would be impossible for any large part of it to
be disposed of in this way.

sheet to Ireland and Wales, where all the evidences of glaciation
are as well marked as in Scotland, though the ice did not, of
course, attain quite so great a thickness.[1]

It is evident that the change of climate requisite to produce
such marvellous effects in the British Isles could not have been
local, and we accordingly find strikingly similar proofs that
Scandinavia and all Northern Europe have also been covered
with a huge ice-sheet; while we have already seen that a similar gigantic glacier buried the Alps, carrying granitic blocks to
the Jura, where it deposited them at a height of 3450 feet above
the sea; while to the south, in the plains of Italy, the terminal
moraines left by the retreating glaciers have formed extensive
hills, those of Ivrea, the work of the great glacier from the Val
d'Aosta, being 15 miles across, and from 700 to 1500 feet high.

Glacial Phenomena in North America.—In North America
the marks of glaciation are even more extensive and striking
than in Europe, stretching over the whole of Canada and to the
south of the Great Lakes as far as latitude 39°. There is in all
these countries a wide-spread deposit like the "till" of Scotland,
produced by the grinding of the great ice-sheet when it was at
its maximum thickness; and also extensive beds of moraine-
matter, true moraines, and travelled blocks, left by the glaciers
as they retreated towards the mountains and finally withdrew
into the upland valleys. There are also, both in Britain, Scandinavia, and North America, proofs of the submersion of the
land beneath the sea to a depth of upwards of a thousand feet;
but this is a subject we need not here enter upon, as our special
object is to show the reality and amount of that wonderful and
comparatively recent change of climate termed the glacial epoch.

Many persons, even among scientific men, who have not given

---

[1] That the ice-sheet was continuous from Scotland to Ireland is proved by the
glacial phenomena in the Isle of Man, where "till" similar to that in Scotland
abounds, and rocks are found in it which must have come from Cumberland and
Scotland, as well as from the North of Ireland. This would show that glaciers from
each of these districts reached the Isle of Man, where they met and flowed southward down the Irish Sea. Ice-marks are traced over the tops of the mountains,
which are nearly 2000 feet high. (See "A Sketch of the Geology of the Isle of
Man," by John Horne, F.G.S., *Transactions of the Edinburgh Geological Society*,
Vol. II., Pt. iii., 1874.)

much attention to the question look upon the whole subject of
the glacial epoch as a geological theory made to explain certain
phenomena which are otherwise a puzzle; and they would not
be much surprised if they were some day told that it was all a
delusion, and that Mr. So-and-so had explained the whole thing
in a much more simple way.   It is to prevent my readers being
imposed upon by any such statements or doubts that I have
given this very brief and imperfect outline of the nature, extent,
and completeness of the evidence on which the existence of the
glacial epoch depends.   There is perhaps no great conclusion
in any science which rests upon a surer foundation than this;
and if we are to be guided by our reason at all in deducing the
unknown from the known, the past from the present, we cannot
refuse our assent to the reality of the glacial epoch of the
Northern Hemisphere in all its more important features.

*Effects of the Glacial Epoch on Animal Life: Warm and
Cold Periods.*—It is hardly necessary to point out what an important effect this great climatal cycle must have had upon all
living things.   When an icy mantle crept gradually over much
of the Northern Hemisphere till large portions of Europe and
North America were reduced to the condition of Greenland
now, the greater part of the animal life must have been driven
southward, causing a struggle for existence which must have led
to the extermination of many forms, and the migration of others into new areas.   But these effects must have been greatly
multiplied and intensified if, as there is very good reason to believe, the glacial epoch itself—or at least the earlier and later
phases of it—consisted of two or more alternations of warm and
cold periods.

The evidence that such was the case is very remarkable.   The
"till," as we have seen, could only have been formed when the
country was entirely buried under a large ice-sheet of enormous
thickness, and when it must therefore have been, in all the parts
so covered, almost entirely destitute of animal and vegetable
life.   But in several places in Scotland fine layers of sand and
gravel, with beds of peaty matter, have been found resting on
"till" and again covered by "till."   Sometimes these intercalated beds are very thin, but in other cases they are twenty or

thirty feet thick, and in them have been found remains of the extinct ox, the Irish elk, the horse, reindeer, and mammoth. Here we have evidence of two distinct periods of intense cold, and an intervening milder period sufficiently prolonged for the country to become covered with vegetation and stocked with animal life. In some districts borings have proved the existence of no less than four distinct formations of "till" separated from each other by beds of sand from two to twenty feet in thickness.[1] Facts of a similar nature have been observed in other parts of our islands. In the East of England, Mr. Skertchly (of the Geological Survey) enumerates four distinct boulder clays with intervening deposits of gravels and sands.[2] Mr. Searles V. Wood, Jr., classes the most recent (Hessle) boulder clay as "postglacial;" but he admits an intervening warmer period, characterized by southern forms of mollusca and insects, after which glacial conditions again prevailed with northern types of mollusca.[3] Elsewhere he says, "Looking at the presence of such fluviatile mollusca as *Cyrena fluminalis* and *Unio littoralis*, and of such mammalia as the hippopotamus and other great pachyderms, and of such a littoral Lusitanian fauna as that of the Selsea bed, where it is mixed up with the remains of some of those pachyderms, as well as of some other features, it has seemed to me that the climate of the earlier part of the postglacial period in England was possibly even warmer than our present climate; and that it was succeeded by a refrigeration sufficiently severe to cause ice to form all round our coasts, and glaciers to accumulate in the valleys of the mountain districts; and that this increased severity of climate was preceded and partially accompanied by a limited submergence, which nowhere apparently exceeded 300 feet, and reached that amount only in the northern counties of England."[4] This decided ad-

---

[1] "The Great Ice Age," p. 177.

[2] These are named, in descending order, Hessle Boulder Clay, Purple Boulder Clay, Chalky Boulder Clay, and Lower Boulder Clay, below which is the Norwich Crag.

[3] "On the Climate of the Postglacial Period," *Geological Magazine*, 1872, pp. 158, 160.

[4] *Geological Magazine*, 1876, p. 396.

mission of an alternation of warm and cold climates since the height of the glacial epoch by so cautious a geologist as Mr. Wood is very important, as is his statement of an accompanying *depression* of the land accompanying the increased cold, because many geologists maintain that a greater elevation of the land is the true and sufficient explanation of glacial periods. Further evidence of this alternation is found both in the Isle of Man and in Ireland, where two distinct boulder clays have been described with intervening beds of gravels and sands.

Palæontological Evidence of Alternate Cold and Warm Periods.—Especially suggestive of a period warmer than the present, immediately following glacial conditions, is the occurrence of the hippopotamus in caves, brick-earths, and gravels of palæolithic age. Entire skeletons of this animal have been found at Leeds in a bed of dark-blue clay overlaid by gravel. Farther north, at Kirkdale cave, in N. lat. 54° 15′, remains of the hippopotamus occur abundantly along with those of the ox, elephant, horse, and other quadrupeds, and with countless remains of the hyenas which devoured them; while it has also been found in cave-deposits in Glamorganshire, at Durham Down, near Bristol, and in the post-Pliocene drifts of Dorsetshire. It is important to note that where it is associated with other mammals in *caves* — which are hyena-dens, and not mere receptacles of water-carried remains—these always imply a mild climate, the elephant and rhinoceros found with it being species characteristic of temperate latitudes (*Elephas antiquus* and *Rhinoceros hemitœchus*). But when it occurs in gravels or in water-borne cave-deposits it is sometimes associated with the mammoth, the woolly rhinoceros, and the reindeer—animals which as certainly imply a cold or even arctic climate. This difference is intelligible if we consider that the hyena, which carried the bones of all these animals into the caves, is itself indicative of a mild climate, and that there is nothing to cause the remains of animals of successive epochs to be intermingled in such caves. In the gravels, however, it is very different. During the warm periods the rivers would be inhabited by hippopotami, and the adjacent plains by elephants and horses, and their remains would be occasionally embedded in deposits formed during floods.

But when the cold period came on, and these had passed southward, the same river-banks would be grazed by mammoths and
reindeer whose remains would sometimes intermingle with those
of the animals which preceded them. It is to be noted, also,
that in many of these river-deposits there are proofs of violent
floods causing much rearrangement of materials, so that the
remains of the two periods would be thus still further intermingled.[1]

The fact of the hippopotamus having lived at 54° N. lat. in
England, quite close to the time of the glacial epoch, is absolutely inconsistent with a mere gradual amelioration of climate
from that time till the present day. The immense quantity
of vegetable food which this creature requires implies a mild
and uniform climate with hardly any severe winter; and no
theory that has yet been suggested renders this possible except
that of alternate cold and warm periods during the glacial epoch
itself. In order that the hippopotamus could have reached
Yorkshire and retired again as the climate changed, we may
suppose it to have been a permanent inhabitant of the Lower
Rhone, between which river and the Rhine there is an easy
communication by means of the Doubs and the Ill, some of
whose tributaries approach within a mile or two of each other
about fifteen miles southwest of Mülhausen. Thence the passage would be easy down the Rhine into the great river which
then flowed up the bed of the North Sea, and thence up the
Humber and Ouse into Yorkshire. By this route there would
be only one water - shed to cross, and this might probably
have been marshy; but we may also suppose the animals to
have ascended the Bristol Channel after passing round a long
extent of French and English coast (which would then have
consisted of vast plains stretching far beyond the Scilly Isles),
in which case they would find an equally easy passage over a
low water-shed from the valley of the Avon to that of the Trent
and Yorkshire Ouse. A consideration of the long and circuitous journey required on any hypothesis will at once convince

---

[1] A. Tylor, on "Quaternary Gravels," *Quarterly Journal of Geological Society
of London*, 1869, pp. 83, 95 (wood-cuts).

us that it could never have been made (as some have supposed) annually during the short hot summer of the glacial period itself; whereas the interglacial warm periods lasting several thousand years would allow for the animals' gradual migration into all suitable river valleys. Thus, the very existence of the hippopotamus in Yorkshire as well as in the South of England, in close association with glacial conditions, must be held to be a strong corroborative argument in favor of the reality of an interglacial warm period.

Evidence of Interglacial Warm Periods on the Continent and in North America.—Besides the evidence already adduced from our own islands, many similar facts have been noted in other countries. In Switzerland two glacial periods are distinctly recognized, between which was a warm period when vegetation was so luxuriant as to form beds of lignite sufficiently thick to be worked for coal. The plants found in these deposits are similar to those now inhabiting Switzerland—pines, oaks, birches, larch, etc.; but numerous animal remains are also found, showing that the country was then inhabited by an elephant (*Elephas antiquus*), a rhinoceros (*Rhinoceros Etruscus*), the urus (*Bos primigenius*), the red deer (*Cervus elephas*), and the cave-bear (*Ursus spelæus*); and there was also abundance of insects.[1]

In Sweden also there are two " tills," the lower one having been in places partly broken up and denuded before the upper one was deposited, but no interglacial deposits have yet been found. In North America more complete evidence has been obtained. On the shores of Lake Ontario sections are exposed showing three separate beds of " till " with intervening stratified deposits, the lower one of which has yielded many plant remains and fresh-water organisms. These deposits are seen to extend continuously for more than nine miles, and the fossiliferous interglacial beds attain a thickness of 140 feet. Similar beds have been discovered near Cleveland, Ohio, consisting, first, of " till " at the lake-level; secondly, of about forty-eight feet of sand and loam; and, thirdly, of unstratified " till " full of striated

---

[1] Heer's " Primæval World of Switzerland," Vol. II., pp. 148–168.

stones, six feet thick.[1] On the other side of the continent, in British Columbia, Mr. G. M. Dawson, geologist to the North American Boundary Commission, has discovered similar evidence of two glaciations divided from each other by a warm period.

This remarkable series of observations, spread over so wide an area, seems to afford ample proof that the glacial epoch did not consist merely of one process of change from a temperate to a cold and arctic climate, which, having reached a maximum, then passed slowly and completely away, but that there were certainly two, and probably several more, alternations of arctic and temperate climates.

It is evident, however, that if there have been, not two only, but a series of such alternations of climate, we could not possibly expect to find more than the most slender indications of them, because each succeeding ice-sheet would necessarily grind down or otherwise destroy much of the superficial deposits left by its predecessors, while the torrents that must always have accompanied the melting of these huge masses of ice would wash away even such fragments as might have escaped the ice itself. It is a fortunate thing, therefore, that we should find any fragments of these interglacial deposits containing animal and vegetable remains; and, just as we should expect, the evidence they afford seems to show that the later phase of the cold period was less severe than the earlier. Of such deposits as were formed on land during the coming-on of the glacial epoch, when it was continually increasing in severity, hardly a trace has been preserved, because each succeeding extension of the ice, being greater and thicker than the last, destroyed what had gone before it till the maximum was reached.

Migrations and Extinction of Organisms Caused by the Glacial Epoch.—Our last glacial epoch was accompanied by at least two considerable submergences and elevations of the land, and there is some reason to think, as we have already explained, that the two classes of phenomena are connected as cause and effect. We can easily see how such repeated submergences and eleva-

---

[1] Dr. James Geikie, in *Geological Magazine*, 1878, p. 77.

tions would increase and aggravate the migrations and extinctions that a glacial epoch is calculated to produce. We can therefore hardly fail to be right in attributing the wonderful changes in animal and vegetable life that have occurred in Europe and North America between the Miocene period and the present day, in part at least, to the two or more cold epochs that have probably intervened. These changes consist, first, in the extinction of a whole host of the higher animal forms; and, secondly, in a complete change of types due to extinction and emigration, leading to a much greater difference between the vegetable and animal forms of the Eastern and Western hemispheres than before existed. Many large and powerful mammalia lived in our own country in Pliocene times, and apparently survived a part of the glacial epoch; but when it finally passed away, they too had disappeared, some having become altogether extinct, while others continued to exist in more southern lands. Among the first class are the sabre-toothed tiger, the extinct Siberian camel (Merycotherium), three species of elephant, two of rhinoceros, a hippopotamus, two bears, five species of deer, and the gigantic beaver; among the latter are the hyena, bear, and lion, which are considered to be only varieties of those which once inhabited Britain. Down to Pliocene times the flora of Europe was very similar to that which now prevails in Eastern Asia and Eastern North America. Hundreds of species of trees and shrubs of peculiar genera which still flourish in those countries are now completely wanting in Europe; and we have good reason to believe that these were exterminated during the glacial period, being cut off from a southern migration, first by the Alps, and then by the Mediterranean; whereas in Eastern America and Asia the mountain-chains run in a north and south direction, and there is nothing to prevent the flora from having been preserved by a southward migration into a milder region.

Our next two chapters will be devoted to a discussion of the causes which brought about the glacial epoch, and that still more extraordinary climatic phenomenon—the mild climate and luxuriant vegetation of the arctic zone. If my readers will follow me with the care and attention so difficult and interesting a prob-

lem requires and deserves, they will find that I have grappled
with all the facts which have to be accounted for, and offered
what I believe is the first complete and sufficient explanation of
them.   The important influence of climatal changes on the dispersal of animals and plants is a sufficient justification for introducing such a discussion into the present volume.

CHAPTER VIII.

THE CAUSES OF GLACIAL EPOCHS.

Various Suggested Causes.—Astronomical Causes of Changes of Climate.—Difference of Temperature Caused by Varying Distance of the Sun.—Properties of Air
and Water, Snow and Ice, in Relation to Climate.—Effects of Snow on Climate.—
High Land and Great Moisture Essential to the Initiation of a Glacial Epoch.—
Perpetual Snow nowhere Exists on Lowlands.—Conditions Determining the Presence or Absence of Perpetual Snow.—Efficiency of Astronomical Causes in Producing Glaciation.—Action of Meteorological Causes in Intensifying Glaciation.
—Summary of Causes of Glaciation.—Effect of Clouds and Fog in Cutting off the
Sun's Heat.—South Temperate America as Illustrating the Influence of Astronomi-
cal Causes on Climate.—Geographical Changes, how far a Cause of Glaciation.—
Land Acting as a Barrier to Ocean Currents.—The Theory of Interglacial Periods
and their Probable Character.—Probable Effect of Winter in Aphelion on the Climate of Britain.—The Essential Principle of Climatal Change Restated.—Prob-
able Date of the Last Glacial Epoch.—Changes of the Sea-level Dependent on
Glaciation.—The Planet Mars as Bearing on the Theory of Eccentricity as a
Cause of Glacial Epochs.

No less than seven different causes have been at various times
advanced to account for the glacial epoch and other changes of
climate which the geological record proves to have taken place.
These, as enumerated by Mr. Searles V. Wood, Jr., are as follows:

1. A decrease in the original heat of our planet.

2. Changes in the obliquity of the ecliptic.

3. The combined effect of the precession of the equinoxes
and of the eccentricity of the earth's orbit.

4. Changes in the distribution of land and water.

5. Changes in the position of the earth's axis of rotation.

6. A variation in the amount of heat radiated by the sun.

7. A variation in the temperature of space.

Of the above, causes 1 and 2 are undoubted realities; but
it is now generally admitted that they are utterly inadequate

to produce the observed effects. Causes 5, 6, and 7 are all purely hypothetical; for, though such changes may have occurred, there is no evidence that they have occurred during geological time, and it is, besides, certain that they would not, either singly or combined, be adequate to explain the whole of the phenomena. There remain causes 3 and 4, which have the advantage of being demonstrated facts, and which are universally admitted to be capable of producing *some* effect of the nature required, the only question being whether, either alone or in combination, they are adequate to produce all the observed effects. It is therefore to these two causes that we shall confine our inquiry, taking first those astronomical causes whose complex and wide-reaching effects have been so admirably explained and discussed by Dr. Croll in numerous papers and in his work " Climate and Time in their Geological Relations."

Astronomical Causes of Changes of Climate.—The earth moves in an elliptical orbit round the sun, which is situated in one of the foci of the ellipse, so that the distance of the sun from us varies during the year to a considerable amount. Strange to say, we are now three millions of miles nearer to the sun in winter than in summer, while the reverse is the case in the Southern Hemisphere; and this must have some effect in making our northern winters less severe than those of the south temperate zone. But the earth moves more rapidly in that part of its orbit which is nearer to the sun, so that our winter is not only milder, but several days shorter, than that of the Southern Hemisphere. The distribution of land and sea and other local causes prevent us from making any accurate estimate of the effects due to these differences; but there can be no doubt that if our winter were now as long as our summer, and we were also three million miles farther from the sun at the former period, a very decided difference of climate would result—our winter would be colder and longer, our summer hotter and shorter. Now there is a combination of astronomical revolutions (the precession of the equinoxes and the motion of the aphelion) which actually brings this change about every 10,500 years, so that after this interval the condition of the two hemispheres is reversed as regards nearness to the sun in summer, and comparative duration of summer and winter; and this change has been going on throughout all geological periods. (See diagram.) The influence of the present phase of precession is perhaps seen in the great extension of the antarctic ice-fields, and the existence of glaciers at the sea-level in the Southern Hemisphere in latitudes corresponding to that of England; but it is not supposed that similar effects would be produced with us at the last cold period, 10,500 years ago, because we are exceptionally favored by the Gulf Stream warming the whole North Atlantic Ocean, and by the prevalence of westerly winds which convey that warmth to our shores; and also by the comparatively small quantity of high land around the North Pole, which does not

N.HEMISPHERE WINTER IN APHELION

S.HEMISPHERE WINTER IN APHELION

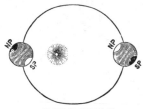

GLACIAL EPOCH IN
N.HEMISPHERE

GLACIAL EPOCH IN
S.HEMISPHERE

DIAGRAM SHOWING THE ALTERED POSITION OF THE POLES AT INTERVALS OF 10,500 YEARS PRODUCED BY THE PRECESSION OF THE EQUINOXES AND THE MOTION OF THE APHELION; AND ITS EFFECT ON CLIMATE DURING A PERIOD OF HIGH ECCENTRICITY.

encourage great accumulations of ice. But the amount of eccentricity itself varies very largely, though very slowly, and it is now nearly at a minimum. It also varies very irregularly; but its amount has been calculated for several million years back. Fifty thousand years ago it was rather less than it is now; but it then increased, and when we come to a hundred thousand years ago, there is a difference of eight and a half millions of miles between our distance from the sun in aphelion and perihelion (as the most distant and nearest points of the earth's orbit are termed). At a hundred and fifty thousand years back it had decreased somewhat—to six millions of miles; but then it increased again, till at two hundred thousand years ago it was ten and a quarter, and at two hundred and ten thousand years ten and a half, millions of miles. By reference to the accompanying diagram, which includes the last great period of eccentricity, we find that for the immense period of a hundred and sixty thousand years (commencing about eighty thousand years ago) the eccentricity was very great, reaching a maximum of three and a half times its present amount at almost the remotest part of this period, at which time the length of summer in one hemisphere and of winter in the other would be nearly twenty-eight days in excess.

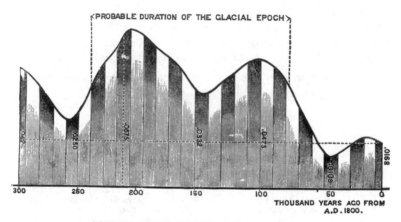

DIAGRAM OF ECCENTRICITY AND PRECESSION.

The dark and light bands mark the phases of precession, the dark showing short mild winters, and the light long cold winters, the contrast being greater as the eccentricity is higher. The horizontal dotted line marks the present eccentricity. The figures show the maxima and minima of eccentricity during the last 300,000 years from Dr. Croll's tables.

Now during all this time our position would change, as above described (and as indicated on the diagram), every ten thousand five hundred years; so that we should have alternate periods of very long and cold winters with short hot summers, and short mild winters with long cool summers. In order to understand the important effects which this would produce, we must ascertain two things—first, what actual difference of temperature would be caused by varying distances of the sun; and, secondly, what are the properties of snow and ice in regard to climate.

Difference of Temperature Caused by Varying Distances of the Sun.—On this subject comparatively few persons have correct ideas, owing to the unscientific manner in which we reckon heat by our thermometers. Our zero is thirty-two degrees below the freezing-point of water, or, in the centigrade thermometer, the freezing-point itself, both of which are equally misleading when applied to cosmical problems. If we say that the mean temperature of a place is 50° F. or 10° C., these figures tell us nothing of how much the sun warms that place, because if the sun were withdrawn the temperature would fall far below either of the zero-points. In the last arctic expedition a temperature of −74° F. was registered, or 106° below the freezing-point of water; and as at the same time the earth, at a depth of two feet, was only −13° F. and the sea-water +28° F., we may be sure that even this intense cold was not near the possible minimum temperature. By various calculations and experiments which cannot be entered upon here, it has been determined that the temperature of space, independent of solar (but not of stellar) influence, is about −239° F., and physicists almost universally adopt this quantity in all estimates of cosmical temperature. It follows that if the mean temperature of the earth's surface at any time is 50° F., it is really warmed by the sun to an amount measured by $50 + 239 = 289°$ F., which is hence termed its absolute temperature. Now during the time of the glacial epoch the greatest distance of the sun in winter was $97\frac{1}{2}$ millions of miles, whereas it is now, in winter, only 91 millions of miles. But the quantity of heat received from the sun is inversely as the square of the distance, so that it would then be in the proportion of 8281 to 9506 now, or nearly one eighth less than its present amount. The mean temperature of England in January is about 39° F., which equals 278° F. of absolute temperature. But the above-named fraction of 278° is 36°, representing the amount which must be deducted to obtain the January temperature during the glacial epoch, which will therefore be 3° F. Our actual temperature at that time might, however, have been very different from this, because the temperature of a place does not depend so much on the amount of heat it receives directly from the sun as on the amount brought to it or carried away from it by warm or cold winds. We often have it bitterly cold in the middle of May when we are receiving as much sun heat as many parts of the tropics, because we get cold winds from the iceberg-laden North Atlantic, and this partially neutralizes the effect of the sun. So we often have it very mild in December if southwesterly winds bring us warm moist air from the Gulf Stream. But though the above method does not give correct results for any one time or place, it is more nearly correct for very large areas, because all the sensible surface heat which produces climates comes from the sun, and its proportionate amount may be very nearly calculated in the manner above described. We may therefore say, generally, that during our northern winter, at the time of the glacial epoch, the Northern Hemisphere was receiving so much less heat from the sun as to lower its surface temperature on an average about 35° F., while during the height of summer of the same period it would be receiving so much more heat as would suffice to raise its mean temperature about 60° F. above what it is now. The winter, however, would be long and the summer short, the difference being twenty-six days.

We have here certainly a superabundant amount of cold in winter to produce a glacial period,[1] especially as this cold would

---

[1] In a letter to *Nature* of October 30, 1879, the Rev. O. Fisher calls attention to a result arrived at by Pouillet, that the temperature which the surface of the ground would assume if the sun were extinguished would be −128° F. instead of −239° F. If this corrected amount were used in our calculations, the January temperature of England during the glacial epoch would come out 17° F., and this Mr. Fisher thinks not low enough to cause any extreme difference from the present climate. In this opinion, however, I cannot agree with him. On the contrary, it would, I think, be a relief to the theory were the amounts of decrease of temperature in winter and increase in summer rendered more moderate, since according to the usual calculation (which I have adopted) the differences are unnecessarily great. I cannot, therefore, think that this modification of the temperatures, should it be ultimately proved to be correct (which is altogether denied by Dr. Croll), would be any serious objection to the adoption of Dr. Croll's theory of the astronomical and physical causes of the glacial epoch.

The reason of the increase of summer heat being 60°, while the decrease of winter cold is only 35°, is because our summer is now *below* and our winter *above* the average. A large part of the 60° increase of temperature would, no doubt, be used up in evaporating water, so that there would be a much less increase of sensible heat; while only a portion of the 35° lowering of temperature in winter would be actually produced, owing to equalizing effect of winds and currents and the storing-up of heat by the earth and ocean.

be long continued; but at the same time we should have almost
tropical heat in summer, although that season would be somewhat shorter.   How, then, it may be asked, could such a climate
have the effect supposed?   Would not the snow that fell in
winter be all melted by the excessively hot summer?   In order
to answer this question, we must take account of certain properties of water and air, snow and ice, to which due weight has not
been given by writers on this subject.

*Properties of Air and Water, Snow and Ice, in Relation to
Climate.*—The great aerial ocean which surrounds us has the
wonderful property of allowing the heat-rays from the sun to
pass through it without its being warmed by them; but when
the earth is heated the air gets warmed by contact with it, and
also to a considerable extent by the heat radiated from the warm
earth, because, although pure dry air allows such dark heat-rays
to pass freely, yet the aqueous vapor and carbonic acid in the
air intercept and absorb them.   But the air thus warmed by
the earth is in continual motion, owing to changes of density.
It rises up and flows off, while cooler air supplies its place; and
thus heat can never accumulate in the atmosphere beyond a
very moderate degree, the excessive sun heat of the tropics being
much of it carried away to the upper atmosphere and radiated
into space.   Water also is very mobile; and although it receives
and stores up a great deal of heat, it is forever dispersing it over
the earth.   The rain, which brings down a certain portion of
heat from the atmosphere, and which often absorbs heat from
the earth on which it falls, flows away in streams to the ocean;
while the ocean itself, constantly impelled by the winds, forms
great currents, which carry off the surplus heated water of the
tropics to the temperate and even to the polar regions, while
colder water flows from the poles to ameliorate the heat of the
tropics.   An immense quantity of sun heat is also used up in
evaporating water, and the vapor thus produced is conveyed by
the aerial currents to distant countries, where, on being condensed into rain, it gives up much of this heat to the earth and
atmosphere.

The power of water in carrying away heat is well exhibited
by the fact of the abnormally high temperature of arid deserts

and of very dry countries generally; while the still more powerful influence of moving air may be appreciated by considering the effects of even our northern sun in heating a tightly closed glass house to far above the temperature produced by the vertical sun of the equator, where the free air and abundance of moisture exert their beneficial influence. Were it not for the large proportion of the sun's heat carried away by air and water, the tropics would become uninhabitable furnaces; as would, indeed, *any* part of the earth where the sun shone brightly throughout a summer's day.

We see, therefore, that the excess of heat derived from the sun at any place cannot be stored up to an important amount owing to the wonderful dispersing agency of air and water; and though some heat does penetrate the ground and is stored up there, this is so little in proportion to the whole amount received, and the larger part of it is so soon given out from the surface layers, that any surplus heat that may be thus preserved during one summer rarely or never remains in sufficient quantity to affect the temperature of the succeeding summer, so that there is no such thing as an accumulation of earth heat from year to year. But though heat cannot, cold can be stored up to an almost unlimited amount, owing to the peculiar property water possesses of becoming solid at a moderately low temperature; and as this is a subject of the very greatest importance to our inquiry—the whole question of the possibility of glacial epochs and warm periods depending on it—we must consider it in some detail.

Effects of Snow on Climate.—Let us, then, examine the very different effects produced by water falling as a liquid in the form of rain, or as a solid in the form of snow, although the two may not differ from each other more than two or three degrees in temperature. The rain, however much of it may fall, runs off rapidly into streams and rivers, and soon reaches the ocean. If cold, it cools the air and the earth somewhat while passing through or over them, but produces no permanent effect on temperature, because a few hours of sunshine restore to the air or the surface soil all the heat they had lost. But if snow falls for a long time, the effect, as we all know, is very different, *because it has no mobility.* It remains where it fell and becomes compacted into a mass, and it then keeps the earth below it and the air above at or near the freezing-point till it is all melted. If the quantity is great, it may take days or weeks to melt; and if snow continues falling, it goes on accumulating all over the surface of a country (which water cannot do), and may thus form such a mass that the warmth of the whole succeeding summer may not be able to melt it. It then produces perpetual snow, such as we find above a certain altitude on all the great mountains of the globe; and when this takes place *cold* is rendered permanent, no amount of sun heat warming the air or the earth much above the freezing-point. This is illustrated by the often-quoted fact that at 80° N. lat. Captain Scoresby had the pitch melted on one side of his ship by the heat of the sun, while water was freezing on the other side owing to the coldness of the air.

The quantity of heat required to melt ice or snow is very great, as we all know by experience of the long time masses of snow will remain unmelted even in warm weather. We shall, however, be better able to appreciate the great effect this has upon climate by a few figures showing what this amount really is. In order to melt one cubic foot of ice, as much heat is required as would heat a cubic foot of water from the freezing-point to 176° F., or two cubic feet to 88° F. To melt a layer of ice a foot thick will therefore use up as much heat as would raise a layer of water two feet thick to the temperature of 88° F.; and the effect becomes still more easily understood if we estimate it as applied to air, for to melt a layer of ice only one and a half inch thick would require as much heat as would raise a stratum of air 800 feet thick from the freezing-point to the tropical heat of 88° F.! We thus obtain a good idea, both of the wonderful power of snow and ice in keeping down temperature, and also of the reason why it requires so long a time to melt away, and is able to go on accumulating to such an extent as to become permanent. These properties would, however, be of no avail if it were liquid, like water; hence it is the state of solidity and almost complete immobility of ice that enables it to produce by its accumulation such extraordinary effects in physical

geography and in climate as we see in the glaciers of Switzerland and the ice-capped interior of Greenland.

*High Land and Great Moisture Essential to the Initiation of
a Glacial Epoch.*—Another point of great importance in connection with this subject is the fact that this permanent storing-
up of cold depends entirely on the annual amount of snow-fall
in proportion to that of the sun and air heat, and not on the actual cold of winter, or even on the average cold of the year. A
place may be intensely cold in winter and may have a short
arctic summer, yet, if so little snow falls that it is quickly melted
by the returning sun, there is nothing to prevent the summer
being hot and the earth producing a luxuriant vegetation. As
an example of this we have great forests in the extreme North
of Asia and America where the winters are colder and the summers shorter than in Greenland, in lat. 62° N., or than in Heard
Island and South Georgia, both in lat. 53° S., in the Southern
Ocean, and almost wholly covered with perpetual snow and ice.
At the "Jardin" on the Mont Blanc range, above the line of
perpetual snow, a thermometer in an exposed situation marked
— 6° F. as the lowest winter temperature; while in many parts
of Siberia mercury freezes several weeks in winter, showing a
temperature below --40° F.; yet here the summers are hot, all
the snow disappears, and there is a luxuriant vegetation. Even
in the very highest latitudes reached by our last arctic expedition there is very little perpetual snow or ice, for Captain Nares
tells us that north of Hayes's Sound, in lat. 79° N., the mountains
were remarkably free from ice-cap, while extensive tracts of
land were free from snow during summer, and covered with a
rich vegetation with abundance of bright flowers. The reason
of this is evidently the scanty snow-fall, which rendered it sometimes difficult to obtain enough to form shelter-banks around
the ships; and this was north of 80° N. lat., where the sun
was absent for a hundred and forty two days.

Perpetual Snow nowhere Exists on Lowlands.—It is a very
remarkable and most suggestive fact that nowhere in the world
at the present time are there any extensive lowlands covered
with perpetual snow. The Tundras of Siberia and the barren
grounds of North America are all clothed with some kind of

summer vegetation;[1] and it is only where there are lofty mountains or plateaus—as in Greenland, Spitzbergen, and Grinnell's
Land—that glaciers, accompanied by perpetual snow, cover the
country, and descend in places to the level of the sea. In the
antarctic regions there are extensive highlands and lofty mountains, and these are everywhere exposed to the influence of moist
sea-air; and it is here, accordingly, that we find the nearest approach to a true ice-cap covering the whole circumference of
the antarctic continent, and forming a girdle of ice-cliffs which
almost everywhere descend to the sea. Such antarctic islands as
South Georgia, South Shetland, and Heard Island are often said
to have perpetual snow at sea-level; but they are all very mountainous, and send down glaciers into the sea, and as they are
exposed to moist sea-air on every side, the precipitation, almost
all of which takes the form of snow even in summer, is of
course unusually large.

That high land in an area of great precipitation is the necessary condition of glaciation is well shown by the general state
of the two polar areas at the present time. The northern part
of the north temperate zone is almost all land, mostly low but
with elevated borders; while the polar area is, with the exception of Greenland and a few other considerable islands, almost
all water. In the Southern Hemisphere the temperate zone is
almost all water, while the polar area is almost all land, or is at
least enclosed by a ring of high and mountainous land. The
result is that in the north the polar area is free from any accumulation of permanent ice (except on the highlands of Greenland and Grinnell's Land), while in the south a complete barrier
of ice of enormous thickness appears to surround the pole. Dr.
Croll shows, from the measured height of numerous antarctic

---

[1] In an account of Professor Nordenskjöld's recent expedition round the northern
coast of Asia. given in *Nature*, November 20, 1879, we have the following passage
fully supporting the statement in the text : "Along the whole coast, from the White
Sea to Behring Strait, no glacier was seen. During autumn the Siberian coast is
nearly free of ice and snow. There are no mountains covered all the year round
with snow, although some of them rise to a height of more than two thousand feet."
It must be remembered that the north coast of Eastern Siberia is in the area of supposed greatest winter cold on the globe.

icebergs (often miles in length) that the ice-sheet from which they are the broken outer fragments must be from a mile to a mile and a half in thickness.[1] As this is the thickness of the outer edge of the ice, it must be far thicker inland; and we thus find that the antarctic continent is at this very time suffering glaciation to quite as great an extent as we have reason to believe occurred in the same latitudes of the Northern Hemisphere during the last glacial epoch.

The accompanying diagrams show the comparative state of the two polar areas both as regards the distribution of land and sea, and the extent of the ice-sheet and floating icebergs. The much greater quantity of ice at the south pole is undoubtedly due to the presence of a large extent of high land, which acts as a condenser, and an unbroken surrounding ocean, which affords a constant supply of vapor; and the effect is intensified by winter being there in aphelion, and thus several days longer than with us, while the whole Southern Hemisphere is at that time farther from the sun, and therefore receives less heat.

We see, however, that with less favorable conditions for the production and accumulation of ice, Greenland is glaciated down to lat. 61°. What, then, would be the effect if the antarctic continent, instead of being confined almost wholly within the south polar circle, were to extend in one or two great mountainous promontories far into the temperate zone? The comparatively small Heard Island, in S. lat. 53°, is even now glaciated down to the sea. What would be its condition were it a northerly extension of a lofty antarctic continent? We may be quite sure that glaciation would then be far more severe, and that an ice-sheet corresponding to that of Greenland might extend to beyond the parallel of 50° S. lat. Even this is probably too low an estimate, for on the west coast of New Zealand, in S. lat. 43° 35′, a glacier even now descends to within seven hundred and five feet of the sea-level; and if those islands were the northern extension of an antarctic continent, we may be pretty sure that they would be nearly in the ice-covered condition of Greenland, although situated in the latitude of Marseilles.

---

[1] "On the Glacial Epoch," by James Croll, *Geological Magazine*, July, August, 1874.

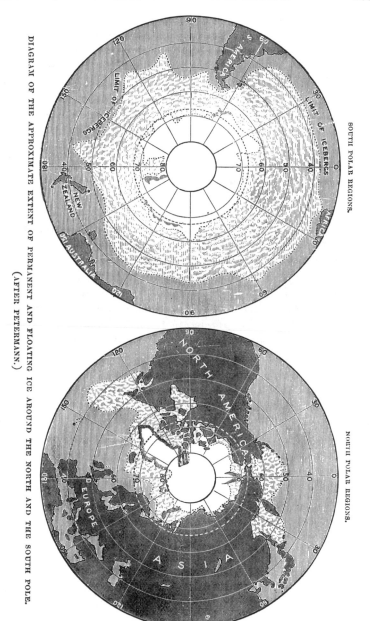

DIAGRAM OF THE APPROXIMATE EXTENT OF PERMANENT AND FLOATING ICE AROUND THE NORTH AND THE SOUTH POLE.
(AFTER PETERMANN.)

SOUTH POLAR REGIONS.

NORTH POLAR REGIONS.

Conditions Determining the Presence or Absence of Perpetual Snow.—It is clear, then, that the vicinity of a sea or ocean to supply moisture, together with high land to serve as a condenser of that moisture into snow, are the prime essentials of a great accumulation of ice; and it is fully in accordance with this view that we find the most undoubted signs of extensive glaciation in the west of Europe and the east of North America, both washed by the Atlantic, and both having abundance of high land to condense the moisture which it supplies. Without these conditions cold alone, however great, can produce no glacial epoch. This is strikingly shown by the fact that in the very coldest portions of the two northern continents—Eastern Siberia and the northwestern shores of Hudson's Bay—there is no perennial covering of snow or ice whatever. No less remarkable is the coincidence of the districts of greatest glaciation with those of greatest rainfall at the present time. Looking at a rain-map of the British Isles, we see that the greatest area of excessive rainfall is the Highlands of Scotland, then follow the West of Ireland, Wales, and the North of England; and these were glaciated pretty nearly in proportion to the area of country over which there is an abundant supply of moisture. So in Europe, the Alps and the Scandinavian mountains have excessive rainfall, and have been areas of excessive glaciation, while the Ural and Caucasian mountains, with less rain, never seem to have been proportionally glaciated. In North America the eastern coast has an abundant rainfall, and New England with Northeastern Canada seems to have been the source of much of the glaciation of that continent.[1]

---

[1] "The general absence of recent marks of glacial action in Eastern Europe is well known; and the series of changes which have been so well traced and described by Professor Szabó as occurring in those districts seems to leave no room for those periodical extensions of 'ice-caps' with which some authors in this country have amused themselves and their readers. Mr. Campbell, whose ability to recognize the physical evidence of glaciers will scarcely be questioned, finds quite the same absence of the proof of extensive ice-action in North America westward of the. meridian of Chicago" (Professor J. W. Judd, in *Geological Magazine*, 1876, p. 535).

The same author notes the diminution of marks of ice-action on going eastward in the Alps; and the Altai Mountains far in Central Asia show no signs of having been largely glaciated. West of the Rocky Mountains, however, in the Sierra Nevada and the coast ranges farther north, signs of extensive old glaciers again appear; all which

The reason why no accumulation of snow or ice ever takes place on arctic lowlands is explained by the observations of Lieutenant Payer of the Austrian Polar Expedition, who found that during the short arctic summer of the highest latitudes the ice-fields diminished four feet in thickness under the influence of the sun and wind. To replace this would require a precipitation of snow equivalent to about forty-five inches of rain, an amount which rarely occurs in lowlands out of the tropics. In Siberia, within and near the Arctic Circle, about six feet of snow covers the country all the winter and spring, and is not sensibly diminished by the powerful sun só long as northerly winds keep the air below the freezing-point and occasional snow-storms occur. But early in June the wind usually changes to southerly, probably the southwestern anti-trades overcoming the northern inflow; and under its influence the snow all disappears in a few days, and the vegetable kingdom bursts into full luxuriance. This is very important as showing the impotence of mere sun heat to get rid of a thick mass of snow so long as the air remains cold, while currents of warm air are in the highest degree effective. If, however, they are not of sufficiently high temperature, or do not last long enough to melt the snow, they are likely to increase it from the quantity of moisture they bring with them, which will be condensed into snow by coming into contact with the frozen surface. We may therefore expect the transition from perpetual snow to a luxuriant arctic vegetation to be very abrupt, depending as it must on a few degrees more or less in the summer temperature of the air; and this is quite in accordance with the fact of corn ripening by the sides of Alpine glaciers.

Efficiency of Astronomical Causes in Producing Glaciation.— Having now collected a sufficient body of facts, let us endeavor to ascertain what would be the state to which the Northern Hemisphere would be reduced by a high degree of eccentricity and a winter in aphelion. When the glacial epoch is supposed to have been at its maximum, about 210,000 years ago, the ec-

phenomena are strikingly in accordance with the theory here advocated of the absolute dependence of glaciation on abundant rainfall and elevated snow condensers and accumulators.

centricity was more than three times as great as it is now; and, according to Dr. Croll's calculations, the mid-winter temperature of the Northern Hemisphere would have been lowered 36° F., while the winter half of the year would have been twenty-six days longer than the summer half. This would bring the January mean temperature of England and Scotland almost down to zero, or about 30° F. of frost, a winter climate corresponding to that of Labrador, or the coast of Greenland on the Arctic Circle. But we must remember that the summer would be just as much hotter than it is now, and the problem to be solved is, whether the snow that fell in winter would accumulate to such an extent that it would not be melted in summer, and so go on increasing year by year till it covered the whole of Scotland, Ireland, and Wales, and much of England. Dr. Croll and Dr. Geikie answer without hesitation that it would. Sir Charles Lyell maintained that it would only do so when geographical conditions were favorable; while the late Mr. Belt has argued that eccentricity alone would not produce the effect unless aided by increased obliquity of the ecliptic, which, by extending the width of the polar regions, would increase the duration and severity of the winter to such an extent that snow and ice would be formed in the arctic and antarctic regions at the same time, whether the winter were in perihelion or aphelion.

The problem we have now to solve is a very difficult one, because we have no case at all parallel to it from which we can draw direct conclusions. It is, however, clear, from the various considerations we have already adduced, that the increased cold of winter, when the eccentricity was great and the sun in aphelion during that season, would not of itself produce a glacial epoch unless the amount of vapor supplied for condensation was also exceptionally great. The greatest quantity of snow falls in the arctic regions in summer and autumn, and with us the greatest quantity of rain falls in the autumnal months. It seems probable, then, that in all northern lands glaciation would commence when autumn occurred in aphelion. All the rain which falls on our mountains at that season would then fall as snow, and, being further increased by the snow of winter, would form accumulations which the summer might not be able to melt.

As time went on, and the aphelion occurred in winter, the perennial snow on the mountains would have accumulated to such an extent as to chill the spring and summer vapors, so that they too would fall as snow, and thus increase the amount of deposition; but it is probable that this would never in our latitude have been sufficient to produce glaciation, were it not for a series of climatal reactions which tend still further to increase the production of snow.

Action of Meteorological Causes in Intensifying Glaciation.— The trade-winds owe their existence to the great difference between the temperature of the equator and that of the poles, which causes a constant flow of air towards the equator. The strength of this flow depends on the difference of temperature and the extent of the cooled and heated masses of air, and this effect is now greatest between the south pole and the equator, owing to the much greater accumulation of ice in the antarctic regions. The consequence is that the southeast trades are stronger than the northeast, the neutral zone or belt of calms between them not being on the equator, but several degrees to the north of it. But just in proportion to the strength of the trade-winds is the strength of the anti-trades, that is, the upper return current which carries the warm moisture-laden air of the tropics towards the poles, descending in the temperate zone as west and southwest winds. These are now strongest in the Southern Hemisphere, and, passing everywhere over a wide ocean, they supply the moisture necessary to produce the enormous quantity of snow which falls in the antarctic area. During the period we are now discussing, however, this state of things would have been partially reversed. The south polar area, having its winter in perihelion, would probably have had less ice, while the north temperate and arctic regions would have been largely ice-clad; and the northeast trades would therefore be stronger than they are now. The southwesterly anti-trades would also be stronger in the same proportion, and would bring with them a greatly increased quantity of moisture, which is the prime necessity to produce a condition of glaciation.

But this is only one half of the effect that would be produced, for the increased force of the trades sets up another action which still further helps on the accumulation of snow and ice. It is now generally admitted that we owe much of our mild climate and our comparative freedom from snow to the influence of the Gulf Stream, which also ameliorates the climate of Scandinavia and Spitzbergen, as shown by the remarkable northward curvature of the isothermal lines, so that Drontheim, in N. lat. 62°, has the same mean temperature as Halifax (Nova Scotia), in N. lat. 45°. The quantity of heat now brought into the North Atlantic by the Gulf Stream depends mainly on the superior strength of the southeast trades. When the northeast trades were the more powerful, the Gulf Stream would certainly be of much less magnitude and velocity; while it is possible, as Dr. Croll thinks, that a large portion of it might be diverted southward, owing to the peculiar form of the east coast of South America, and so go to swell the Brazilian current and ameliorate the climate of the Southern Hemisphere.

That effects of this nature would follow from any increase of the arctic and decrease of the antarctic ice may be considered certain; and Dr. Croll has clearly shown that in this case cause and effect act and react on each other in a remarkable way. The increase of snow and ice in the Northern Hemisphere is the cause of an increased supply of moisture being brought by the more powerful anti-trades; and this greater supply of moisture leads to an extension of the ice, which reacts in still further increasing the supply of moisture. The same increase of snow and ice, by causing the northeast to be stronger than the southeast trade-winds, diminishes the force of the Gulf Stream, and this diminution lowers the temperature of the North Atlantic both in summer and winter, and thus helps on still further the formation and perpetuation of the icy mantle. It must also be remembered that these agencies are at the same time acting in a reverse way in the Southern Hemisphere, diminishing the supply of the moisture carried by the anti-trades, and increasing the temperature by means of more powerful southward ocean currents; and all this again reacts on the Northern Hemisphere, increasing yet further the supply of moisture by the more powerful southwesterly winds, while still further lowering the temperature by the southward diversion of the Gulf Stream.

Summary of Principal Causes of Glaciation.—I have now sufficiently answered the question why the short hot summer would not melt the snow which accumulated during the long cold winter (produced by high eccentricity and winter in aphelion), although the annual amount of heat received from the sun was exactly the same as it is now, and equal in the two hemispheres. It may be well, before going further, briefly to summarize the essential causes of this apparent paradox. These are—primarily, the fact that solar heat cannot be stored up, owing to its being continually carried away by air and water, while cold can be so stored up, owing to the comparative immobility of snow and ice ; and, in the second place, because the two great heat-distributing agencies, the winds and the ocean currents, are so affected by an increase of the snow and ice towards one pole and its diminution towards the other as to help on the process when it has once begun, and by their action and reaction produce a maximum of effect which, without their aid, would be altogether unattainable.

But even this does not exhaust the causes at work all tending in one direction. Snow and ice reflect heat to a much greater degree than does land or water. The heat, therefore, of the short summer would have far less effect than is due to its calculated amount in melting the snow, because so much of it would be lost by reflection. A portion of the reflected heat would, no doubt, warm the vapor in the atmosphere ; but this heat would be carried off to other parts of the earth, while a considerable portion of the whole would be lost in space. It must also be remembered that an enormous quantity of heat is used up in melting snow and ice, without raising its temperature ; each cubic foot of ice requiring as much heat to melt it as would raise nearly six cubic feet of water 30° F. It has, however, been argued that because when water is frozen it evolves just as much heat as it requires to melt it again, there is no loss of heat on the whole ; and, as this is adduced as a valid argument over and over again in every criticism of Dr. Croll's theory, it may be well to consider it a little more closely. In the act of freezing, no doubt, water gives up some of its heat to the surrounding air ; but *that air still remains below the freezing-point*, or freezing would not take place. The heat liberated by freezing is, therefore, what may be termed low-grade heat—heat incapable of melting snow or ice; while the heat absorbed while ice or snow is melting is high-grade heat, such as is capable of melting snow and supporting vegetable growth. Moreover, the low-grade heat liberated in the formation of snow is usually liberated high up in the atmosphere, where it may be carried off by winds to more southern latitudes; while the heat absorbed in melting the surface of snow and ice is absorbed close to the earth, and is thus prevented from warming the lower atmosphere, which is in contact with vegetation. The two phenomena, therefore, by no means counterbalance or counteract each other, as it is so constantly and superficially asserted that they do.

Effect of Clouds and Fog in Cutting off the Sun's Heat.—Another very important cause of diminution of heat during summer in a glaciated country would be the intervention of clouds and fogs, which would reflect or absorb a large proportion of the sun heat and prevent it reaching the surface of the earth; and such a cloudy atmosphere would be a necessary result of large areas of high land covered with snow and ice. That such a prevalence of fogs and cloud is an actual fact in all ice-clad countries has been shown by Dr. Croll most conclusively, and he has further shown that the existence of perpetual snow often depends upon it. South Georgia, in the latitude of Yorkshire, is almost, and Sandwich Land, in the latitude of the North of Scotland, is entirely, covered with perpetual snow; yet in their summer the sun is three million miles nearer the earth than it is in our summer, and the heat actually received from the sun must be sufficient to raise the temperature 20° F. higher than in the same latitudes in the Northern Hemisphere, were the conditions equal—instead of which their summer temperature is probably fully 20° lower. The chief cause of this can only be that the heat of the sun does not reach the surface of the earth; and that this is the fact is testified by all antarctic voyagers. Darwin notes the cloudy sky and constant moisture of the southern part of Chili, and in his remarks on the climate and productions of the antarctic islands he says, "In the Southern Ocean the winter is not so excessively cold, but the summer is far less hot (than in the north), *for the clouded sky seldom allows the sun to warm the ocean,* itself a bad absorbent of heat; and hence the main temperature of the year, which regulates the zone of perpetually congealed under-soil, is low." Sir James Ross, Lieutenant Wilkes, and other antarctic voyagers speak of the snow-storms, the absence of sunshine, and the freezing temperature in the height of summer; and Dr. Croll shows that this is a constant phenomenon accompanying the presence of large masses of ice in every part of the world.[1]

In reply to the objections of a recent critic, Dr. Croll has given a new proof of this important fact by comparing the known amount of snow-fall with the equally well-known melting power of direct sun heat in different latitudes. He says, " The annual precipitation on Greenland in the form of snow and rain, according to Dr. Rink, amounts to only twelve inches, and two inches of this he considers is never melted, but is carried away in the form of icebergs. The quantity of heat received at the equator from sunrise to sunset, if none were cut off by the atmosphere, would melt three and a third inches of ice, or a hundred feet in a year. The quantity received between latitude 60° and 80°, which is that of Greenland, is, according to Meech, one half that received at the equator. The heat received by Greenland from the sun, if none were cut off by the atmosphere, would therefore melt fifty feet of ice per annum, or fifty times the amount of snow which falls on that continent. What, then, cuts off the ninety-eight per cent. of the sun's heat ?" The only possible answer is that it is the clouds and fog during a great part of the summer, and reflection from the surface of the snow and ice when these are absent.

South Temperate America as Illustrating the Influence of Astronomical Causes on Climate.—Those persons who still doubt the effect of winter in aphelion with a high degree of eccentricity in producing glaciation should consider how the condition of south temperate America at the present day is explicable if

---

[1] For numerous details and illustrations, see the paper "On Ocean Currents in Relation to the Physical Theory of Secular Changes of Climate," in the *Philosophical Magazine*, 1870.

they reject this agency. The line of perpetual snow in the Southern Andes is so low as 6000 feet in the same latitude as the Pyrenees. In the latitude of the Swiss Alps, mountains only 6200 feet high produce immense glaciers which descend to the sea-level; while, in the latitude of Cumberland, mountains only from 3000 to 4000 feet high have every valley filled with streams of ice descending to the sea-coast and giving off abundance of huge icebergs.[1]   Here we have exactly the condition of things to which England and Western Europe were subjected during the latter portion of the glacial epoch, when every valley in Wales, Cumberland, and Scotland had its glacier; and to what can this state of things be imputed if not to the fact that there is now a moderate amount of eccentricity, and the winter of the Southern Hemisphere is in aphelion? The mere geographical position of the southern extremity of America does not seem especially favorable to the production of such a state of glaciation. The land narrows from the tropics southward, and terminates altogether in about the latitude of Edinburgh; the mountains are of moderate height; while during summer the sun is three millions of miles nearer, and the heat received from it is equivalent to a rise of 20° F. as compared with the same season in the Northern Hemisphere. The only important differences are the open Southern Ocean, the longer and colder winter, and the general low temperature caused by the south polar ice. But the great accumulation of south polar ice is itself due to the great extent of high land within the Antarctic Circle acted upon by the long cold winter and furnished with moisture by the surrounding wide ocean. These conditions of high land and open ocean we know did not prevail to so great an extent in the Northern Hemisphere during the glacial epoch as they do in the Southern Hemisphere at the present time; but the other acting cause—the long cold winter—existed in a far higher degree, owing to the eccentricity being about three times as much as it is now. It is, so far as we know or are justified in believing, the *only* efficient cause of glaciation which was undoubtedly much more powerful at that time; and

---

[1] See Darwin's "Naturalist's Voyage round the World," 2d ed., pp. 244–251.

we are therefore compelled to accept it as the most probable cause of the much greater glaciation which then prevailed.

Geographical Changes, how far a Cause of Glaciation.— Messrs. Croll and Geikie have both objected to the views of Sir Charles Lyell as to the preponderating influence of the distribution of land and sea on climate; and they maintain that if the land were accumulated almost wholly in the equatorial regions, the temperature of the earth's surface as a whole would be lowered, not raised, as Sir Charles Lyell maintained. The reason given is that the land being heated heats the air, which rises and thus gives off much of the heat to space, while the same area covered with water would retain more of the heat, and by means of currents carry it to other parts of the earth's surface. But although the mean temperature of the whole earth might be somewhat lowered by such a disposition of the land, there can be little doubt that it would render all extremes of temperature impossible, and that even during a period of high eccentricity there would be no glacial epochs, and perhaps no such thing as ice anywhere produced. This would result from there being no land near the poles to retain snow, while the constant interchange of water by means of currents between the polar and tropical regions would most likely prevent ice from ever forming in the sea. On the other hand, were all the land accumulated in the polar and temperate regions, there can be little doubt that a state of almost perpetual glaciation of much of the land would result, notwithstanding that the whole earth should theoretically be at a somewhat higher temperature. Two main causes would bring about this glaciation. A very large area of elevated land in high latitudes would act as a powerful condenser of the enormous quantity of vapor produced by the whole of the equatorial and much of the temperate regions being areas of evaporation, and thus a greater accumulation of snow and ice would take place around both poles than would be possible under any other conditions. In the second place, there would be little or no check to this accumulation of ice, because, owing to the quantity of land around the polar areas, warm oceanic currents could not reach them, while the warm winds would necessarily bring so much moisture that they would help on instead of checking the process of ice-accumulation. If we suppose the continents to be of the same total area and to have the same extent and altitude of mountain-ranges as the present ones, these mountains must necessarily offer an almost continuous barrier to the vapor-bearing winds from the south, and the result would probably be that three fourths of the land would be in the ice-clad condition of Greenland, while a comparatively narrow belt of the more southern lowlands would alone afford habitable surfaces or produce any woody vegetation.

Notwithstanding, therefore, the criticism above referred to, I believe that Sir Charles Lyell was substantially right, and that the two ideal maps given in the "Principles of Geology" (11th ed., Vol. I., p. 270), if somewhat modified so as to allow a freer passage of currents in the tropics, do really exhibit a condition of the earth which by geographical changes alone would bring about a perpetual summer or an almost universal winter. But we have seen in our sixth chapter that there is the strongest cumulative evidence, almost amounting to demonstration, that for all known geological periods our continents and oceans have occupied the same general position they do now, and that no such radical changes in the distribution of sea and land as imagined, by way of hypothesis, by Sir Charles Lyell have ever occurred. Such an hypothesis, however, is not without its use in our present inquiry; for if we obtain thereby a clear conception of the influence of such great changes on climate, we are the better able to appreciate the tendency of lesser changes, such as have undoubtedly often occurred.

Land as a Barrier to Ocean Currents.—We have seen already the great importance of elevated land to serve as condensers and ice-accumulators; but there is another and hardly less important effect that may be produced by an extension of land in high latitudes, which is, to act as a barrier to the flow of ocean currents. In the region with which we are more immediately interested it is easy to see how a comparatively slight alteration of land and sea, such as has undoubtedly occurred, would produce an enormous effect on climate. Let us suppose, for instance, that the British Isles again became continental, and that this continental land extended across the Faroe Islands and Iceland to Greenland. The whole of the warm waters of the Atlantic, with the Gulf Stream, would then be shut out from Northern Europe, and the result would almost certainly be that snow would accumulate on the high mountains of Scandinavia till they became glaciated to as great an extent as Greenland, and the cold thus produced would react on our own country and cover the Grampians with perpetual snow, like mountains of the same height at even a lower latitude in South America.

If a similar change were to occur on the opposite side of the Atlantic, very different effects would be produced. Suppose, for instance, the east side of Greenland were to sink considerably, while on the west the sea-bottom were to rise in Davis's Strait so as to unite Greenland with Baffin's Land, thus stopping altogether the cold arctic current with its enormous stream of icebergs from the west coast of Greenland. Such a change might cause a great accumulation of ice in the higher polar latitudes, but it would certainly produce a wonderful ameliorating effect on the climate of the east coast of North America, and might raise the temperature of Labrador to that of Scotland. Now these two changes have almost certainly occurred, either together or separately, during the Tertiary period, and they must have had a considerable effect either in aiding or checking the terrestrial and astronomical causes affecting climate which were then in operation.

It would be easy to suggest other probable changes which would produce a marked effect on climate; but we will only refer to the subsidence of the Isthmus of Panama, which has certainly happened more than once in Tertiary times. If this subsidence were considerable, it would have allowed much of the accumulated warm water which initiates the Gulf Stream to pass into the Pacific; and if this occurred while astronomical causes were tending to bring about a cold period in the Northern Hemisphere, the resulting glaciation might be exceptionally severe. The effect of this change would, however, be neutralized if at the same epoch the Lesser and Greater Antilles formed a connected land.

Now, as such possible and even probable geographical changes are very numerous, they must have produced important effects;

and though we may admit that the astronomical causes already explained were the most important in determining the last glacial epoch, we must also allow that geographical changes must often have had an equally important and perhaps even a preponderating influence on climate. We must also remember that changes of land and sea are almost always accompanied by elevation or depression of the pre-existing land; and whereas the former produces its chief effect by diverting the course of warm or cold oceanic currents, the latter is of not less importance in adding to or diminishing those areas of condensation and ice-accumulation which, as we have seen, are the most efficient agents in producing glaciation.

If, then, Sir Charles Lyell may have somewhat erred in attaching too exclusive an importance to geographical changes as bringing about mutations of climate, his critics have, I think, attached far too little importance to these changes. We know that they have always been in progress to a sufficient extent to produce important climatal effects; and we shall probably be nearest the truth if we consider that great extremes of cold have only occurred when astronomical and geographical causes were acting in the same direction, and thus produced a cumulative result; while, through the agency of warm oceanic currents, the latter alone have been the chief cause of mild climates in high latitudes, as we shall prove in our next chapter.[1]

---

[1] The influence of geographical changes on climate is now held by many geologists who oppose what they consider the extravagant hypotheses of Dr. Croll. Thus, Professor Dana imputes the glacial epoch chiefly, if not wholly, to elevation of the land caused by the lateral pressure due to shrinking of the earth's crust that has caused all other elevations and depressions. He says, "Now that elevation of the land over the higher latitudes which brought on the glacial era is a natural result of the same agency, and a natural and almost necessary counterpart of the coral-island subsidence which must have been then in progress. The accumulating, folding, solidification, and crystallization of rocks attending all the rock-making and mountain-making through the Palæozoic, Mesozoic, and Cenozoic eras had greatly stiffened the crust in these parts; and hence, in after-times, the continental movements resulting from the lateral pressure necessarily appeared over the more northern portions of the continent, where the accumulations and other changes had been relatively small. To the subsidence which followed the elevation, the weight of the ice-cap may have contributed in some small degree. But the great balancing movements of the crust of the continental and oceanic areas then going forward must have had a greatly pre-

On the Theory of Interglacial Periods and their Probable Character.—The theory by which the glacial epoch is here explained is one which apparently necessitates repeated changes from glacial to warm periods, with all the consequences and modifications both of climate and physical geography which follow or accompany such changes. It is essentially a theory of alternation; and it is certainly remarkable in how many cases geologists have independently deduced some alternations of climate as probable. Such are the interglacial deposits indicating a mild climate, both in Europe and America; an early phase of very severe glaciation when the "till" was deposited, with later less extensive glaciation when moraines were left in the valleys; several successive periods of submergence and elevation, the later ones becoming less and less in amount, as indicated by the raised beaches slightly elevated above our present coast-line; and, lastly, the occurrence in the same deposits of animal remains indicating both a warm and a cold climate, and especially the existence of the hippopotamus in Yorkshire soon after the period of extreme glaciation.

But although the evidence of *some* alternations of climate seems indisputable, and no suggestion of any adequate cause for them other than the alternating phases of precession during high eccentricity has been made, it by no means follows that these changes were always very great—that is to say, that the ice completely disappeared and a warm climate prevailed throughout the whole year. It is quite evident that during the height of the glacial epoch there was a combination of causes at work which led to a large portion of Northwestern Europe and Eastern America being buried in ice to a greater extent even than

---

ponderating effect in the oscillating agency of all time—lateral pressure within the crust."—*American Journal of Science and Arts*, 3d Series, Vol. IX., p. 318.

In the second edition of his "Manual of Geology," Professor Dana suggests elevation of arctic lands sufficient to exclude the Gulf Stream as a source of cold during glacial epochs. This, he thinks, would have made an epoch of cold at any era of the globe. A deep submergence of Behring Strait, letting in the Pacific warm current to the polar area, would have produced a mild arctic climate like that of the Miocene period. When the warm current was shut out from the polar area, it would yet reach near to it, and bring with it that abundant moisture necessary for glaciation.— Pp. 541, 755, 756.

Greenland is now, since it certainly extended beyond the land and filled up all the shallow seas between our islands and Scandinavia. Among these causes we must reckon a diminution of the force of the Gulf Stream, or its being diverted from the northwestern coasts of Europe; and what we have to consider is, whether the alteration from a long cold winter and short hot summer to a short mild winter and long cool summer would greatly affect the amount of ice *if the ocean currents remained the same.* The force of these currents are, it is true, by our hypothesis, modified by the increase or diminution of the ice in the two hemispheres alternately, and they then react upon climate; but they cannot be thus changed till after the ice-accumulation has been considerably affected by other causes. Their direction may indeed be greatly changed by slight alterations in the outline of the land, while they may be barred out altogether by other alterations of not very great amount; but such changes as these have no relation to the alteration of climates caused by the changing phases of precession.

Now the existence at the present time of an ice-clad Greenland is an anomaly in the Northern Hemisphere only to be explained by the fact that cold currents from the polar area flow down both sides of it. In Eastern Asia we have the lofty Stanovoi Mountains in the same latitude as the southern part of Greenland, which, though their summits are covered with perpetual snow, give rise to no ice-sheet, and, apparently, even to no important glaciers—a fact undoubtedly connected with the warm Japan current flowing partially into the Sea of Ochotsk. So in Northwest America we have the lofty coast range culminating in Mt. St. Elias, nearly 15,000 feet high, and an extensive tract of high land to the north and northwest, with glaciers comparable in size with those of New Zealand, although situated in latitude 60° instead of in latitude 45°. Here, too, we have the main body of the Japan current turning east and south, and thus producing a mild climate, little inferior to that of Norway, warmed by the Gulf Stream. We thus have it made clear that could the two arctic currents be diverted from Greenland, that country would become free from ice, and might even be completely forest-clad and inhabitable; while if the Japan current were to be diverted from the coast of North America and a cold current come out of Behring Strait, the entire northwestern extremity of America would even now become buried in ice.

Now it is the opinion of the best American geologists that during the height of the glacial epoch Northeastern America was considerably elevated.[1] This elevation would bring the wide area of the banks of Newfoundland far above water, causing the American coast to stretch out in an immense curve to a point more than 600 miles east of Halifax ; and this would certainly divert much of the greatly reduced Gulf Stream straight across to the coast of Spain. The consequence of such a state of things would probably be that the southward-flowing arctic currents would be much reduced in velocity ; and the enormous quantity of icebergs continually produced by the ice-sheets of all the lands bordering the North Atlantic would hang about their shores and the adjacent seas, filling them with a dense ice-pack far surpassing that of the antarctic regions, and chilling the atmosphere so as to produce constant clouds and fog with almost perpetual snow-storms, even at midsummer, such as now prevail in the worst portions of the Southern Ocean.

But when such was the state of the North Atlantic (and, however caused, such *must* have been its state during the height of the glacial epoch), can we suppose that the mere change from the distant sun in winter and near sun in summer to the reverse could bring about any important alteration—*the physical and geographical causes of glaciation remaining unchanged ?* For, certainly, the less powerful sun of summer, even though lasting somewhat longer, could not do more than the much more powerful sun did during the phase of summer in perihelion, while during the less severe winters the sun would have far less power than when it was equally near and at a very much greater altitude in summer. It seems to me, therefore, quite certain that whenever *extreme* glaciation has been brought about by high eccentricity combined with favorable geographical and physical causes (and without this combination it is doubtful whether *extreme* glaciation would ever occur), then the ice-sheet will *not* be

---

[1] Dana's " Manual of Geology," 2d ed., p. 540.

removed during the alternate phases of precession so long as these geographical and physical causes remain unaltered. It is true that the warm and cold oceanic currents, which are the most important agents in increasing or diminishing glaciation, depend for their strength and efficiency upon the comparative extents of the northern and southern ice-sheets; but these ice-sheets cannot, I believe, increase or diminish to any important extent unless some geographical or physical change first occurs.[1]

If this argument is valid, then it would follow that, so long as eccentricity was high, whatever condition of climate was brought about by it in combination with geographical causes would persist through several phases of precession; but this would not necessarily be the case when the eccentricity itself changed and became more moderate. It would then depend upon the proportionate effect of climatal and geographical causes in producing glaciation as to what change would be produced by the changing phases of precession; and we can best examine this question by considering the probable effect of the change in precession during the next period of 10,500 years, with the present moderate degree of eccentricity.

Probable Effect of Winter in Aphelion on the Climate of Britain. — Let us then suppose the winters of the Northern

[1] In reply to an objection of a somewhat similar nature to this, Dr. Croll has recently stated (*Geological Magazine*, October, 1879) that he "has not assumed that the comparative disappearance of the ice on the warm hemisphere during the period of high eccentricity is due to any additional heat derived from the sun in consequence of the greater length of the summer," but that "the real and effective cause of the disappearance of the ice was the enormous transferrence of equatorial heat to temperate and polar regions by means of ocean currents." But this is surely arguing in a circle; for the ocean currents are mainly due to the difference of temperature of the polar and equatorial areas combined with the peculiar form and position of the continents, and some one or more of these factors must be altered *before* the ocean currents towards the north pole can be increased. The only factor available is the antarctic ice; and if this were largely increased, the northward-flowing currents might be so increased as to melt some of the arctic ice. But the very same argument applies to both poles. Without some geographical change the antarctic ice could not materially diminish during its winter in perihelion, nor increase to any important extent during the opposite phase. We therefore seem to have no available agency by which to get rid of the ice over a glaciated country *so long as the geographical conditions remained unchanged and the eccentricity continued high.*

Hemisphere to become longer and much colder, the summers being proportionately shorter and hotter, without any other change whatever. The long cold winter would certainly bring down the snow-line considerably, covering large areas of high land with snow during the winter months, and extending all glaciers and ice-fields. This would chill the superincumbent atmosphere to such an extent that the warm sun and winds of spring and early summer would bring clouds and fog, so that the sun heat would be cut off and much vapor be condensed as snow. The greater sun heat of summer would, no doubt, considerably reduce the snow and ice; but it is, I think, quite certain that the extra accumulation would not be all melted, and that therefore the snow-line would be permanently lowered. This would be a necessary result, because the greater part of the increased cold of winter would be stored up in snow and ice, while the increased heat of summer could not be in any way stored up, but would be largely prevented from producing any effect, by reflection from the surface of the snow, and by the intervention of clouds and fog which would carry much of the heat they received to other regions. It follows that 10,000 years hence, when our winter occurs in aphelion (instead of, as now, in perihelion), there will be produced a colder climate, independently of any change of land and sea, of heights of mountains, or the force of currents.

But if this is true, then the reverse change, bringing the sun back into exactly the same position with regard to us as it is in now (all geographical and physical conditions remaining unchanged), would certainly bring back again our present milder climate. The change either way would not probably be very great, but it might be sufficient to bring the snow-line down to 3000 feet in Scotland, so that all the higher mountains had their tops covered with perpetual snow. This perpetual snow, down to a fixed line, would be kept up by the necessary supply of snow falling during autumn, winter, and spring, and this would, as we have seen, depend mainly on the increased length and greatly increased cold of the winter. As both the duration and the cold of winter decreased, the amount of snow would certainly decrease; and of this lesser quantity of snow a larger proportion would be melted by the longer, though somewhat cooler, summer. This would follow because the total amount of sun heat received during the summer would be the same as before, while it would act on a less quantity of snow; there would thus be a smaller surface to reflect the heat, and a smaller condensing area to produce fogs, while the diminished intensity of the sun would produce a less dense canopy of clouds, which have been shown to be of prime importance in checking the melting of snow by the sun. We have considered this case, for simplicity of reasoning, on the supposition that all geographical and physical causes remained unchanged. But if an alteration of the climate of the whole north temperate and arctic zones occurred, as here indicated, this would certainly affect both the winds and currents, in the manner already explained (see p. 135), so as to react upon climate and increase the differences produced by phases of precession. How far that effect would be again increased by corresponding but opposite changes in the Southern Hemisphere it is impossible to say. It may be that existing geographical and physical conditions are there such potent agents in producing a state of glaciation that no change in the phases of precession would materially affect it. Still, as the climate of the whole Southern Hemisphere is dominated by the great mass of ice within the Antarctic Circle, it seems probable that if the winter were shorter and the summer longer the quantity of ice would slightly diminish; and this would again react on the northern climate as already fully explained.

The Essential Principle of Climatal Change Restated.—The preceding discussion has been somewhat lengthy, owing to the varied nature of the facts and arguments adduced and the extreme complexity of the subject. But if, as I venture to hope, the principle here laid down is a sound one, it will be of the greatest assistance in clearing away some of the many difficulties that beset the whole question of geological climates. This principle is, briefly, that the great features of climate are determined by a combination of causes, of which geographical conditions and the degree of eccentricity of the earth's orbit are by far the most important; that when these combine to produce a severe glacial epoch, the changing phases of precession every 10,500 years have very little, if any, effect on the character of the climate, as mild or glacial, though it may modify the seasons; but when the eccentricity becomes moderate and the resulting climate less severe, then the changing phases of precession bring about a considerable alteration, and even a partial reversal, of the climate.

The reason of this may perhaps be made clearer by considering the stability of either very cold or very mild conditions, and the comparative instability of an intermediate state of climate. When a country is largely covered with ice, we may look upon it as possessing the accumulated or stored-up cold of a long series of preceding winters; and, however much heat is poured upon it, its temperature cannot be raised above the freezing, point till that store of cold is got rid of—that is, till the ice is all melted. But the ice itself, when extensive, tends to its own preservation, even under the influence of heat; for the chilled atmosphere becomes filled with fog, and this keeps off the sun heat; and then snow falls even during summer, and the stored-up cold does not diminish during the year. When, however, only a small portion of the surface is covered with ice, the exposed earth becomes heated by the hot sun, this warms the air, and the warm air melts the adjacent ice. It follows that, towards the equatorial limits of a glaciated country, alternations of climate may occur during a period of high eccentricity; while nearer the pole, where the whole country is completely ice-clad, no amelioration may take place. Exactly the same thing will occur inversely with mild arctic climates; but this is a subject which will be discussed in the next chapter.

This view of the character of the last glacial epoch strictly corresponds with the facts adduced by geologists. The interglacial deposits never exhibit any indication of a climate whose warmth corresponded to the severity of the preceding cold, but rather of a partial amelioration of that cold; while it is only the very latest of them, which we may suppose to have occurred when the eccentricity was considerably diminished, that exhibit any indications of a climate at all warmer than that which now prevails.[1]

---

[1] In a recent number of the *Geological Magazine* (April, 1880) Mr. Searles V.

Probable Date of the Glacial Epoch.—The state of extreme glaciation in the Northern Hemisphere, of which we gave a general description at the commencement of the preceding chapter, is a fact of which there can be no doubt whatever, and it occurred at a period so recent, geologically, that all the mollusca were the same as species still living. There is clear geological proof, however, that considerable changes of sea and land, and a large amount of valley-denudation, took place during and since the glacial epoch; while, on the other hand, the surface-markings produced by the ice have been extensively preserved; and, taking all these facts into consideration, the period of about 200,000 years since it reached its maximum, and about 80,000 years since it passed away, is generally considered by geologists to be

---

Wood adduces what he considers to be the "conclusive objection" to Dr. Croll's eccentricity theory, which is, that during the last glacial epoch Europe and North America were glaciated very much in proportion to their respective climates now, which are generally admitted to be due to the distribution of oceanic currents. But Dr. Croll admits his theory "to be baseless unless there was a complete diversion of the warm ocean currents from the hemisphere glaciated," in which case there ought to be no difference in the extent of glaciation in Europe and North America. Whether or not this is a correct statement of Dr. Croll's theory, the above objection certainly does not apply to the views here advocated; but as I also hold the "eccentricity theory" in a modified form, it may be as well to show why it does not apply. In the first place, I do not believe that the Gulf Stream was "completely diverted" during the glacial epoch, but that it was diminished in force, and (as described on p. 136) *partly* diverted southward. A portion of its influence would, however, still remain to cause a difference between the climates of the two sides of the Atlantic; and to this must be added two other causes—the far greater penetration of warm sea-water into the European than into the North American continent, and the proximity to America of the enormous ice-producing mass of Greenland. We have thus three distinct causes, all combining to produce a more severe winter climate on the west than on the east of the Atlantic during the glacial epoch; and though the first of these —the Gulf Stream—was not nearly so powerful as it is now, neither is the difference indicated by the ice-extension in the two countries so great as the present difference of winter temperature, which is the essential point to be considered. The ice-sheet of the United States is usually supposed to have extended about ten, or, at most, twelve, degrees farther south than it did in Western Europe, whereas we must go twenty degrees farther south in the former country to obtain the same mean winter temperature we find in the latter, as may be seen by examining any map of winter isothermals. This difference very fairly corresponds to the difference of conditions existing during the glacial epoch and the present time, so far as we are able to estimate them, and it certainly affords no grounds of objection to the theory by which the glaciation is here explained.

ample. There seems, therefore, to be little doubt that in increased eccentricity we have found one of the chief exciting causes of the glacial epoch, and that we are therefore able to fix its date with a considerable probability of being correct. The enormous duration of the glacial epoch itself (including its interglacial mild or warm phases) as compared with the lapse of time since it finally passed away is a consideration of the greatest importance, and has not yet been taken fully into account in the interpretation given by geologists of the physical and biological changes that were coincident with and probably dependent on it.

Changes of the Sea-level Dependent on Glaciation.—It has been pointed out by Dr. Croll that many of the changes of level of sea and land which occurred about the time of the glacial epoch may be due to an alteration of the sea-level caused by a shifting of the earth's centre of gravity; and physicists have generally admitted that the cause is a real one, and must have produced some effect of the kind indicated. It is evident that if ice-sheets several miles in thickness were removed from one polar area and placed on the other, the centre of gravity of the earth would shift towards the heavier pole, and the sea would necessarily follow it, and would rise accordingly. Extreme glacialists have maintained that during the height of the glacial epoch an ice-cap extended from about 50° N. lat. in Europe, and 40° N. lat. in America, continually increasing in thickness till it reached at least six miles thick at the pole; but this view is now generally given up. A similar ice-cap is, however, believed to exist on the antarctic pole at the present day, and its transferrence to the Northern Hemisphere would, it is calculated, produce a rise of the ocean to the extent of 800 or 1000 feet. We have, however, shown that the production of any such ice-cap is improbable, if not impossible, because snow and ice can only accumulate where precipitation is greater than melting and evaporation, and this is never the case except in areas exposed to the full influence of the vapor-bearing winds. The outer rim of the ice-sheet would inevitably exhaust the air of so much of its moisture that what reached the inner parts would produce far less snow than would be melted by the long hot days of summer. The accumulations of ice were therefore probably confined, in the Northern Hemisphere, to the coasts exposed to moist winds, and where elevated land and mountain-ranges afforded condensers to initiate the process of glaciation; and we have already seen that the evidence strongly supports this view. Even with this limitation, however, the mass of accumulated ice would be enormous, as indeed we have positive evidence that it was, and might have caused a sufficient shifting of the centre of gravity of the earth to produce a submergence of about 150 or 200 feet.

But this would only be the case if the accumulation of ice on one pole was accompanied by a diminution on the other, and this may have occurred to a limited extent during the earlier stages of the glacial epoch, when alternations of warmer and colder periods would be caused by winter occurring in perihelion or aphelion. If, however, as we maintain, no such alternations occurred when the eccentricity was near its maximum, then the ice would accumulate in the Southern Hemisphere at the same time as in the Northern, unless changed geographical conditions, of which we have no evidence whatever, prevented such accumulations. That there was such a greater accumulation of ice is shown by the traces of ancient glaciers in the Southern Andes and in New Zealand, and also, according to several writers, in South Africa; and the indications in all these localities point to a period so recent that it must almost certainly have been contemporaneous with the glacial period of the Northern Hemisphere.[1] This greater accumulation of ice in both hemispheres

---

[1] The recent extensive glaciation of New Zealand is generally imputed by the local geologists to a greater elevation of the land; but I cannot help believing that the high phase of eccentricity which caused our own glacial epoch was, at all events, an assisting cause. This is rendered more probable if taken in connection with the following very definite statement of glacial markings in South Africa. Captain Aylward, in his "Transvaal of To-day" (p. 171), says, "It will be interesting to geologists and others to learn that the entire country, from the summits of the Quathlamba to the junction of the Vaal and Orange rivers, shows marks of having been swept over, and that at no very distant period, by vast masses of ice from east to west. The striations are plainly visible, scarring the older rocks, and marking the hill-sides — getting lower and lower and less visible as, descending from the mountains, the kopjies (small hills) stand wider apart; but, wherever the hills narrow

would lower the whole ocean by the quantity of water abstracted
from it, while any want of perfect synchronism between the
decrease of the ice at the two poles would cause a movement of
the centre of gravity of the earth, and a slight rise of the sea-
level at one pole and depression at the other. It is also generally believed that a great accumulation of ice might cause subsidence by its pressure on the flexible crust of the earth, and we
thus have a very complex series of agents leading to elevations
and subsidences of limited amount, such as seem always to have
accompanied glaciation. This complexity of the causes at work

---

towards each other, again showing how the vast ice-fields were checked, thrown up,
and raised against their eastern extremities."

This passage is evidently written by a person familiar with the phenomena of
glaciation ; and as Captain Aylward's preface is dated from Edinburgh, he has probably seen similar markings in Scotland. The country described consists of the most
extensive and lofty plateau in South Africa, rising to a mountain-knot with peaks
more than 10,000 feet high, thus offering an appropriate area for the condensation
of vapor and the accumulation of snow. At present, however, the mountains do not
reach the snow-line, and there is no proof that they have been much higher in recent
times, since the coast of Natal is now said to be rising. It is evident that no slight
elevation would now lead to the accumulation of snow and ice in these mountains,
situated as they are between 27° and 30° S. lat. ; since the Andes, which in 32°
S. lat. reach 23,300 feet high, and in 28° S. lat. 20,000, with far more extensive
plateaus, produce no ice-fields. We cannot, therefore, believe that a few thousand
feet of additional elevation, even if it occurred so recently as indicated by the presence of striations, would have produced the remarkable amount of glaciation above
described ; while from the analogy of the Northern Hemisphere we may well believe
that it was mainly due to the same high eccentricity that led to the glaciation of
Western and Central Europe and Eastern North America.

These observations confirm those of Mr. G. W. Stow, who, in a paper published
in the *Quarterly Journal of the Geological Society* (Vol. XXVII., p. 539), describes
similar phenomena in the same mountains, and also mounds and ridges of unstratified
clay packed with angular boulders; while farther south the Stormberg Mountains
are said to be similarly glaciated, with immense accumulations of morainic matter in
all the valleys. We have here *all* the chief surface phenomena characteristic of a
glaciated country only a few degrees south of the tropic; and, taken in connection
with the evidence of Professor Hartt, who describes true moraines near Rio Ja-
neiro, situated on the tropic itself, we can hardly doubt the occurrence of some
general and wide-spread cause of glaciation in the Southern Hemisphere at a period
so recent that the superficial phenomena are as well preserved as in Europe. Such
evidences of recent glaciation in the Southern Hemisphere are quite inexplicable
without calling in the aid of the recent phase of high eccentricity; and they may be
fairly claimed as adding another link to the long chain of argument in favor of the
theory here advocated.

may explain the somewhat contradictory evidence as to rise and
fall of land, some authors maintaining that it stood higher, and
others lower, during the glacial period.

*The State of the Planet Mars, as Bearing on the Theory of
Eccentricity, as a Cause of Glacial Periods.*—It is well known
that the polar regions of the planet Mars are covered with white
patches or disks, which undergo considerable alterations of size
according as they are more or less exposed to the sun's rays.
They have therefore been generally considered to be snow or
ice caps, and to prove that Mars is now undergoing something
like a glacial period. It must always be remembered, however,
that we are very ignorant of the exact physical conditions of
the surface of Mars. It appears to have a cloudy atmosphere
like our own, but the gaseous composition of that atmosphere
may be different, and the clouds may be formed of other matter
besides aqueous vapor. Its much smaller mass and attractive
power must have an effect on the nature and extent of these
clouds, and the heat of the sun may consequently be modified
in a way quite different from anything that obtains upon our
earth. Bearing these difficulties and uncertainties in mind, let
us see what are the actual facts connected with the supposed
polar snows of Mars.[1]

Mars offers an excellent subject for comparison with the earth
as regards this question, because its eccentricity is now a little
greater than the maximum eccentricity of the earth during the
last million years (Mars eccentricity, 0.0931; earth eccentricity,
850,000 years back, 0.0707); the inclination of its axis is also a
little greater than ours (Mars, 28° 51′; earth, 23° 27′), and both
Mars and the earth are so situated that they now have the winter of their Northern hemispheres in perihelion, that of their
Southern hemispheres being in aphelion. If, therefore, the physical condition of Mars were the same, or nearly the same, as that
of the earth, all circumstances combine, according to Mr. Croll's

---

[1] The astronomical facts connected with the motions and appearance of the planet
are taken from a paper by Mr. Edward Carpenter, M.A., in the *Geological Magazine* of March, 1877, entitled "Evidence Afforded by Mars on the Subject of Glacial
Periods," but I arrive at somewhat different conclusions from those of the writer of
the paper.

hypothesis, to produce a severe glacial epoch in its *southern*, with a perpetual spring or summer in its *northern*, hemisphere; while, on the hypothesis here advocated, we should expect glaciation at both poles. As a matter of fact, Mars has two snow-caps of nearly equal magnitude at their maximum in winter, but varying very unequally. The northern cap varies slowly and little, the southern varies rapidly and largely.

In the year 1830 the *southern* snow was observed, during the midsummer of Mars, to diminish to half its former diameter in a fortnight (the duration of such phenomena on Mars being reckoned in Martian months equivalent to one twelfth of a Martian year). Thus on June 23 it was 11° 30′ in diameter, and on July 9 had diminished to 5° 46′, after which it rapidly increased again. In 1837 the same cap was observed near its maximum in winter, and was found to be about 35° in diameter.

In the same year the *northern* snow-cap was observed during its summer, and was found to vary as follows:

May 4, Diameter of spot,		31° 24′
June 4, "	"	28° 0′
" 17, "	"	22° 54′
July 4, "	"	18° 24′
" 12, "	"	15° 20′
" 20, "	"	18° 0′

We thus see that Mars has two permanent snow-caps of nearly equal size in winter, but diminishing very unequally in summer, when the southern cap is reduced to nearly one third the size of the northern; and this fact is held by Mr. Carpenter, as it was by the late Mr. Belt, to be opposed to the view of the hemisphere which has winter in aphelion (as the southern now has both in the earth and Mars) having been alone glaciated during periods of high eccentricity.[1]

---

[1] In an article in *Nature* of January 1, 1880, the Rev. T. W. Webb states that in 1877 the pole of Mars (? the south pole) was, according to Schiaparelli, entirely free of snow. He remarks also on the regular contour of the supposed snows of Mars as offering a great contrast to ours, and also the strongly marked dark border which has often been observed. On the whole, Mr. Webb seems to be of opinion that there can be no really close resemblance between the physical condition of the earth and Mars, and that any arguments founded on such supposed similarity are therefore untrustworthy.

Before, however, we can draw any conclusion from the case of Mars, we must carefully scrutinize the facts and the conditions they imply. In the first place, there is evidently this radical difference between the state of Mars now and of the earth during a glacial period—that Mars has no great ice-sheets spreading over her temperate zone as the earth undoubtedly had. This we know from the fact of the *rapid* disappearance of the white patches over a belt three degrees wide in a fortnight (equal to a width of about 100 miles of our measure), and in the Northern Hemisphere of eight degrees wide (about 280 miles) between May 4 and July 12. Even with our much more powerful sun, which gives us more than twice as much heat as Mars receives, no such diminution of an ice-sheet, or of glaciers of even moderate thickness, could possibly occur; but the phenomenon is, on the contrary, exactly analogous to what actually takes place on the plains of Siberia in summer. These, as I am informed by Mr. Seebohm, are covered with snow during winter and spring to a depth of six or eight feet, which diminishes very little even under the hot suns of May, till warm winds combine with the sun in June, when in about a fortnight the whole of it disappears, and a little later the whole of Northern Asia is free from its winter covering. As, however, the sun of Mars is so much less powerful than ours, we may be sure that the snow (if it is real snow) is much less thick—a mere surface-coating in fact, such as occurs in parts of Russia where the precipitation is less, and the snow accordingly does not exceed two or three feet in thickness.

We now see the reason why the *southern* pole of Mars parts with its white covering so much quicker and to so much greater an extent than the *northern*, for the south pole during summer is nearest the sun, and, owing to the great eccentricity of Mars, would have about one third more heat than during the summer of the Northern Hemisphere; and this greater heat would cause the winds from the equator to be both warmer and more powerful, and able to produce the same effects on the scanty Martian snows as they produce on our northern plains. The reason why both poles of Mars are almost equally snow-covered in winter is not difficult to understand. Owing to the greater obliquity of the ecliptic, and the much greater length of the year, the polar regions will be subject to winter darkness fully twice as long as with us, and the fact that one pole is nearer the sun during this period than the other at a corresponding period will therefore make no perceptible difference. It is also probable that the two poles of Mars are approximately alike as regards their geographical features, and that neither of them is surrounded by very high land on which ice may accumulate. With us at the present time, on the other hand, geographical conditions completely mask and even reverse the influence of eccentricity, and that of winter in perihelion in the Northern and summer in perihelion in the Southern Hemisphere. In the north we have a preponderance of sea within the Arctic Circle, and of lowlands in the temperate zone. In the south exactly opposite conditions prevail, for there we have a preponderance of land (and much of it high land) within the Antarctic Circle, and of sea in the temperate zone. Ice, therefore, accumulates in the south, while a thin coating of snow, easily melted in summer, is the prevalent feature in the north; and these contrasts react upon climate to such an extent that, in the Southern Ocean, islands in the latitude of Ireland have glaciers descending to the level of the sea, and constant snow-storms in the height of summer, although the sun is then actually nearer the earth than it is during our northern summer!

It is evident, therefore, that the phenomena presented by the varying polar snows of Mars are in no way opposed to that modification of Dr. Croll's theory of the conditions which brought about the glacial epochs of our Northern Hemisphere which is here advocated, but are perfectly explicable on the same general principles, if we keep in mind the distinction between an ice-sheet—which a summer's sun cannot materially diminish, but may even increase by bringing vapor to be condensed into snow—and a thin snowy covering which may be annually melted and annually renewed with great rapidity and over large areas. Except within the small circles of perpetual polar snow, there can at the present time be no ice-sheets in Mars; and the reason why this permanent snowy area is more extensive around the northern than around the southern pole may be partly due to higher land at the north, but is perhaps sufficiently explained by the diminished power of the summer sun, owing to its greatly increased distance at that season in the Northern Hemisphere, so that it is not able to melt so much of the snow which has accumulated during the long night of winter.

CHAPTER IX.

ANCIENT GLACIAL EPOCHS, AND MILD CLIMATES IN THE ARC-TIC REGIONS.

Dr. Croll's Views on Ancient Glacial Epochs.—Effects of Denudation in Destroying the Evidence of Remote Glacial Epochs. — Rise of Sea-level Connected with Glacial Epochs a Cause of Further Denudation.—What Evidence of Early Glacial Epochs may be Expected.—Evidences of Ice-action during the Tertiary Period. —The Weight of the Negative Evidence.—Temperate Climates in the Arctic Regions.—The Miocene Arctic Flora.—Mild Arctic Climates of the Cretaceous Period.—Stratigraphical Evidence of Long-continued Mild Arctic Conditions.— The Causes of Mild Arctic Climates.—Geographical Conditions Favoring Mild Northern Climates in Tertiary Times.—The Indian Ocean as a Source of Heat in Tertiary Times.—Condition of North America during the Tertiary Period.— Effect of High Eccentricity on Warm Polar Climates.—Evidences as to Climate in the Secondary and Palæozoic Epochs.—Warm Arctic Climates in Early Secondary and Palæozoic Times.—Conclusions as to the Climates of Secondary and Tertiary Periods.—General View of Geological Climates as Dependent on the Physical Features of the Earth's Surface.—Estimate of the Comparative Effects of Geographical and Physical Causes in Producing Changes of Climate.

If we adopt the view set forth in the preceding chapter as to the character of the glacial epoch and of the accompanying alternations of climate, it must have been a very important agent in producing changes in the distribution of animal and vegetable life. The intervening mild periods, which almost certainly occurred during its earlier and later phases, were sometimes more equable than even our present insular climate, and severe frosts were probably then unknown. During the eight or ten thousand years that each such mild period lasted, some portions of the north temperate zone which had been buried in snow or ice would become again clothed with vegetation and stocked with animal life, both of which, as the cold again came on, would be driven southward, or perhaps partially exterminated. Forms usually separated would thus be crowded together, and a struggle for existence would follow which must have led

to the modification or the extinction of many species. When
the survivors in the struggle had reached a state of equilibrium,
a fresh field would be opened to them by the later ameliorations
of climate; the more successful of the survivors would spread
and multiply; and after this had gone on for thousands of generations, another change of climate, another southward migration, another struggle of northern and southern forms, would
take place.

But if the last glacial epoch has coincided with, and has been
to a considerable extent caused by, a high eccentricity of the
earth's orbit, we are naturally led to expect that earlier glacial
epochs would have occurred whenever the eccentricity was unusually large. Dr. Croll has published tables showing the varying amounts of eccentricity for three million years back; and
from these it appears there have been many periods of high eccentricity, which has often been far greater than at the time of
the last glacial epoch.[1] The accompanying diagram has been
drawn from these tables, and it will be seen that the highest
eccentricity occurred 850,000 years ago, at which time the difference between the sun's distance at aphelion and perihelion
was thirteen and a half millions of miles, whereas during the
last glacial period the maximum difference was ten and a half
million miles.

Now, judging by the amount of organic and physical change
that occurred during and since the glacial epoch, and that which
has occurred since the Miocene period, it is considered probable
that this maximum of eccentricity coincided with some part of
the latter period; and Dr. Croll maintains that a glacial epoch
must then have occurred surpassing in severity that of which
we have such convincing proofs, and consisting like it of alternations of cold and warm phases every 10,500 years. The
diagram also shows us another long-continued period of high
eccentricity from 1,750,000 to 1,950,000 years ago, and yet
another almost equal to the maximum 2,500,000 years back.
These may perhaps have occurred during the Eocene and Cre-

---

[1] *London, Edinburgh, and Dublin Philosophical Magazine*, Vol. XXXVI., pp. 144–
150 (1868).

taceous epochs respectively, or all may have been included within the limits of the Tertiary period. As two of these high eccentricities greatly exceed that which caused our glacial epoch, while the third is almost equal to it and of longer duration, they seem to afford us the means of testing rival theories of the causes of glaciation. If, as Dr. Croll argues, high eccentricity is the great and dominating agency in bringing on glacial epochs, geographical changes being subordinate, then there must have been glacial epochs of great severity at all these three periods; while if he is also correct in supposing that the alternate phases of precession would inevitably produce glaciation in one hemisphere and a proportionately mild and equable climate in the opposite hemisphere, then we should have to look for evidence of exceptionally warm and exceptionally cold periods occurring alternately, and with several repetitions, within a space of time which, geologically speaking, is very short indeed.

Let us, then, inquire first into the character of the evidence we should expect to find of such changes of climate, if they have occurred; we shall then be in a better position to estimate at its proper value the evidence that actually exists, and, after giving it due weight, to arrive at

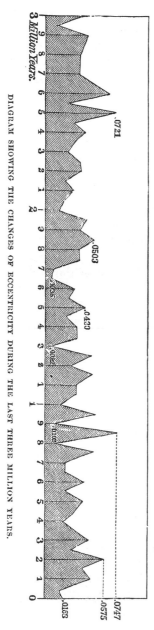

DIAGRAM SHOWING THE CHANGES OF ECCENTRICITY DURING THE LAST THREE MILLION YEARS.

some conclusion as to the theory that best explains and har-
monizes it.

*Effects of Denudation in Destroying the Evidence of Remote
Glacial Epochs.*—It may be supposed that if earlier glacial
epochs than the last did really occur, we ought to meet with
some evidence of the fact corresponding to that which has
satisfied us of the extensive recent glaciation of the Northern
Hemisphere; but Dr. Croll and other writers have ably argued
that no such evidence is likely to be found. It is now generally
admitted that subaerial denudation is a much more powerful
agent in lowering and modifying the surface of a country than
was formerly supposed. It has, in fact, been proved to be so
powerful that the difficulty now felt is, not to account for the
denudation which can be proved to have occurred, but to explain the apparent persistence of superficial features which
ought long ago to have been destroyed.

A proof of the lowering and eating-away of the land surface
which every one can understand is to be found in the quantity
of solid matter carried down to the sea and to low grounds by
rivers. This is capable of pretty accurate measurement, and has
been so measured for several rivers, large and small, in different
parts of the world. The details of these measurements will be
given in a future chapter, and it is only necessary here to state
that the average of them all gives us this result—that one foot
must be taken off the entire surface of the land each 3000 years,
in order to produce the amount of sediment and matter in solution which is actually carried into the sea. To give an idea of
the limits of variation in different rivers, it may be mentioned
that the Mississippi is one which denudes its valley at a slow
rate, taking 6000 years to remove one foot; while the Po is the
most rapid, taking only 729 years to do the same work in its
valley. The cause of this difference is very easy to understand.
A large part of the area of the Mississippi basin consists of the
almost rainless prairie and desert regions of the west, while its
sources are in comparatively arid mountains with scanty snow-
fields or in a low forest-clad plateau. The Po, on the other
hand, is wholly in a district of abundant rainfall, while its
sources are spread over a great amphitheatre of snowy Alps

nearly 400 miles in extent, where the denuding forces are at a maximum. As Scotland is a mountain region of rather abundant rainfall, the denuding power of its rains and rivers is probably rather above than under the average; but, to avoid any possible exaggeration, we will take it at a foot in 4000 years.

Now, if the end of the glacial epoch be taken to coincide with the termination of the last period of high eccentricity, which occurred about 80,000 years ago (and no geologist will consider this too long for the changes which have since taken place), it follows that the entire surface of Scotland must have been since lowered an average amount of twenty feet. But over large areas of alluvial plains, and wherever the rivers have spread during floods, the ground will have been raised instead of lowered; and on all nearly level ground and gentle slopes there will have been comparatively little denudation; so that proportionally much more must have been taken away from mountain-sides and from the bottoms of valleys having a considerable downward slope. One of the very highest authorities on the subject of denudation, Mr. Archibald Geikie, estimates the area of these more rapidly denuded portions as only one tenth of the comparatively level grounds, and he further estimates that the former will be denuded about ten times as fast as the latter. It follows that the valleys will be deepened and widened on the average about five feet in the 4000 years instead of one foot; and thus many valleys must have been deepened and widened one hundred feet, and some even more, since the glacial epoch, while the more level portions of the country will have been lowered, on the average, only about two feet.

Now, Dr. Croll gives us the following account of the present aspect of the surface of a large part of the country :

" Go where one will in the lowlands of Scotland and he shall hardly find a single acre whose upper surface bears the marks of being formed by the denuding agents now in operation. He will observe everywhere mounds and hollows which cannot be accounted for by the present agencies at work. . . . In regard to the general surface of the country, the present agencies may be said to be just beginning to carve a new line of features out of the old glacially formed surface. But so little progress has yet been made that the kames, gravel-mounds, knolls of boulder clay, etc., still retain in most cases their original form." [1]

The facts here seem a little inconsistent, and we must suppose that Dr. Croll has somewhat exaggerated the universality and complete preservation of the glaciated surface. The amount of average denudation, however, is not a matter of opinion, but of measurement; and its consequences can in no way be evaded. They are, moreover, strictly proportionate to the time elapsed; and if so much of the old surface of the country has certainly been remodelled or carried into the sea since the last glacial epoch, it becomes evident that any surface phenomena produced by still earlier glacial epochs *must* have long since entirely disappeared.

Rise of the Sea-level Connected with Glacial Epochs a Cause of Further Denudation.—There is also another powerful agent that must have assisted in the destruction of any such surface deposits or markings. During the last glacial epoch itself there were several oscillations of the land, one at least of considerable extent, during which shell-bearing gravels were deposited on the flanks of the Welsh and Irish mountains, now 1300 feet above sea-level; and there is reason to believe that other subsidences of the same area, though perhaps of less extent, may have occurred at various times during the Tertiary period. Many writers, as we have seen, connect this subsidence with the glacial period itself, the unequal amount of ice at the two poles causing the centre of gravity of the earth to be displaced, when, of course, the surface of the ocean will conform to it, and appear to rise in the one hemisphere and sink in the other. If this is the case, subsidences of the land are natural concomitants of a glacial period, and will powerfully aid in removing all evidence of its occurrence. We have seen reason to believe, however, that during the height of the glacial epoch the extreme cold persisted through the successive phases of precession; and if so, both polar areas would probably be glaciated at once. This would cause the abstraction of a large quantity of water from the ocean, and a proportionate elevation of the land, which would react on the

---

[1] "Climate and Time in their Geological Relations," p. 341.

accumulation of snow and ice, and thus add another to that wonderful series of physical agents which act and react on each
other so as to intensify glacial epochs.

But whether or not these causes would produce any important fluctuations of the sea-level is of comparatively little importance to our present inquiry, because the wide extent of marine
Tertiary deposits in the Northern Hemisphere, and their occurrence at considerable elevations above the present sea-level, afford the most conclusive proofs that great changes of sea and
land have occurred throughout the entire Tertiary period; and
these repeated submergences and emergences of the land, combined with subaerial and marine denudation, would undoubtedly
destroy all those superficial evidences of ice-action on which we
mainly depend for proofs of the occurrence of the last glacial
epoch.

What Evidence of Early Glacial Epochs may be Expected.—
Although we may admit the force of the preceding argument as
to the extreme improbability of our finding any clear evidence
of the superficial action of ice during remote glacial epochs,
there is, nevertheless, one kind of evidence that we ought to
find, because it is both wide-spread and practically indestruc-
tible.

One of the most constant of all the phenomena of a glaciated
country is the abundance of icebergs produced by the breaking-
off of the ends of glaciers which terminate in arms of the sea, or
of the terminal face of the ice-sheet which passes beyond the
land into the ocean. In both these cases abundance of rocks
and débris, such as form the terminal moraines of glaciers on
land, are carried out to sea and deposited over the sea-bottom of
the area occupied by icebergs. In the case of an ice-sheet it is
almost certain that much of the ground-moraine, consisting of
mud and embedded stones, similar to that which forms the
"till" when deposited on land, will be carried out to sea with
the ice, and form a deposit of marine "till" near the shore.

It has, indeed, been objected that when an ice-sheet covered an
entire country there would be no moraines, and that rocks or
débris are very rarely seen on icebergs. But during every glacial epoch there will be a southern limit to the glaciated area,

and everywhere near this limit the mountain-tops will rise far above the ice and deposit on it great masses of débris; and as the ice-sheet spreads, and again as it passes away, this moraine-forming area will successively occupy the whole country. But even such an ice-clad country as Greenland is now known to have protruding peaks and rocky masses which give rise to moraines on its surface;[1] and, as rocks from Cumberland and Ireland were carried by the ice-sheet to the Isle of Man, there must have been a very long period during which the ice-sheets of Britain and Ireland terminated in the ocean and sent off abundance of rock-laden bergs into the surrounding seas; and the same thing must have occurred along all the coasts of Northern Europe and Eastern America.

We cannot, therefore, doubt that throughout the greater part of the duration of a glacial epoch the seas adjacent to the glaciated countries would receive continual deposits of large rocks, rock-fragments, and gravel similar to the material of modern and ancient moraines, and analogous to the drift and the numerous travelled blocks which the ice has undoubtedly scattered broadcast over every glaciated country; and these rocks and boulders would be embedded in whatever deposits were then forming, either from the matter carried down by rivers or from the mud ground off the rocks and carried out to sea by the glaciers themselves. Moreover, as icebergs float far beyond the limits of the countries which gave them birth, these ice-borne materials would be largely embedded in deposits forming from the denudation of countries which had never been glaciated, or from which the ice had already disappeared.

But if every period of high eccentricity produced a glacial epoch of greater or less extent and severity, then, on account of the frequent occurrence of a high phase of eccentricity during the 3,000,000 years for which we have the tables, these boulder and rock-strewn deposits would be both numerous and extensive. 400,000 years ago the eccentricity was almost exactly the same as it is now, and it continually increased from that time up to the glacial epoch. Now, if we take double the present eccen-

---

[1] *Nature*, Vol. XXI., p. 345, " The Interior of Greenland."

tricity as being sufficient to produce some glaciation in the temperate zone, we find (by drawing out the diagram at p. 163 on a larger scale) that during 1,150,000 years out of the 2,400,000 years immediately preceding the last glacial epoch, the eccentricity reached or exceeded this amount, consisting of sixteen separate epochs, divided from each other by periods varying from 30,000 to 200,000 years. But if the last glacial epoch was at its maximum 200,000 years ago, a space of 3,000,000 years will certainly include much, if not all, of the Tertiary period; and even if it does not, we have no reason to suppose that the character of the eccentricity would suddenly change beyond the 3,000,000 years.

It follows, therefore, that if periods of high eccentricity, like that which appears to have been synchronous with our last glacial epoch, and is generally admitted to have been one of its efficient causes, always produced glacial epochs (with or without alternating warm periods), then the whole of the Tertiary deposits in the north temperate and arctic zones should exhibit constantly alternating boulder and rock-bearing beds, or coarse rock-strewn gravels analogous to our existing glacial drift, and with some corresponding change of organic remains. Let us, then, see what evidence can be adduced of the existence of such deposits, and whether it is adequate to support the theory of repeated glacial epochs during the Tertiary period.

Evidences of Ice-action during the Tertiary Period.—The Tertiary fossils both of Europe and North America indicate throughout warm or temperate climates, except those of the more recent Pliocene deposits which merge into the earlier glacial beds. The Miocene deposits of Central and Southern Europe, for example, contain marine shells of some genera now only found farther south, while the fossil plants often resemble those of Madeira and the Southern States of North America. Large reptiles, too, abounded; and man-like apes lived in the south of France and in Germany. Yet in Northern Italy, near Turin, there are beds of sandstone and conglomerate full of characteristic Miocene shells, but containing in an intercalated deposit angular blocks of serpentine and greenstone often of enormous size, one being fourteen feet long, and another twen-

ty-six feet. Some of the blocks were observed by Sir Charles
Lyell to be faintly striated and partly polished on one side, and
they are scattered through the beds for a thickness of nearly one
hundred and fifty feet. It is interesting that the particular bed in
which the blocks occur yields no organic remains, though these
are plentiful both in the underlying and overlying beds, as if
the cold of the icebergs had driven away the organisms adapted
to live only in a comparatively warm sea. Rock similar in kind
to these erratics occurs about twenty miles distant in the Alps.

The Eocene period is even more characteristically tropical in
its flora and fauna, since palms and Cycadaceæ, turtles, snakes,
and crocodiles then inhabited England. Yet on the north side
of the Alps, extending from Switzerland to Vienna, and also
south of the Alps near Genoa, there is a deposit of finely stratified sandstone several thousand feet in thickness, quite destitute
of organic remains, but containing, in several places in Switzerland, enormous blocks either angular or partly rounded, and
composed of oolitic limestone or of granite. Near the Lake of
Thun some of the granite blocks found in this deposit are of
enormous size, one of them being one hundred and five feet
long, ninety feet wide, and forty-five feet thick! The granite
is red, and of a peculiar kind which cannot be matched anywhere
in the Alps, or indeed elsewhere. Similar erratics have also been
found in beds of the same age in the Carpathians and in the
Apennines, indicating probably an extensive inland European
sea into which glaciers descended from the surrounding mountains, depositing these erratics, and cooling the water so as to destroy the mollusca and other organisms which had previously inhabited it. It is to be observed that wherever these erratics occur they are always in the vicinity of great mountain-ranges;
and although these can be proved to have been in great part elevated during the Tertiary period, we must also remember that
they must have been since very much lowered by denudation,
of the amount of which the enormously thick Eocene and Miocene beds now forming portions of them are in some degree a
measure as well as a proof. It is not, therefore, at all improbable that during some part of the Tertiary period these mountains may have been far higher than they are now, and this we

know might be sufficient for the production of glaciers descending to the sea-level, even were the climate of the lowlands somewhat warmer than at present.[1]

The Weight of the Negative Evidence.—But when we proceed to examine the Tertiary deposits of other parts of Europe, and especially of our own country, for evidence of this kind, not only is such evidence completely wanting, but the facts are of so definite a character as to satisfy most geologists that it can never have existed; and the same may be said of temperate North America and of the arctic regions generally.

In his carefully written paper on " The Climate Controversy," Mr. Searles V. Wood, Jr., remarks on this point as follows: "Now the Eocene formation is complete in England, and is exposed in continuous section along the north coast of the Isle of Wight from its base to its junction with the Oligocene (or Lower Miocene according to some), and along the northern coast of Kent from its base to the Lower Bagshot Sand. It has been intersected by railway and other cuttings in all directions and at all horizons, and pierced by wells innumerable; while from its strata in England, France, and Belgium the most extensive collections of organic remains have been made of any formation yet explored, and from nearly all its horizons, for at one place or another in these three countries nearly every horizon may be said to have yielded fossils of some kind. These fossils, however,

---

[1] Professor J. W. Judd says, "In the case of the Alps, I know of no glacial phenomena which are not capable of being explained, like those of New Zealand, by a great extension of the area of the tracts above the snow-line which would collect more ample supplies for the glaciers protruded into surrounding plains. And when we survey the grand panoramas of ridges, pinnacles, and peaks produced, for the most part, by subaerial action, we may well be prepared to admit that before the intervening ravines and valleys were excavated, the glaciers shed from the elevated plateaus must have been of vastly greater magnitude than at present " ("Contributions to the Study of Volcanoes," *Geological Magazine*, 1876, p. 536). Professor Judd applies these remarks to the last as well as to previous glacial periods in the Alps; but surely there has been no such extensive alteration and lowering of the surface of the country since the erratic blocks were deposited on the Jura and the great moraines formed in North Italy, as this theory would imply. We can hardly suppose wide areas to have been lowered thousands of feet by denudation, and yet have left other adjacent areas apparently untouched: and it is even very doubtful whether such an extension of the snow-fields would alone suffice for the effects which were certainly produced.

whether they be the remains of a flora such as that of Sheppey, or of a vertebrate fauna containing the crocodile and alligator, such as is yielded by beds indicative of terrestrial conditions, or of a molluscan assemblage such as is present in marine or fluvio-marine beds of the formation, are of unmistakably tropical or subtropical character throughout; and no trace whatever has appeared of the intercalation of a glacial period, much less of successive intercalations indicative of more than one period of 10,500 years' glaciation. Nor can it be urged that the glacial epochs of the Eocene in England were intervals of dry land, and so have left no evidence of their existence behind them, because a large part of the continuous sequence of Eocene deposits in this country consists of alternations of fluviatile, fluvio-marine, and purely marine strata; so that it seems impossible that during the accumulation of the Eocene formation in England a glacial period could have occurred without its evidences being abundantly apparent. The Oligocene of Northern Germany and Belgium, and the Miocene of those countries and of France, have also afforded a rich molluscan fauna, which, like that of the Eocene, has as yet presented no indication of the intrusion of anything to interfere with its uniformly subtropical character." [1]

This is sufficiently striking; but when we consider that this enormous series of deposits, many thousand feet in thickness, consists wholly of alternations of clays, sands, marls, shales, or limestones, with a few beds of pebbles or conglomerate, not one of the whole series containing irregular blocks of foreign material, boulders, or gravel such as we have seen to be the essential characteristic of a glacial epoch; and when we find that this very same general character pervades all the extensive Tertiary deposits of temperate North America—we shall, I think, be forced to the conclusion that no general glacial epochs could have occurred during their formation. It must be remembered that the "imperfection of the geological record" will not help us here, because the series of Tertiary deposits is unusually complete, and we must suppose some destructive agency to have selected all the intercalated glacial beds and to have so completely

---

[1] *Geological Magazine*, 1876, p. 392.

made away with them that not a fragment remains, while preserving all, or almost all, the *interglacial* beds; and to have acted thus capriciously, not in one limited area only, but over the whole Northern Hemisphere, with the local exceptions on the flanks of great mountain-ranges already referred to.

Temperate Climates in the Arctic Regions.—As we have just seen, the geological evidence of the persistence of subtropical or warm climates in the north temperate zone during the greater part of the Tertiary period is almost irresistible; and we have now to consider the still more extraordinary series of observations which demonstrate that this amelioration of climate extended into the arctic zone, and into countries now almost wholly buried in snow and ice. These warm arctic climates have been explained by Dr. Croll as due to periods of high eccentricity with winter in perihelion, a theory which implies alternating epochs of glaciation far exceeding what now prevails; and it is therefore necessary to examine the evidence pretty closely in order to see if this view is more tenable in the case of the north polar regions than we have found it to be in that of the north temperate zone.

The most recent of these milder climates is perhaps indicated by the abundant remains of large mammalia—such as the mammoth, woolly rhinoceros, bison, and horse—in the icy alluvial plains of Northern Siberia, and especially in the Liakhov Islands, in the same latitude as the North Cape of Asia. These remains occur not in one or two spots only, as if collected by eddies at the mouth of a river, but along the whole borders of the Arctic Ocean; and it is generally admitted that the animals must have lived upon the adjacent plains, and that a considerably milder climate than now prevails could alone have enabled them to do so. At what period this occurred we do not know, but one of the last intercalated mild periods of the glacial epoch itself seems to offer all the necessary conditions. Again, Sir Edward Belcher discovered on the dreary shores of Wellington Channel, in $75\frac{1}{2}°$ N. lat., the trunk and root of a fir-tree which had evidently grown where it was found. It appeared to belong to the species *Abies alba*, or white fir, which now reaches 68° N. lat., and is the most northerly conifer known. Similar trees, one

four feet in circumference and thirty feet long, were found by
Lieutenant Mecham in Prince Patrick's Island, in lat. 76° 12′ N.;
and other arctic explorers have found remains of trees in high
latitudes which may all probably be referred to the same mild
period as that of the ice-preserved arctic mammalia.

Similar indications of a recent milder climate are found in
Spitzbergen. Professor Nordenskjöld says, "At various places
on Spitzbergen, at the bottom of Lomme Bay, at Cape Thordsen, in Blomstrand's strata in Advent Bay, there are found
large and well-developed shells of a bivalve, *Mytilus edulis*,
which is not now found living on the coasts of Spitzbergen,
though on the west coast of Scandinavia it everywhere covers
the rocks near the sea-shore. These shells occur most plentifully in the bed of a river which runs through Reindeer Valley
at Cape Thordsen. They are probably washed out of a thin
bed of sand at a height of about twenty or thirty feet above the
present sea-level, which is intersected by the river. The geological age of this bed cannot be very great, and it has clearly been
formed since the present basin of the Ice Sound, or at least the
greater part of it, has been hollowed out by glacial action." [1]

The Miocene Arctic Flora.—One of the most startling and important of the scientific discoveries of the last twenty years has
been that of the relics of a luxuriant Miocene flora in various
parts of the arctic regions. It is a discovery that was totally unexpected, and is even now considered by many men of science to
be completely unintelligible; but it is so thoroughly established,
and it has such a direct and important bearing on the subjects we
are discussing in the present volume, that it is necessary to lay a
tolerably complete outline of the facts before our readers.

The Miocene flora of temperate Europe was very like that of
Eastern Asia, Japan, and the warmer part of Eastern North
America of the present day. It is very richly represented in
Switzerland by well-preserved fossil remains, and after a close
comparison with the flora of other countries Professor Heer
concludes that the Swiss Lower Miocene flora indicates a climate
corresponding to that of Louisiana, North Africa, and South

---

[1] *Geological Magazine*, 1876, "Geology of Spitzbergen," p. 267.

China, while the Upper Miocene climate of the same country would correspond to that of the south of Spain, Southern Japan, and Georgia (U. S. of America). Of this latter flora, found chiefly at Oeninghen, in the northern extremity of Switzerland, 465 species are known, of which 166 species are trees or shrubs, half of them being evergreens. They comprise sequoyas like the California giant trees, camphor-trees, cinnamons, sassafras, bignonias, cassias, gleditschias, tulip-trees, and many other American genera, together with maples, ashes, planes, oaks, poplars, and other familiar European trees represented by a variety of extinct species. If we now go to the west coast of Greenland, in 70° N. lat., we find abundant remains of a flora of the same general type as that of Oeninghen, but of a more northern character. We have a sequoya identical with one of the species found at Oeninghen, a chestnut, salisburia, liquidambar, and sassafras, and even a magnolia. We have also seven species of oaks, two planes, two vines, three beeches, four poplars, two willows, a walnut, a plum, and several shrubs supposed to be evergreens—altogether 137 species, mostly well and abundantly preserved!

But even farther north, in Spitzbergen, in 78° and 79° N. lat., and one of the most barren and inhospitable regions on the globe, an almost equally rich fossil flora has been discovered, including several of the Greenland species, and others peculiar, but mostly of the same genera. There seem to be no evergreens here except coniferæ, one of which is identical with the swamp-cypress (*Taxodium distichum*) now found living in the Southern United States! There are also eleven pines, two Libocedrus, two sequoyas, with oaks, poplars, birches, planes, limes, a hazel, an ash, and a walnut; also water-lilies, pond-weeds, and an iris—altogether about a hundred species of flowering plants. Even in Grinnell Land, within eight and a quarter degrees of the pole, a similar flora existed, twenty-five species of fossil plants having been collected by the last arctic expedition, of which eighteen were identical with the species from other arctic localities. This flora comprised poplars, birches, hazels, elms, viburnums, and eight species of conifers, including the swamp-cypress, and the Norway spruce (*Pinus abies*) which does not now extend beyond 69½° N. lat.

Fossil plants closely resembling those just mentioned have been found at many other arctic localities, especially in Iceland, on the Mackenzie River in 65° N. lat., and in Alaska. As an intermediate station we have, in the neighborhood of Dantzic, in lat. 55° N., a similar flora, with the swamp-cypress, sequoyas, oaks, poplars, and some cinnamons, laurels, and figs. A little farther south, near Breslau, north of the Carpathians, a rich flora has been found allied to that of Oeninghen, but wanting in some of the more tropical forms. Again, in the Isle of Mull, in Scotland, in about 56½° N. lat., a plant-bed has been discovered containing a hazel, a plane, and a sequoya, apparently identical with a Swiss Miocene species.

We thus find one well-marked type of vegetation spread from Switzerland and Vienna to North Germany, Scotland, Iceland, Greenland, Alaska, and Spitzbergen, some few of the species even ranging over the extremes of latitude between Oeninghen and Spitzbergen; but the great majority being distinct, and exhibiting decided indications of a decrease of temperature according to latitude, though much less in amount than now exists. Some writers have thought that the great similarity of the floras of Greenland and Oeninghen is a proof that they were not contemporaneous, but successive; and that of Greenland has been supposed to be as old as the Eocene. But the arguments yet adduced do not seem to prove such a difference of age, because there is only that amount of specific and generic diversity between the two which might be produced by distance and difference of temperature, under the exceptionally equable climate of the period. We have even now examples of an equally wide range of well-marked types; as in temperate South America, where many of the genera and some of the species range from the Strait of Magellan to Valparaiso—places differing as much in latitude as Switzerland and West Greenland; and the same may be said of North Australia and Tasmania, where, at a greater latitudinal distance apart, closely allied forms of Eucalyptus, Acacia, Casuarina, Stylidium, Goodenia, and many other genera would certainly form a prominent feature in any fossil flora now being preserved.

Mild Arctic Climates of the Cretaceous Period.—In the Upper

Cretaceous deposits of Greenland (in a locality not far from those of the Miocene age last described) another remarkable flora has been discovered, agreeing generally with that of Europe and North America of the same geological age. Sixty-five species of plants have been identified, of which there are fifteen ferns, two cycads, eleven coniferæ, three monocotyledons, and thirty-four dicotyledons. One of the ferns is a tree-fern with thick stems, which has also been found in the Upper Greensand of England. Among the conifers the giant sequoyas are found, and among the dicotyledons the genera Populus, Myrica, Ficus, Sassafras, Andromeda, Diospyros, Myrsine, Panax, as well as magnolias, myrtles, and leguminosæ. Several of these groups occur also in the much richer deposits of the same age in North America and Central Europe; but all of them evidently afford such fragmentary records of the actual flora of the period that it is impossible to say that any genus found in one locality was absent from the other merely because it has not yet been found there. On the whole, there seems to be less difference between the floras of arctic and temperate latitudes in Upper Cretaceous than in Miocene times.

In the same locality in Greenland (70° 33′ N. lat. and 52° W. long.), and also in Spitzbergen, a more ancient flora, of Lower Cretaceous age, has been found; but it differs widely from the other in the great abundance of cycads and conifers and the scarcity of exogens, which latter are represented by a single poplar. Of the thirty-eight ferns, fifteen belong to the genus Gleichenia, now almost entirely tropical. There are four genera of cycads, and three extinct genera of conifers, besides Glyptostrobus and Torreya (now found only in China and California), six species of true pines, and five of the genus Sequoya, one of which occurs also in Spitzbergen. The European deposits of the same age closely agree with these in their general character; conifers, cycads, and ferns forming the mass of the vegetation, while exogens are entirely absent, the above-named Greenland poplar being the oldest known dicotyledonous plant.[1]

---

[1] The preceding account is mostly derived from Professor Heer's great work "Flora Fossilis Arctica."

If we take these facts as really representing the flora of the period, we shall be forced to conclude that, measured by the change effected in its plants, the lapse of time between the Lower and Upper Cretaceous deposits was far greater than between the Upper Cretaceous and the Miocene—a conclusion quite opposed to the indications afforded by the mollusca and the higher animals of the two periods. It seems probable, therefore, that these Lower Cretaceous plants represent local peculiarities of vegetation such as now sometimes occur in tropical countries. On sandy or coralline islands in the Malay Archipelago there will often be found a vegetation consisting almost wholly of cycads, pandani, and palms; while a few miles off, on moderately elevated land, not a single specimen of either of these families may be seen, but a dense forest of dicotyledonous trees covering the whole country. A lowland vegetation such as that above described might be destroyed and its remains preserved by a slight depression, allowing it to be covered up by the detritus of some adjacent river; while not only would the subsidence of high land be a less frequent occurrence, but when it did occur the steep banks would be undermined by the waves, and the trees falling down would be floated away, and would either be cast on some distant shore, or slowly decay on the surface or in the depths of the ocean.

From the remarkable series of facts now briefly summarized, we learn that whenever plant-remains have been discovered within the arctic regions, either in Tertiary or Cretaceous deposits, they show that the climate was one capable of supporting a rich vegetation of trees, shrubs, and herbaceous plants, similar in general character to that which prevailed in the temperate zone at the same periods, but showing the influence of a less congenial climate. These deposits belong to at least four distinct geological horizons, and have been found widely scattered within the Arctic Circle; yet nowhere has any proof been obtained of intercalated cold periods such as would be indicated by the remains of a stunted vegetation, or a molluscan fauna similar to that which now prevails there.

Stratigraphical Evidence of Long-continued Mild Arctic Conditions.—Let us now turn to the stratigraphical evidence, which, as we have already shown, offers a crucial test of the occurrence or non-occurrence of glaciation during any extensive geological period; and here we have the testimony of perhaps the greatest living authority on arctic geology—Professor Nordenskjöld. In his lecture on the "Former Climate of the Polar Regions" he says, "The character of the coasts in the arctic regions is especially favorable to geological investigations. While the valleys are, for the most part, filled with ice, the sides of the mountains in summer, even in the 80th degree of latitude, and to a height of 1000 or 1500 feet above the level of the sea, are almost wholly free from snow. Nor are the rocks covered with any amount of vegetation worth mentioning, and, moreover, the sides of the mountains on the shore itself frequently present perpendicular sections which everywhere expose their bare surfaces to the investigator. The knowledge of a mountain's geognostic character, at which one, in the more southerly countries, can only arrive after long and laborious researches, removal of soil and the like, is here gained almost at the first glance; and as we have never seen in Spitzbergen nor in Greenland, in these sections, often many miles in length, and including, one may say, all formations from the Silurian to the Tertiary, any boulders even as large as a child's head, there is not the smallest probability that strata of any considerable extent containing boulders are to be found in the polar tracts previous to the middle of the Tertiary period. Since, then, both an examination of the geognostic condition and an investigation of the fossil flora and fauna of the polar lands show no signs of a glacial era having existed in those parts before the termination of the Miocene period, we are fully justified in rejecting, on the evidence of actual observation, the hypotheses founded on purely theoretical speculations which assume the many-times-repeated alternation of warm and glacial climates between the present time and the earliest geological ages."[1] And again, in his "Sketch of the Geology of Spitzbergen," after describing the various formations down to the Miocene, he says, "All the fossils found in the foregoing strata show that Spitzbergen, during former geo-

---

[1] *Geological Magazine*, 1875, p. 531.

logical ages, enjoyed a magnificent climate, which, indeed, was
somewhat colder during the Miocene period, but was still favorable for an extraordinarily abundant vegetation, much more
luxuriant than that which now occurs even in the southern part
of Scandinavia; and I have in these strata sought in vain for
any sign that, as some geologists have of late endeavored to
render probable, these favorable climatic conditions have been
broken off by intervals of ancient glacial periods. The profiles
I have had the opportunity to examine during my various Spitzbergen expeditions would certainly, if laid down on a line, occupy an extent of *a thousand English miles;* and if any former
glacial period had existed in this region there ought to have
been some trace to be observed of erratic blocks, or other formations which distinguish glacial action. But this has not been
the case. In the strata, whose length I have reckoned alone, I
have not found a single fragment of a foreign rock so large as
a child's head." [1]

Now it is quite impossible to ignore or evade the force of
this testimony as to the continuous warm climates of the north
temperate and polar zones throughout Tertiary times. The
evidence extends over a vast area, both in space and time; it is
derived from the work of the most competent living geologists;
and it is absolutely consistent in its general tendency. We have
in the Lower Cretaceous period an almost tropical climate in
France and England, a somewhat lower temperature in the
United States, and a mild insular climate in the arctic regions.
In each successive period the climate becomes somewhat less
tropical; but down to the Upper Miocene it remains warm
temperate in Central Europe, and cold temperate within the
polar area, with not a trace of any intervening periods of arctic
cold. It then gradually cools down and merges through the
Pliocene into the glacial epoch in Europe, while in the arctic
zone there is a break in the record between the Miocene and
the recent glacial deposits. [2]

---

[1] *Geological Magazine*, 1876, p. 266.

[2] It is interesting to observe that the Cretaceous flora of the United States (that
of the Dakota group) indicates a somewhat cooler climate than that of the following
Eocene period. Mr. De Rance (in the geological appendix to Captain Sir G. Nares's

Accepting this as a substantially correct account of the general climatic aspect of the Tertiary period in the Northern Hemisphere, let us see whether the principles we have already laid down will enable us to give a satisfactory explanation of its causes.

The Causes of Mild Arctic Climates.—In his remarkable series of papers on " Ocean Currents," Dr. James Croll has proved, with a wealth of argument and illustration whose cogency is irresistible, that the very habitability of our globe is due to the equalizing climatic effects of the waters of the ocean; and that it is to the same cause that we owe, either directly or indirectly, almost all the chief diversities of climate between places situated in the same latitude. Owing to the peculiar distribution of land and sea upon the globe, more than its fair proportion of the warm equatorial waters is directed towards the western shores of Europe, the result being that the British Isles, Norway, and Spitzbergen have all a milder climate than any other parts of the globe in corresponding latitudes. A very small portion of the arctic regions, however, obtains this benefit, and it thus remains, generally speaking, a land of snow and ice, with too short a summer to nourish more than a very scanty and fugitive vegetation. The only other opening than that between Iceland and Britain by which warm water penetrates within the Arctic Circle is through Behring Strait; but this is both shallow and limited in width, and the consequence is that the larger part of the warm currents of the Pacific turns back along the shores of the Aleutian Islands and Northwest America, while a very small quantity enters the icy ocean.

But if there were other and wider openings into the Arctic Ocean, a vast quantity of the heated water which is now turned

---

"Narrative of a Voyage to the Polar Sea ") remarks as follows : " In the overlying American Eocenes occur types of plants occurring in the European Miocenes and still living, proving the truth of Professor Lesquereux's postulate, that the plant types appear in America a stage in advance of their advent in Europe. These plants point to a far higher mean temperature than those of the Dakota group, to a dense atmosphere of vapor, and a luxuriance of ferns and palms." This is very important as adding further proof to the view that the climates of former periods are not due to any general refrigeration, but to causes which were subject to change and alternation in former ages as now.

backward would enter it, and would produce an amelioration of the climate of which we can hardly form a conception. A great amelioration of climate would also be caused by the breaking-up or the lowering of such arctic highlands as now favor the accumulation of ice; while the interpenetration of the sea into any part of the great continents in the tropical or temperate zones would again tend to raise the winter temperature, and render any long continuance of snow in their vicinity almost impossible.

Now geologists have proved, quite independently of any such questions as we are here discussing, that changes of the very kinds above referred to have occurred during the Tertiary period; and that there has been, speaking broadly, a steady change from a comparatively fragmentary and insular condition of the great north temperate lands in early Tertiary times to that more compact and continental condition which now prevails. It is, no doubt, difficult and often impossible to determine how long any particular geographical condition lasted, or whether the changes in one country were exactly coincident with those in another; but it will be sufficient for our purpose briefly to indicate those more important changes of land and sea during the Tertiary period which must have produced a decided effect on the climate of the Northern Hemisphere.

Geographical Changes Favoring Mild Northern Climates in Tertiary Times.—The distribution of the Eocene and Miocene formations shows that during a considerable portion of the Tertiary period an inland sea, more or less occupied by an archipelago of islands, extended across Central Europe between the Baltic and the Black and Caspian seas, and thence by narrower channels southeastward to the valley of the Euphrates and the Persian Gulf, thus opening a communication between the North Atlantic and the Indian Ocean. From the Caspian also a wide arm of the sea extended during some part of the Tertiary epoch northward to the Arctic Ocean, and there is nothing to show that this sea may not have been in existence during the whole Tertiary period. Another channel probably existed over Egypt[1] into the eastern basin of the Mediterranean

---

[1] Mr. S. B. J. Skertchley informs me that he has himself observed thick Tertiary deposits, consisting of clays and anhydrous gypsum, at Berenice, on the borders of

and the Black Sea; while it is probable that there was a communication between the Baltic and the White Sea, leaving
Scandinavia as an extensive island. Turning to India, we find
that an arm of the sea of great width and depth extended from
the Bay of Bengal to the mouths of the Indus; while the enormous depression indicated by the presence of marine fossils of
Eocene age at a height of 16,500 feet in Western Thibet renders
it not improbable that a more direct channel across Afghanistan
may have opened a communication between the West Asiatic
and Polar seas.

It may be said that the changes here indicated are not warranted by an actual knowledge of continuous Tertiary deposits
over the situations of the alleged marine channels; but it is no
less certain that the seas in which any particular strata were deposited were *always* more extensive than the fragments of those
strata now existing, and *often* immensely more extensive. The
Eocene deposits of Europe, for example, have certainly undergone enormous denudation, both marine and subaerial, and may
have once covered areas where we now find older deposits (as
the chalk once covered the weald), while they certainly exist
concealed under some Miocene, Pliocene, or recent beds. We
find them widely scattered over Europe and Asia, and often
elevated into lofty mountain-ranges; and we should certainly
err far more seriously in confining the Eocene seas to the exact
areas where we now find Eocene rocks than in liberally extending them so as to connect the several detached portions of the
formation whenever there is no valid argument against our
doing so. Considering, then, that some one or more of the sea-
communications here indicated almost certainly existed during
Eocene and Miocene times, let us endeavor to estimate the probable effect such communications would have upon the climate
of the Northern Hemisphere.

The Indian Ocean as a Source of Heat in Tertiary Times.—
If we compare the Indian Ocean with the South Atlantic, we
shall see that the position and outline of the former are both

---

Egypt and Nubia, at a height of about 600 feet above the sea-level; but these may
have been of fresh-water origin.

very favorable for the accumulation of a large body of warm
water moving northward. Its southern opening between
South Africa and Australia is very wide, and the tendency of
the trade-winds would be to concentrate the currents towards
its northwestern extremity, just where the two great channels
above described formed an outlet to the northern seas. As will
be shown in our nineteenth chapter, there were probably, during
the earlier portion of the Tertiary period at least, several large
islands in the space between Madagascar and South India ; but
these had wide and deep channels between them, and their effect
would probably have been favorable to the conveyance of heated water northward by concentrating the currents, and thus
producing massive bodies of moving water analogous to the
Gulf Stream of the Atlantic.[1] Less heat would thus be lost by
evaporation and radiation in the tropical zone, and an impulse
would be acquired which would carry the warm water into the
north polar area. About the same period Australia was divided
into two islands, separated by a wide channel in a north and
south direction (see Chapter XXII.), and through this another
current would almost certainly set northward, and be directed
to the northwest by the southern extension of Malayan Asia.
The more insular condition at this period of Australia, India,
and North Africa, with the depression and probable fertility of
the Central Asiatic plateau, would lead to the Indian Ocean
being traversed by regular trade-winds instead of by variable
monsoons, and thus the constant *vis a tergo*, which is so efficient
in the Atlantic, would keep up a steady and powerful current
towards the northern parts of the Indian Ocean, and thence
through the midst of the European archipelago to the northern
seas.

Now it is quite certain that such a condition as we have here
sketched out would produce a wonderful effect on the climate
of Central Europe and Western and Northern Asia. Owing to
the warm currents being concentrated in inland seas, instead of

---

[1] By referring to our map of the Indian Ocean showing the submarine banks indicating ancient islands (Chap. XIX.), it will be evident that the southeast trade-
winds, then exceptionally powerful, would cause a vast body of water to enter the
deep Arabian Sea.

being dispersed over a wide ocean like the North Atlantic, much more heat would be conveyed into the Arctic Ocean, and this would altogether prevent the formation of ice on the northern shores of Asia, which continent did not then extend nearly so far north and was probably deeply interpenetrated by the sea. This open ocean to the north, and the warm currents along all the northern lands, would so equalize temperature that even the northern parts of Europe might then have enjoyed a climate fully equal to that of the warmer parts of New Zealand at the present day, and might have well supported the luxuriant vegetation of the Miocene period, even without any help from similar changes in the Western Hemisphere.[1]

Condition of North America during the Tertiary Period.— But changes of a somewhat similar character have also taken place in America and the Pacific. An enormous area west of the Mississippi, extending over much of the Rocky Mountains, consists of marine Cretaceous beds 10,000 feet thick, indicating great and long-continued subsidence, and an insular condition of Western America with a sea probably extending northward to the Arctic Ocean. As marine Tertiary deposits are found conformably overlying these Cretaceous strata, Professor Dana is of opinion that the great elevation of this part of America did not begin till early Tertiary times. Other Tertiary beds in California, Alaska, Kamtschatka, the Mackenzie River, the Parry Islands, and Greenland indicate partial submergence of all these lands with the possible influx of warm water from the Pacific;

---

[1] In his recently published "Lectures on Physical Geography," Professor Haughton calculates that more than half the solar heat of the torrid zone is carried to the temperate zones by ocean currents. The Gulf Stream itself carries one twelfth of the total amount, but it is probable that a very small fraction of this quantity of heat reaches the polar seas, owing to the wide area over which the current spreads in the North Atlantic. The corresponding stream of the Indian Ocean in Miocene times would have been fully equal to the Gulf Stream in heating power, while, owing to its being so much more concentrated, a large proportion of its heat may have reached the polar area. But the Arctic Ocean occupies less than one tenth of the area of the tropical seas; so that whatever proportion of the heat of the tropical zone was conveyed to it would, by being concentrated into one tenth of the surface, produce an enormously increased effect. Taking this into consideration, we can hardly doubt that the opening of a sufficient passage from the Indian Ocean to the arctic seas would produce the effects above indicated.

and the considerable elevation of some of the Miocene beds in
Greenland and Spitzbergen renders it probable that these countries were then much less elevated, in which case only their
higher summits would be covered with perpetual snow, and no
glaciers would descend to the sea.

In the Pacific there was probably an elevation of land coun-
terbalancing, to some extent, the great depression of so much of
the northern continents. Our map in Chapter XV. shows the
islands that would be produced by an elevation of the great
shoals under a thousand fathoms deep, and it is seen that these
all trend in a southeast and northwest direction, and would thus
facilitate the production of definite currents impelled by the
southeast trades towards the Northwest Pacific, where they would
gain access to the polar seas through Behring Strait, which was,
perhaps, sometimes both wider and deeper than at present.

Effect of these Changes on the Climate of the Arctic Regions.
—These various changes of sea and land, all tending towards a
transferrence of heat from the equator to the north temperate
zone, were not improbably still further augmented by the existence of a great inland South American sea occupying what are
now the extensive valleys of the Amazon and Orinoco, and forming an additional reservoir of superheated water to add to the
supply poured into the North Atlantic.

It is not, of course, supposed that all the modifications here indicated coexisted at the same time. We have good reason to
believe, from the known distribution of animals in the Tertiary
period, that land communications have at times existed between
Europe or Asia and North America, either by way of Behring
Strait, or by Iceland, Greenland, and Labrador. But the same
evidence shows that these land communications were the exception rather than the rule, and that they occurred only at long intervals and for short periods, so as at no time to bring about anything like a complete interchange of the productions of the two
continents.[1] We may therefore admit that the communication
between the tropical and arctic oceans was occasionally inter-

---

[1] For an account of the resemblances and differences of the mammalia of the two
continents during the Tertiary epoch, see my " Geographical Distribution of Animals," Vol. I., pp. 140, 156.

rupted in one or other direction; but if we look at a globe instead of a Mercator's chart of the world, we shall see that the disproportion between the extent of the polar and tropical seas is so enormous that a single wide opening, with an adequate impulse to carry in a considerable stream of warm water, would be amply sufficient for the complete abolition of polar snow and ice, when aided by the absence of any great areas of high land within the polar circle, such high land being, as we have seen, essential to the production of perpetual snow even at the present time.

Those who wish to understand the effect of oceanic currents in conveying heat to the north temperate and polar regions should study the papers of Dr. Croll already referred to. But the same thing is equally well shown by the facts of the actual distribution of heat due to the Gulf Stream. The difference between the mean annual temperatures of the opposite coasts of Europe and America is well known and has been already quoted; but the difference of their mean *winter* temperature is still more striking, and it is this which concerns us as more especially affecting the distribution of vegetable and animal life. Our mean winter temperature in the West of England is the same as that of the Southern United States, as well as that of Shanghai in China, both about twenty degrees of latitude farther south ; and as we go northward the difference increases, so that the winter climate of Nova Scotia in lat. 45° is found within the Arctic Circle on the coast of Norway; and if the latter country did not consist almost wholly of precipitous snow-clad mountains, it would be capable of supporting most of the vegetable products of the American coast in the latitude of Bordeaux.[1]

With these astounding facts before us, due wholly to the tranferrence of a portion of the warm currents of the Atlantic to the shores of Europe, even with all the disadvantages of an icy sea to the northeast and ice-covered Greenland to the north-

---

[1] Professor Haughton has made an elaborate calculation of the difference between existing climates and those of Miocene times, for all the places where a Miocene flora has been discovered, by means of the actual range of corresponding species and genera of plants. Although this method is open to the objection that the ranges of plants and animals are not determined by temperature only, yet the results may be

west, how can we doubt the enormously greater effect of such a
condition of things as has been shown to have existed during
the Tertiary epoch? Instead of *one* great stream of warm water
spreading widely over the North Atlantic and thus losing the
greater part of its store of heat *before* it reaches the arctic seas,
we should have *several* streams conveying the heat of far more
extensive tropical oceans by comparatively narrow inland channels, thus being able to transfer a large proportion of their heat
into the northern and arctic seas. The heat that they gave out
during the passage, instead of being widely dispersed by winds
and much of it lost in the higher atmosphere, would directly
ameliorate the climate of the continents they passed through,
and prevent all accumulation of snow except on the loftiest
mountains. The formation of ice in the arctic seas would then
be impossible; and the mild winter climate of the latitude of
North Carolina, which, by the Gulf Stream, is transferred 20°
northward to our islands, might certainly, under the favorable
conditions which prevailed during the Cretaceous, Eocene, and
Miocene periods, have been carried another 20° north to Greenland and Spitzbergen; and this would bring about exactly the
climate indicated by the fossil arctic vegetation. For it must
be remembered that the arctic summers are, even now, really
hotter than ours; and if the winter's cold were abolished and all

---

approximately correct, and are very interesting. The following table, which summa-
rizes these results, is taken from his "Lectures on Physical Geography," p. 344 :

	Latitude.	Present Temperature.	Miocene Temperature.	Difference.
1. Switzerland...........	47°.00	53°.6 F.	69°.8 F.	16°.2 F.
2. Dantzic...............	54°.21	45°.7	62°.6	16°.9
3. Iceland...............	65°.30	35°.6	48°.2	12°.6
4. Mackenzie River.......	65°.00	19°.4	48°.2	28°.8
5. Disco (Greenland).....	70°.00	19°.6	55°.6	36°.0
6. Spitzbergen...........	78°.00	16°.5	51°.8	35°.3
7. Grinnell Land.........	81°.44	1°.7	42°.3	44°.0

It is interesting to note that Iceland, which is now exposed to the full influence of
the Gulf Stream, was only 12°.6 F. warmer in Miocene times ; while Mackenzie River,
now totally removed from its influence, was 28° warmer. This, as well as the greater
increase of temperature as we go northward and the polar area becomes more limited,
is quite in accordance with the view of the causes which brought about the Miocene
climate which is here advocated.

ice-accumulation prevented, the high northern lands would be able to support a far more luxuriant summer vegetation than is possible in our unequal and cloudy climate.[1]

Effect of High Eccentricity on the Warm Polar Climates.— If the explanation of the cause of the glacial epoch given in the last chapter is a correct one, it will, I believe, follow that changes in the amount of eccentricity will produce no important alteration of the climates of the temperate and arctic zones so long as favorable geographical conditions, such as have been now sketched out, render the accumulation of ice impossible. The effect of a high eccentricity in producing a glacial epoch was shown to be due to the capacity of snow and ice for storing up cold, and its singular power (when in large masses) of preserving itself unmelted under a hot sun by itself causing the interposition of a protective covering of cloud and vapor. But mobile currents of warm water have no such power of accumulating and storing up heat or cold from one year to another, though they do in a pre-eminent degree possess the power of equalizing the temperature of winter and summer, and of conveying the superabundant heat of the tropics to ameliorate the rigor of the arctic winters. However great was the difference between the amount of heat received from the sun in winter and summer in the arctic zone during a period of high eccentricity and winter in aphelion, the inequality would be greatly diminished by the free ingress of warm currents to the polar area; and if this was

---

[1] The objection has been made that the long polar night would of itself be fatal to the existence of such a luxuriant vegetation as we know to have existed as far as 80° N. lat., and that there must have been some alteration of the position of the pole, or diminution of the obliquity of the ecliptic, to permit such plants as magnolias and large-leaved maples to flourish. But there appear to be really no valid grounds for such an objection. Not only are numbers of Alpine and arctic evergreens deeply buried in the snow for many months without injury, but a variety of tropical and subtropical plants are preserved in the hot-houses of St. Petersburg and other northern cities which are closely matted during winter, and are thus exposed to as much darkness as the night of the arctic regions. We have, besides, no proof that any of the arctic trees or large shrubs were evergreens, and the darkness would certainly not be prejudicial to deciduous plants. With a suitable temperature there is nothing to prevent a luxuriant vegetation up to the pole, and the long-continued day is known to be highly favorable to the development of foliage, which in the same species is larger and better developed in Norway than in the South of England.

sufficient to prevent any accumulation of ice, the summers would be warmed to the full extent of the powers of the sun during the long polar day, which is such as to give the pole at midsummer more heat during the twenty-four hours than the equator receives during its day of twelve hours. The only difference, then, that would be directly produced by the changes of eccentricity and precession would be that the summers would be at one period almost tropical, at the other of a more mild and uniform temperate character; while the winters would be at one time somewhat longer and colder, but never, probably, more severe than they are now in the West of Scotland.

But though high eccentricity would not directly modify the mild climates produced by the state of the Northern Hemisphere which prevailed during Cretaceous, Eocene, and Miocene times, it might indirectly affect it by increasing the mass of antarctic ice, and thus increasing the force of the trade-winds and the resulting northward-flowing warm currents. Now there are many peculiarities in the distribution of plants and of some groups of animals in the Southern Hemisphere which render it almost certain that there has sometimes been a greater extension of the antarctic lands during Tertiary times; and it is therefore not improbable that a more or less glaciated condition may have been a long-persistent feature of the Southern Hemisphere, due to the peculiar distribution of land and sea which favors the production of ice-fields and glaciers. And as we have seen that during the last three million years the eccentricity has been almost always much higher than it is now, we should expect that the quantity of ice in the Southern Hemisphere will usually have been greater, and will thus have tended to increase the force of those oceanic currents which produce the mild climates of the Northern Hemisphere.

Evidences of Climate in the Secondary and Palæozoic Epochs.
—We have already seen that so far back as the Cretaceous period there is the most conclusive evidence of the prevalence of a very mild climate not only in temperate but also in arctic lands, while there is no proof whatever, or even any clear indication, of early glacial epochs at all comparable in extent and severity with that which has so recently occurred; and we have seen

reason to connect this state of things with a distribution of land and sea highly favorable to the transferrence of warm water from equatorial to polar latitudes. So far as we can judge by the plant-remains of our own country, the climate appears to have been almost tropical in the Lower Eocene period; and as we go further back we find no clear indications of a higher, but often of a lower, temperature, though always warmer or more equable than our present climate. The abundant corals and reptiles of the Oolite and Lias indicate equally tropical conditions; but further back, in the Trias, the flora and fauna become poorer, and there is nothing incompatible with a climate no warmer than that of the Upper Miocene. This poverty is still more marked in the Permian formation, and it is here that clear indications of ice-action are found in the Lower Permian conglomerates of the West of England. These beds contain abundant fragments of various rocks, often angular and sometimes weighing half a ton, while others are partially rounded, and have polished and striated surfaces, just like the stones of the "till." They lie confusedly bedded in a red unstratified marl, and some of them can be traced to the Welsh hills from twenty to fifty miles distant. This remarkable formation was first pointed out as proving a remote glacial period by Professor Ramsay; and Sir Charles Lyell agreed that this is the only possible explanation that, with our present knowledge, we can give of them.

Permian breccias are also found in Ireland, containing blocks of Silurian and Old Red Sandstone rocks, which Professor Hull believes could only have been carried by floating ice. Similar breccias occur in the South of Scotland, and these are stated to be "overlaid by a deposit of glacial age, so similar to the breccia below as to be with difficulty distinguished from it."[1]

These numerous physical indications of ice-action over a considerable area during the same geological period, coinciding with just such a poverty of organic remains as might be produced by a very cold climate, are very important, and seem clearly to indicate that at this remote period geographical conditions were such as to bring about a glacial epoch in our part of the world.

---

[1] *Geological Magazine*, 1873, p. 320.

Boulder-beds also occur in the Carboniferous formation, both in Scotland, on the continent of Europe, and in North America; and Professor Dawson considers that he has detected true glacial deposits of the same age in Nova Scotia. Boulder-beds also occur in the Silurian rocks of Scotland and North America, and, according to Professor Dawson, even in the Huronian, older than our Cambrian. None of these indications are, however, so satisfactory as those of Permian age, where we have the very kind of evidence we looked for in vain throughout the whole of the Tertiary and Secondary periods. Its presence in several localities in such ancient rocks as the Permian is not only most important as indicating a glacial epoch of some kind in Palæozoic times, but confirms us in the validity of our conclusion that the *total* absence of any such evidence throughout the Tertiary and Secondary epochs demonstrates the absence of recurring glacial epochs in the Northern Hemisphere, notwithstanding the frequent recurrence of periods of high eccentricity.

Warm Arctic Climates in Early Secondary and Palæozoic Times.—The evidence we have already adduced of the mild climates prevailing in the arctic regions throughout the Miocene, Eocene, and Cretaceous periods is supplemented by a considerable body of facts relating to still earlier epochs.

In the Jurassic period, for example, we have proofs of a mild arctic climate in the abundant plant-remains of East Siberia and Amoorland, with less productive deposits in Spitzbergen, and at Andöen in Norway, just within the Arctic Circle. But even more remarkable are the marine remains found in many places in high northern latitudes, among which we may especially mention the numerous ammonites and the vertebræ of huge reptiles of the genera Ichthyosaurus and Teleosaurus found in the Jurassic deposits of the Parry Islands, in 77° N. lat.

In the still earlier Triassic age, nautili and ammonites inhabited the seas of Spitzbergen, where their fossil remains are now found.

In the Carboniferous formation we again meet with plant-remains and beds of true coal in the arctic regions. Lepidodendrons and Calamites, together with large spreading ferns, are found at Spitzbergen, and at Bear Island, in the extreme north of Eastern Siberia; while marine deposits of the same age contain abundance of large stony corals.

Lastly, the ancient Silurian limestones, which are widely spread in the high arctic regions, contain abundance of corals and cephalopodous mollusca resembling those from the same deposits in more temperate lands.

Conclusions as to the Climates of Tertiary and Secondary Periods.—If now we look at the whole series of geological facts as to the animal and vegetable productions of the arctic regions in past ages, it is certainly difficult to avoid the conclusion that they indicate a climate of a uniformly temperate or warm character. Whether in Miocene, Upper or Lower Cretaceous, Jurassic, Triassic, Carboniferous, or Silurian times, and in all the numerous localities extending over more than half the polar regions, we find one uniform climatic aspect in the fossils. This is quite inconsistent with the theory of alternate cold and mild epochs during phases of high eccentricity, and persistent cold epochs when the eccentricity was as low as it is now, or lower, for that would imply that the duration of cold conditions was *greater* than that of warm. Why, then, should the fauna and flora of the cold epochs *never* be preserved? Mollusca and many other forms of life are abundant in the arctic seas, and there is often a luxuriant dwarf woody vegetation on the land, yet in no one case has a single example of such a fauna or flora been discovered of a date anterior to the last glacial epoch. And this argument is very much strengthened when we remember that an exactly analogous series of facts is found over all the temperate zones. Everywhere we have abundant floras and faunas indicating warmer conditions than such as now prevail, but never in a single instance one which as clearly indicates colder conditions. The fact that drift with arctic shells was deposited during the last glacial epoch, as well as gravels and crag with the remains of arctic animals and plants, shows us that there is nothing to prevent such deposits being formed in cold as well as in warm periods; and it is quite impossible to believe that in every place and at all epochs all records of the former have been destroyed, while in a considerable number of instances those of the latter have been preserved. When to this uniform testimony of the palæontological evidence we add the equally uniform absence of any indication of those ice-borne rocks, boulders, and drift which are the constant and necessary accompaniment of every period of glaciation, and which must inevitably pervade all the marine deposits formed over a wide area so long as the state of glaciation continues, we are driven to the conclusion that the last glacial epoch of the Northern Hemisphere was exceptional, and was not preceded by numerous similar glacial epochs throughout Tertiary and Secondary time.

But, although glacial epochs (with the one or two exceptions already referred to) were certainly absent, considerable changes of climate may have frequently occurred, and these would lead to important changes in the organic world. We can hardly doubt that some such change occurred between the Lower and Upper Cretaceous periods, the floras of which exhibit such an extraordinary contrast in general character. We have also the testimony of Mr. J. S. Gardner, who has long worked at the fossil floras of the Tertiary deposits, and who states that there is strong negative and some positive evidence of alternating warmer and colder conditions, not glacial, contained not only in English Eocene, but all Tertiary beds throughout the world.[1] In the case of marine faunas it is more difficult to judge, but the numerous changes in the fossil remains from bed to bed, only a few feet and sometimes a few inches apart, may be sometimes due to change of climate; and when it is recognized that such changes have probably occurred at all geological epochs, and their effects are systematically searched for, many peculiarities in the distribution of organisms through the different members of one deposit may be traced to this cause.

General View of Geological Climates as Dependent on the Physical Features of the Earth's Surface.—In the preceding chapters I have earnestly endeavored to arrive at an explanation of geological climates in the temperate and arctic zones which should be in harmony with the great body of geological facts now available for their elucidation. If my conclusions as here set forth diverge considerably from those of Dr. Croll, it is not

---

[1] *Geological Magazine*, 1877, p. 137.

from any want of appreciation of his facts and arguments, since for many years I have upheld and enforced his views to the best of my ability. But a careful re-examination of the whole question has now convinced me that an error has been made in estimating the comparative effect of geographical and astronomical causes on changes of climate, and that, while the latter have undoubtedly played an important part in bringing about the glacial epoch, it is to the former that the mild climates of the arctic regions are almost entirely due. If I have now succeeded in approaching to a true solution of this difficult problem, I owe it mainly to the study of Dr. Croll's writings, since my theory is entirely based on the facts and principles so clearly set forth in his admirable papers on " Ocean Currents in Relation to the Distribution of Heat over the Globe." The main features of this theory, as distinct from that of Dr. Croll, I will now endeavor to summarize.

Looking at the subject broadly, we see that the climatic condition of the Northern Hemisphere is the result of the peculiar distribution of land and water upon the globe ; and the general permanence of the position of the continental and oceanic areas —which we have shown to be proved by so many distinct lines of evidence—is also implied by the general stability of climate throughout long geological periods. The land surface of our earth appears to have always consisted of three great masses in the north temperate zone, narrowing southward, and terminating in three comparatively narrow extremities represented by Southern America, South Africa, and Australia. Towards the north these masses have approached each other, and have sometimes become united, leaving beyond them a considerable area of open polar sea. Towards the south they have never been much farther prolonged than at present; but far beyond their extremities an extensive mass of land has occupied the south polar area.

This arrangement is such as would cause the Northern Hemisphere to be always (as it is now) warmer than the Southern, and this would lead to the preponderance of northward winds and ocean currents, and would bring about the concentration of the latter in three great streams carrying warmth to the north polar regions. These streams would, as Dr. Croll has so well shown, be greatly increased in power by the glaciation of the south polar land; and whenever any considerable portion of this land was elevated, such a condition of glaciation would certainly be brought about, and would be heightened whenever a high degree of eccentricity prevailed.

It appears to be the general opinion of geologists that the great continents have undergone a process of development from earlier to later times. Professor Dana says, " The North American continent, which since early time had been gradually expanding in each direction from the Northern Azoic, eastward, westward, and southward, and which, after the Palæozoic, was finished in its rocky foundation, excepting on the borders of the Atlantic and Pacific and the area of the Rocky Mountains, had reached its full expansion at the close of the Tertiary period. The progress from the first was uniform and systematic: the land was at all times simple in outline; and its enlargement took place with almost the regularity of an exogenous plant." [1]

A similar development undoubtedly took place in the European area, which was apparently never so compact and so little interpenetrated by the sea as it is now, while Europe and Asia have only become united into one unbroken mass since late Tertiary times.

If, however, the greater continents have become more compact and massive from age to age, and have received their chief extensions northward at a comparatively recent period, while the antarctic lands had a corresponding but somewhat earlier development, we have all the conditions requisite to explain the persistence, with slight fluctuations, of warm climates far into the north polar area throughout Palæozoic, Mesozoic, and Tertiary times. At length, during the latter part of the Tertiary epoch, a considerable elevation took place, closing up several of the water passages to the north, and raising up extensive areas in the arctic regions to become the receptacle of snow and ice fields. This elevation is indicated by the abundance of Miocene and the absence of Pliocene deposits in the arctic zone, and the

---

[1] " Manual of Geology," 2d ed., p. 525.

considerable altitude of many Miocene rocks in Europe and North America; and the occurrence at this time of a long-continued period of high eccentricity necessarily brought on the glacial epoch in the manner already described in our last chapter.

We thus see that the last glacial epoch was the climax of a great process of continental development which has been going on throughout long geological ages; and that it was the direct consequence of the north temperate and polar land having attained a great extension and a considerable altitude just at the time when a phase of very high eccentricity was coming on. Throughout earlier Tertiary and Secondary times an equally high eccentricity often occurred, but it never produced a glacial epoch, because the north temperate and polar areas had less high land, and were more freely open to the influx of warm oceanic currents. But wherever great plateaus with lofty mountains occurred in the temperate zone a considerable *local* glaciation might be produced, which would be specially intense during periods of high eccentricity; and it is to such causes we must impute the indications of ice-action in the vicinity of the Alps during the Tertiary period. The Permian glaciation appears to have been more extensive, and it is quite possible that at this remote epoch a sufficient mass of high land existed in our area and northward towards the pole to have brought on a true glacial period comparable with that which has so recently passed away.

Estimate of the Comparative Effects of Geographical and Astronomical Causes in Producing Changes of Climate.—It appears, then, that while geographical and physical causes alone, by their influence on ocean currents, have been the main agents in producing the mild climates which for such long periods prevailed in the arctic regions, the concurrence of astronomical causes—high eccentricity with winter in aphelion—was necessary to the production of the great glacial epoch. If we reject this latter agency, we shall be obliged to imagine a concurrence of geographical changes at a very recent period of which we have no evidence. We must suppose, for example, that a large part of the British Isles—Scotland, Ireland, and Wales at all events—were simultaneously elevated so as to bring extensive areas above the line of perpetual snow; that about the same time Scandinavia, the Alps, and the Pyrenees received a similar increase of altitude; and that, almost simultaneously, Eastern North America, the Sierra Nevada of California, the Caucasus, Lebanon, the southern mountains of Spain, the Atlas range, and the Himalayas were each some thousands of feet higher than they are now; for all these mountains present us with indications of a recent extension of their glaciers, in superficial phenomena so similar to those which occur in our own country and in Western Europe that we cannot suppose them to belong to a different epoch. Such a supposition is rendered more difficult by the general concurrence of scientific testimony to a partial submergence during the glacial epoch, not only in all parts of Britain, but in North America, Scandinavia, and, as shown by the wide extension of the drift, in Northern Europe; and when to this we add the difficulty of understanding how any probable addition to the altitude of our islands could have brought about the extreme amount of glaciation which they certainly underwent, and when, further, we know that a phase of very high eccentricity did occur at a period which is generally admitted to agree well with physical evidence of the time elapsed since the cold passed away, there seems no sufficient reason why such an agency should be ignored.

No doubt a prejudice has been excited against it in the minds of many geologists, by its being thought to lead *necessarily* to frequently recurring glacial epochs throughout all geological time. But I have here endeavored to show that this is *not* a necessary consequence of the theory, because a concurrence of favorable geographical conditions is essential to the initiation of a glaciation, which, when once initiated, has a tendency to maintain itself throughout the varying phases of precession occurring during a period of high eccentricity. When, however, geographical conditions favor warm arctic climates—as it has been shown they have done throughout the larger portion of geological time—then changes of eccentricity, to however great an extent, have no tendency to bring about a state of glaciation, because warm oceanic currents have a preponderating influence, and without very large areas of high northern land to act as condensers, no perpetual snow is possible, and hence the initial process of glaciation does not occur.

The theory as now set forth should commend itself to geologists, since it shows the direct dependence of climate on physical processes which are guided and modified by those changes in the earth's surface which geology alone can trace out. It is in perfect accord with the most recent teachings of the science as to the gradual and progressive development of the earth's crust from the rudimentary formations of the Azoic age, and it lends support to the view that no important departure from the great lines of elevation and depression originally marked out on the earth's surface have ever taken place.

It also shows us how important an agent in the production of a habitable globe, with comparatively small extremes of climates over its whole area, is the great disproportion between the extent of the land and the water surfaces. For if these proportions had been reversed, large areas of land would necessarily have been removed from the beneficial influence of aqueous currents or moisture-laden winds; and slight geological changes might easily lead to half the land surface becoming covered with perpetual snow and ice or being exposed to extremes of summer heat and winter cold, of which our water-permeated globe at present affords no example. We thus see that what are usually regarded as geographical anomalies—the disproportion of land and water, the gathering of the land mainly into one hemisphere, and the singular arrangement of the land in three great southward-pointing masses—are really facts of the greatest significance and importance, since it is to these very anomalies that the universal spread of vegetation and the adaptability of so large a portion of the earth's surface for human habitation are directly due.

CHAPTER X.

THE EARTH'S AGE, AND THE RATE OF DEVELOPMENT OF ANIMALS AND PLANTS.

Various Estimates of Geological Time.—Denudation and Deposition of Strata as a Measure of Time.—How to Estimate the Thickness of the Sedimentary Rocks.— How to Estimate the Average Rate of Deposition of the Sedimentary Rocks.— The Rate of Geological Change probably Greater in very Remote Times.—Value of the Preceding Estimate of Geological Time.—Organic Modification Dependent on Change of Conditions.—Geographical Mutations as a Motive Power in Bringing about Organic Changes.—Climatal Revolutions as an Agent in Producing Organic Changes.—Present Condition of the Earth one of Exceptional Stability as regards Climate.—Date of Last Glacial Epoch, and its Bearing on the Measurement of Geological Time.—Concluding Remarks.

THE subjects discussed in the last three chapters introduce us to a difficulty which has hitherto been considered a very formidable one—that the maximum age of the habitable earth, as deduced from physical considerations, does not afford sufficient time either for the geological or the organic changes of which we have evidence. Geologists continually dwell on the slowness of the processes of upheaval and subsidence, of denudation of the earth's surface, and of the formation of new strata; while on the theory of development as expounded by Mr. Darwin the variation and modification of organic forms is also a very slow process, and has usually been considered to require an even longer series of ages than might satisfy the requirements of physical geology alone.

As an indication of the periods usually contemplated by geologists, we may refer to Sir Charles Lyell's calculation in the tenth edition of his " Principles of Geology" (omitted in later editions), by which he arrived at two hundred and forty millions of years as having probably elapsed since the Cambrian period —a very moderate estimate in the opinion of most geologists. This calculation was founded on the rate of modification of the

species of mollusca; but much more recently Professor Haughton has arrived at nearly similar figures from a consideration of
the rate of formation of rocks and their known maximum thickness, whence he deduces a maximum of two hundred millions of
years for the whole duration of geological time as indicated by
the series of stratified formations.[1] But in the opinion of all our
first naturalists and geologists, the period occupied in the formation of the known stratified rocks only represents a portion, and
perhaps a small portion, of geological time. In the last edition
of the " Origin of Species " (p. 286), Mr. Darwin says, " Consequently, if the theory be true, it is indisputable that before the
lowest Cambrian stratum was deposited long periods elapsed—as
long as, or probably far longer than, the whole interval from
the Cambrian age to the present day; and that during these
vast periods the world swarmed with living creatures." Professor Huxley, in his anniversary address to the Geological Society in 1870, adduced a number of special cases showing that,
on the theory of development, almost all the higher forms of
life must have existed during the Palæozoic period. Thus, from
the fact that almost the whole of the Tertiary period has been
required to convert the ancestral Orohippus into the true horse,
he believes that, in order to have time for the much greater
change of the ancestral Ungulata into the two great odd-toed
and even-toed divisions (of which change there is no trace even
among the earliest Eocene mammals), we should require a large
portion, if not the whole, of the Mesozoic or Secondary period.
Another case is furnished by the bats and whales, both of which
strange modifications of the mammalian type occur perfectly developed in the Eocene formation. What countless ages back must
we then go for the origin of these groups—the whales from some
ancestral carnivorous animal, and the bats from the insectivora!
And even then we have to seek for the common origin of carnivora, insectivora, ungulata, and marsupials at a far earlier period; so that, on the lowest estimate, we must place the origin
of the mammalia very far back in Palæozoic times. Similar
evidence is afforded by reptiles, of which Professor Huxley says,

---

[1] *Nature*, Vol. XVIII. (July, 1878), p. 268.

"If the very small differences which are observable between the crocodiles of the older Secondary formations and those of the present day furnish any sort of an approximation towards an estimate of the average rate of change among reptiles, it is almost appalling to reflect how far back in Palæozoic times we must go before we can hope to arrive at that common stock from which the crocodiles, lizards, *Ornithoscelida*, and *Plesiosauria*, which had attained so great a development in the Triassic epoch, must have been derived." Professor Ramsay has expressed similar views, derived from a general study of the whole series of geological formations and their contained fossils. He says, speaking of the abundant, varied, and well-developed fauna of the Cambrian period, "In this earliest known *varied* life we find no evidence of its having lived near the beginning of the zoological series. In a broad sense, compared with what must have gone before, both biologically and physically, all the phenomena connected with this old period seem, to my mind, to be of quite a recent description ; and the climates of seas and lands were of the very same kind as those the world enjoys at the present day."[1]

These opinions, and the facts on which they are founded, are so weighty that we can hardly doubt that, if the time since the Cambrian epoch is correctly estimated at two hundred millions of years, the date of the commencement of life on the earth cannot be much less than five hundred millions ; while it may not improbably have been longer, because the reaction of the organism under changes of the environment is believed to have been less active in low and simple than in high and complex forms of life, and thus the processes of organic development may for countless ages have been excessively slow.

But, according to the physicists, no such periods as are here contemplated can be granted. From a consideration of the possible sources of the heat of the sun, as well as from calculations of the period during which the earth can have been cooling to bring about the present rate of increase of temperature as we

---

[1] "On the Comparative Value of Certain Geological Ages Considered as Items of Geological Time," *Proceedings of the Royal Society*, 1874, p. 334.

descend beneath the surface, Sir William Thomson concludes
that the crust of the earth cannot have been solidified much longer than one hundred million years (the maximum possible being four hundred millions), and this conclusion is held by Dr.
Croll and other men of eminence to be almost indisputable.[1] It
will therefore be well to consider on what data the calculations
of geologists have been founded, and how far the views here set
forth as to frequent changes of climate throughout all geological time may affect the rate of biological change.

  Denudation and Deposition of Strata as a Measure of Time.
—The materials of all the stratified rocks of the globe have been
obtained from the dry land. Every point of the surface is exposed to the destructive influences of sun and wind, frost, snow,
and rain, which break up and wear away the hardest rocks as
well as the softer deposits, and by means of rivers convey the
worn material to the sea. The existence of a considerable depth
of soil over the greater part of the earth's surface; of vast heaps
of rocky débris at the foot of every inland cliff; of enormous
deposits of gravel, sand, and loam; as well as the shingle, pebbles, sand, or mud of every sea-shore, alike attest the universality of this destructive agency. It is no less clearly shown by
the way in which almost every drop of running water—whether
in gutter, brooklet, stream, or large river — becomes discolored
after each heavy rainfall, since the matter which causes this dis-
coloration must be derived from the surface of the country, must
always pass from a higher to a lower level, and must ultimately
reach the sea, unless it is first deposited in some lake, or by the
overflowing of a river goes to form an alluvial plain. The universality of this subaerial denudation, both as regards space and
time, renders it certain that its cumulative effects must be very
great; but no attempt seems to have been made to determine
the magnitude of these effects till Mr. Alfred Tylor, in 1853,[2]
pointed out that by measuring the quantity of solid matter
brought down by rivers (which can be done with considerable

---

  [1] *Transactions of the Royal Society of Edinburgh*, Vol. XXIII., p. 161; *Quarterly
Journal of Science*, 1877 (Croll, on the "Probable Origin and Age of the Sun").
  [2] *Philosophical Magazine*, April, 1853.

accuracy), we may obtain the amount of lowering of the land area, and also the rise of the ocean-level, owing to the quantity of matter deposited on its floor. A few years later Dr. Croll applied the same method in more detail to an estimate of the amount by which the land is lowered in a given period; and the validity of this method has been upheld by Professor Geikie, Sir Charles Lyell, and all our best geologists, as affording a means of actually determining with some approach to accuracy the time occupied by one important phase of geological change.

The quantity of matter carried away from the land by a river is greater than at first sight appears, and is more likely to be under- than over-estimated. By taking samples of water near the mouth of a river (but above the influence of the tide) at a sufficient number of points in its channel and at different depths, and repeating this daily or at other short intervals throughout the year, it is easy to determine the quantity of solid matter held in suspension and solution; and if corresponding observations determine the quantity of water that is discharged, the total amount of solid matter brought down annually may be calculated. But, besides this, a considerable quantity of sand, or even gravel, is carried along the bottom or bed of the river, and this has rarely been estimated, so that the figures hitherto obtained are usually under the real quantities. There is also another source of error caused by the quantity of matter the river may deposit in lakes or in flooded lands during its course; for this adds to the amount of denudation performed by the river, although the matter so deposited does not come down to the sea. After a careful examination of all the best records, Professor A. Geikie arrives at the following results as to the quantity of matter removed by seven rivers from their basins, estimated by the number of years required to lower the whole surface an average of one foot:

The Mississippi removes one foot in 6000 years.
The Ganges   "   "   2358   "
The Hoang Ho   "   "   1464   "
The Rhone   "   "   1528   "
The Danube   "   "   6846   "
The Po   "   "   729   "
The Nith   "   "   4723   "

Here we see an intelligible relation between the character of the river basin and the amount of denudation. The Mississippi has a large portion of its basin in an arid country, and its sources are either in forest-clad plateaus or in mountains free from glaciers and with a scanty rainfall. The Danube flows through Eastern Europe, where the rainfall is considerably less than in the West, while comparatively few of its tributaries rise among the loftiest Alps. The proportionate amounts of denudation being, then, what we might expect, and as all are probably under rather than over the truth, we may safely take the average of them all as representing an amount of denudation which, if not true for the whole land surface of the globe, will certainly be so for a very considerable proportion of it. This average is almost exactly one foot in three thousand years.[1] The mean altitude of the several continents has been estimated to be as follows: Europe, 671 feet; Asia, 1132 feet; Africa, 900 feet; North America, 748 feet; and South America, 1151 feet. At the rate of denudation above given, it results that, were no other forces at work, Europe would be planed down to the sea-level in about two million years; while, if we take a somewhat slower rate for North America, that continent might last about three million years.[2] This also implies that the mean height of these conti-

---

[1] It has usually been the practice to take the amount of denudation in the Mississippi valley, or one foot in six thousand years, as a measure of the rate of denudation in Europe, from an idea apparently of being on the "safe side," and of not overestimating the rate of change. But this appears to me a most unphilosophical mode of proceeding, and unworthy of scientific inquiry. What should we think of astronomers if they always took the lowest estimates of planetary or stellar distances, instead of the mean results of observation, "in order to be on the safe side?" As if error in one direction were any worse than error in another. Yet this is what geologists do systematically. Whenever any calculations are made involving the antiquity of man, it is those that give the *lowest* results that are always taken, for no reason apparently except that there was, for so long a time, a prejudice, both popular and scientific, against the great antiquity of man; and now that a means has been found of measuring the rate of denudation, they take the slowest rate instead of the mean rate, apparently only because there is now a scientific prejudice in favor of extremely slow geological change. I take the mean of the whole; and, as this is almost exactly the same as the mean of the three great European rivers—the Rhone, Danube, and Po—I cannot believe that this will not be nearer the truth for Europe than taking one North American river as the standard.

[2] These figures are merely used to give an idea of the rate at which denudation is

nents would have been double what it is now two million and
three million years ago respectively ; and, as we have no reason
to suppose this to have been the case, we are led to infer the
constant action of that upheaving force which the presence of
sedimentary formations even on the highest mountains also
demonstrates.

We have already discussed the unequal rate of denudation on
hills, valleys, and lowlands in connection with the evidence of
remote glacial epochs (p. 164); what we have now to consider
is, what becomes of all this denuded matter, and how far the
known rate of denudation affords us a measure of the rate of
deposition, and thus gives us some indication of the lapse of
geological time from a comparison of this rate with the observed
thickness of stratified rocks on the earth's surface.

How to Estimate the Thickness of the Sedimentary Rocks.—
The sedimentary rocks, of which the earth's crust is mainly composed, consist, according to Sir Charles Lyell's classification, of
fourteen great formations, of which the most ancient is the Laurentian, and the most recent the Post-tertiary; with thirty important subdivisions, each of which again consists of a more or
less considerable number of distinct beds or strata.   Thus, the
Silurian formation is divided into Upper and Lower Silurian,
each characterized by a distinct set of fossil remains; and the
Upper Silurian again consists of a large number of separate
beds, such as the Wenlock Limestone, the Upper Llandovery
Sandstone, the Lower Llandovery Slates, etc., each usually characterized by a difference of mineral composition or mechanical
structure as well as by some peculiar fossils.   These beds and
formations vary greatly in extent, both above and beneath the

---

actually going on now; but if no elevatory forces were at work, the rate of denudation would certainly diminish as the mountains were lowered and the slope of the
ground everywhere rendered flatter.   This would follow not only from the diminished
power of rain and rivers, but because the climate would become more uniform, the
rainfall probably less, and no rocky peaks would be left to be fractured and broken up
by the action of frosts.   It is certain, however, that no continent has ever remained
long subject to the influences of denudation alone; for, as we have seen in our sixth
chapter, elevation and depression have always been going on in one part or other of
the surface.

surface, and are also of very various thicknesses in different localities. A thick bed or series of beds often thins out in a
given direction, and sometimes disappears altogether, so that
two beds which were respectively above and beneath it may
come into contact. As an example of this thinning-out, American geologists adduce the Palæozoic formations of the Appa-
lachian Mountains, which have a total thickness of 42,000 feet,
but as they are traced westward thin out till they become only
4000 feet in total thickness. In like manner, the Carboniferous grits and shales are 1800 feet thick in Yorkshire and Lancashire, but they thin out southward, so that in Leicestershire
they are only 3000 feet thick; and similar phenomena occur
in all strata and in every part of the world. It must be observed that this thinning-out has nothing to do with denudation (which acts upon the surface of a country so as to produce
great irregularities of contour), but is a regular attenuation of
the layers of rock, due to a deficiency of sediment in certain
directions at the original formation of the deposit. Owing to
this thinning-out of stratified rocks, they are, on the whole, of far
less extent than is usually supposed. When we see a geological
map showing successive formations following each other in long
irregular belts across the country (as is well seen in the case of
the Secondary rocks of England), and a corresponding section
showing each bed dipping beneath its predecessor, we are apt
to imagine that beneath the uppermost bed we shall find all the
others following in succession like the coats of an onion. But
this is far from being the case, and a remarkable proof of the
narrow limitation of these formations has been recently obtained
by a boring at Ware through the Chalk and Gault clay, which
latter immediately rests on the Upper Silurian Wenlock Limestone, full of characteristic fossils, at a depth of only 800 feet.
Here we have an enormous gap, showing that none of earlier
Secondary or late Palæozoic formations extend to this part of
England, unless, indeed, they had been all once elevated and
entirely swept away by denudation.[1]

---

[1] The following statement of the depths at which the Palæozoic formations have
been reached in various localities in and around London was given by Mr. H. B.
Woodward in his address to the Norwich Geological Society in 1879 :

But if we consider how such deposits are now forming, we shall find that the thinning-out of the beds of each formation, and their restriction to irregular bands and patches, is exactly what we should expect. The enormous quantity of sediment continually poured into the sea by rivers gradually subsides to the bottom as soon as the motion of the water is checked. All the heavier material must be deposited near the shore or in those areas over which it is first spread by the tides or currents of the ocean; while only the very fine mud and clay are carried out to considerable distances. Thus all stratified deposits will form most quickly near the shores, and will thin out rapidly at greater distances, little or none being formed in the depths of the great oceans. This important fact was demonstrated by the specimens of sea-bottom examined during the voyage of the *Challenger*, all the "shore-deposits" being usually confined within a distance of one hundred or one hundred and fifty miles from the coast, while the "deep-sea" deposits are either purely organic, being formed of the calcareous or siliceous skeletons of Globigerinæ, radiolarians, and Diatomaceæ, or are clays formed of undissolved portions of these, together with the disintegrated or dissolved materials of pumice and volcanic dust, which, being very light, are carried by wind or by water over the widest oceans.

From the preceding considerations we shall be better able to appreciate the calculations as to the thickness of stratified deposits made by geologists. Professor Ramsay has calculated that the sedimentary rocks of Britain alone have a total maximum thickness of 72,600 feet; while Professor Haughton, from a survey of the whole world, estimates the maximum thickness

---

Deep Wells through the Tertiary and Cretaceous Formations.

Harwich	at 1022 feet reached		Carboniferous Rock.
Kentish Town.........	" 1114 "	"	Old Red Sandstone.
Tottenham Court Road	" 1064 "	"	Devonian.
Blackwall	" 1004 "	"	Devonian or Old Red Sandstone.
Ware	" 800 "	"	Silurian (Wenlock Shale).

We thus find that over a wide area, extending from London to Ware and Harwich, the whole of the formations from the Oolite to the Permian are wanting, the Cretaceous resting on the Carboniferous or older Palæozoic rocks; and the same deficiency extends across to Belgium, where the Tertiary beds are found resting on Carboniferous at a depth of less than 400 feet.

of the known stratified rocks at 177,200 feet. Now these maximum thicknesses of each deposit will have been produced only
where the conditions were exceptionally favorable, either in deep
water near the mouths of great rivers, or in inland seas, or in
places to which the drainage of extensive countries was conveyed by ocean currents; and this great thickness will necessarily be accompanied by a corresponding thinness, or complete
absence of deposit, elsewhere. How far the series of rocks found
in any extensive area, as Europe or North America, represents
the whole series of deposits which have been made there we
cannot tell; but there is no reason to think that it is a very inadequate representation of their maximum thickness, though it
undoubtedly is of their extent and bulk. When we see in how
many distinct localities patches of the same formation occur, it
seems improbable that the whole of the deposits formed during
any one period should have been destroyed, even in such an area
as Europe, while it is still more improbable that they should be
so destroyed over the whole world; and if any considerable portion of them is left, that portion may give a fair idea of their
average, or even of their maximum, thickness. In his admirable paper on " The Mean Thickness of the Sedimentary Rocks," [1]
Dr. James Croll has dwelt on the extent of denudation in diminishing the mean thickness of the rocks that have been
formed, remarking, " Whatever the present mean thickness of
all the sedimentary rocks of our globe may be, it must be small
in comparison to the mean thickness of all the sedimentary
rocks which have been formed. This is obvious from the fact
that the sedimentary rocks of one age are partly formed from
the destruction of the sedimentary rocks of former ages. From
the Laurentian age down to the present day the stratified rocks
have been undergoing constant denudation." This is perfectly
true, and yet the mean thickness of that portion of the sedimentary rocks which remains may not be very different from that of
the entire mass, because denudation acts only on those rocks
which are exposed on the surface of a country, and most largely
on those that are upheaved; while, except in the rare case of an

---

[1] *Geological Magazine*, Vol. VIII., March, 1871.

extensive formation being *quite horizontal,* and wholly exposed
to the sea or to the atmosphere, denudation can have no tendency to diminish the thickness of any entire deposit.[1] Unless,
therefore, a formation is completely destroyed by denudation in
every part of the world (a thing very improbable), we may have
in existing rocks a not very inadequate representation of the
mean thickness of all that have been formed, and even of the
maximum thickness of the larger portion. This will be the
more likely because it is almost certain that many rocks con-
temporaneously formed are counted by geologists as distinct
formations whenever they differ in lithological character or in
organic remains. But we know that limestones, sandstones, and
shales are always forming at the same time; while a great difference in organic remains may arise from comparatively slight
changes of geographical features, or from difference in the depth
or purity of the water in which the animals lived.[2]

  *How to Estimate the Average Rate of Deposition of the Sed-
imentary Rocks.* — But if we take the estimate of Professor
Haughton (177,200 feet), which, as we have seen, is probably excessive, for the maximum thickness of the sedimentary rocks of
the globe of all known geological ages, can we arrive at any estimate of the rate at which they were formed? Dr. Croll has
attempted to make such an estimate, but he has taken for his
basis the mean thickness of the rocks, which we have no means
whatever of arriving at, and which he guesses, allowing for denudation, to be equal to the maximum thickness as measured by
geologists. The land area of the globe is, according to Dr. Croll,

---

  [1] Mr. C. Lloyd Morgan has well illustrated this point by comparing the generally
tilted-up strata denuded on their edges to a library in which a fire had acted on the
exposed edges of the books, destroying a great mass of literature, but leaving a portion of each book in its place, which portion represents the thickness, but not the size,
of the book (*Geological Maagzine,* 1878, p. 161).

  [2] Professor J. Young thinks it highly probable that "the Lower Greensand is
contemporaneous with part of the Chalk, so were parts of the Wealden; nay, even
of the Purbeck a portion must have been forming while the Cretaceous sea was gradually deepening southward and westward." Yet these deposits are always arranged
successively, and their several thicknesses added together to obtain the total thickness of the formations of the country.   (See Presidential Address, Sect. C, British
Association, 1876.)

57,000,000 square miles, and he gives the coast-line as 116,000 miles. This, however, is, for our purpose, rather too much, as it allows for bays, inlets, and the smaller islands. An approximate measurement on a globe shows that 100,000 miles will be nearer the mark, and this has the advantage of being an easily remembered even number. The distance from the coast to which shore-deposits usually extend may be reckoned at about one hundred or one hundred and fifty miles, but by far the larger portion of the matter brought down from the land will be deposited comparatively close to the shore; that is, within twenty or thirty miles. If we suppose the portion deposited beyond thirty miles to be added to the deposits within that distance, and the whole reduced to a uniform thickness in a direction at right angles to the coast, we should probably include all areas where deposits of the maximum thickness are forming at the present time, along with a large but unknown proportion of surface where the deposits were far below the maximum thickness. This follows, if we consider that deposit must go on very unequally along different parts of a coast, owing to the distance from each other of the mouths of great rivers and the limitations of ocean currents; and because, compared with the areas over which a thick deposit is forming annually, those where there is little or none are probably at least twice as extensive. If, therefore, we take a width of thirty miles along the whole coast-line of the globe, as representing the area over which deposits are forming, corresponding to the maximum thickness as measured by geologists, we shall certainly over- rather than under-estimate the possible rate of deposit.[1]

---

[1] As by far the larger portion of the denuded matter of the globe passes to the sea through comparatively few great rivers, the deposits must often be confined to very limited areas. Thus the denudation of the vast Mississippi basin must be almost all deposited in a limited portion of the Gulf of Mexico, that of the Nile within a small area of the Eastern Mediterranean, and that of the great rivers of China—the Hoang Ho and Yang-tse-kiang—in a small portion of the Eastern Sea. Enormous lengths of coast, like those of Western America and Eastern Africa, receive very scanty deposits; so that thirty miles in width along the whole of the coasts of the globe will probably give an area greater than that of the area of average deposit, and certainly greater than that of maximum deposit, which is the basis on which I have here made my estimates. In the case of the Mississippi, it is stated by Count Pourtales that

Now a coast-line of 100,000 miles with a width of 30 gives an area of 3,000,000 square miles, on which the denuded matter of the whole land area of 57,000,000 square miles is deposited. As these two areas are as 1 to 19, it follows that deposition, as measured by maximum thickness, goes on at least nineteen times as fast as denudation—probably very much faster. But the mean rate of denudation over the whole earth is about one foot in three thousand years; therefore the rate of maximum deposition will be at least nineteen feet in the same time; and as the total maximum thickness of all the stratified rocks of the globe is, according to Professor Haughton, 177,200 feet, the time required to produce this thickness of rock at the present rate of denudation and deposition is only 28,000,000 years.[1]

The Rate of Geological Change Probably Greater in very Remote Times.—The opinion that denudation and deposition went on more rapidly in early times owing to the frequent occurrence of vast convulsions and cataclysms was strenuously opposed by Sir Charles Lyell, who so well showed that causes of the very same nature as those now in action were sufficient to account for all the phenomena presented by the rocks throughout the whole series of geological formations. But while upholding the sound-

---

along the plateau between the mouth of the river and the southern extremity of Florida, for two hundred and fifty miles in width, the bottom consists of clay with some sand and but few Rhizopods; but beyond this distance the soundings brought up either Rhizopod shells alone, or these mixed with coral sand, Nullipores, and other calcareous organisms (Dana's "Manual of Geology," 2d ed., p. 671). It is probable, therefore, that a large proportion of the entire mass of sediment brought down by the Mississippi is deposited on the limited area above indicated.

Professor Dana further remarks, "Over interior oceanic basins, as well as off a coast in quiet depths, fifteen or twenty fathoms and beyond, the deposits are mostly of fine silt, fitted for making fine argillaceous rocks, as shales or slates. When, however, the depth of the ocean falls off below a hundred fathoms, the deposition of silt in our existing oceans mostly ceases, unless in the case of a great bank along the border of a continent."

[1] From the same data Professor Haughton estimates a minimum of two hundred million years for the duration of geological time; but he arrives at this conclusion by supposing the products of denudation to be uniformly spread over the *whole sea-bottom* instead of over a narrow belt near the coasts—a supposition entirely opposed to all the known facts, and which had been shown by Dr. Croll five years previously to be altogether erroneous. (See *Nature*, Vol. XVIII., p. 268, where Professor Haughton's paper is given as read before the Royal Society.)

ness of the views of the "uniformitarians" as opposed to the
"convulsionists," we must yet admit that there is reason for believing in a gradually increasing intensity of all telluric action
as we go back into past time. This subject has been well treated
by Mr. W. J. Sollas,[1] who shows that if, as all physicists maintain, the sun gave out perceptibly more heat in past ages than
now, this alone would cause an increase in almost all the forces
that have brought about geological phenomena. With greater
heat there would be a more extensive aqueous atmosphere, and
a greater difference between equatorial and polar temperatures ;
hence, more violent winds, heavier rains and snows, and more
powerful oceanic currents, all producing more rapid denudation.
At the same time, the internal heat of the earth being greater,
it would be cooling more rapidly, and thus the forces of con-
traction—which cause the upheaving of mountains, the eruption
of volcanoes, and the subsidence of extensive areas—would be
more powerful and would still further aid the process of denudation. Yet again, the earth's rotation was certainly more rapid
in very remote times, and this would cause more impetuous
tides and still further add to the denuding power of the ocean.
It thus appears that as we go back into the past, *all* the forces
tending to the continued destruction and renewal of the earth's
surface would be in more powerful action, and must, therefore,
tend to reduce the time required for the deposition and upheaval of the various geological formations. It may be true, as many
geologists assert, that the changes here indicated are so slow
that they would produce comparatively little effect within the
time occupied by the known sedimentary rocks; yet, whatever
effect they did produce would certainly be in the direction here
indicated, and as several causes are acting together, their combined effect may have been by no means unimportant. It must
also be remembered that such an increase of the primary forces
on which all geologic change depends would act with great effect in still further intensifying those alternations of cold and
warm periods in each hemisphere, or, more frequently, of excessive and equable seasons, which have been shown to be the re-

---

[1] See *Geological Magazine* for 1877, p. 1.

sult of astronomical combined with geographical revolutions; and this would again increase the rapidity of denudation and deposition, and thus still further reduce the time required for the production of the known sedimentary rocks. It is evident, therefore, that these various considerations all combine to prove that, in supposing that the rate of denudation has been on the average only what it is now, we are almost certainly overestimating the *time* required to have produced the whole series of formations from the Cambrian upwards.

Value of the Preceding Estimate of Geological Time.—It is not, of course, supposed that the calculation here given makes any approach to accuracy, but it is believed that it does indicate the *order* of magnitude of the time required. We have a certain number of data which are not guessed, but the result of actual measurement; such are, the amount of solid matter carried down by rivers, the width of the belt within which this matter is mainly deposited, and the maximum thickness of the known stratified rocks.[1] A considerable but unknown amount of denudation is effected by the waves of the ocean eating away coast-lines. This was once thought to be of more importance than subaerial denudation, but it is now believed to be comparatively slow in its action.[2] Whatever it may be, however, it adds to the rate of formation of new strata, and its omission from the calculation is again on the side of making the lapse of time greater rather than less than the true amount. Even if a considerable modification

---

[1] In his reply to Sir W. Thomson, Professor Huxley *assumed* one foot in a thousand years as a not improbable rate of deposition. The above estimate indicates a far higher rate; and this follows from the well-ascertained fact that the area of deposition is many times smaller than the area of denudation.

[2] Dr. Croll and Professor Geikie have shown that marine denudation is very small in amount as compared with subaerial, since it acts only locally on the *edge* of the land, whereas the latter acts over every foot of the *surface*. Mr. W. T. Blanford argues that the difference is still greater in tropical than in temperate latitudes, and arrives at the conclusion that "if over British India the effects of marine to those of fresh-water denudation in removing the rocks of the country be estimated at 1 to 100, I believe that the result of marine action will be greatly overstated" ("Geology and Zoology of Abyssinia," p. 158, note). Now, as our estimate of the rate of subaerial denudation cannot pretend to any precise accuracy, we are justified in neglecting marine denudation altogether, especially as we have no method of estimating it for the whole earth with any approach to correctness.

should be needed in some of the assumptions it has been necessary to make, the result must still show that, so far as the time required for the formation of the known stratified rocks, the hundred million years allowed by physicists is not only ample, but
will permit of even more than an equal period anterior to the
lowest Cambrian rocks, as demanded by Mr. Darwin—a demand
supported and enforced by the arguments, taken from independent standpoints, of Professor Huxley and Professor Ramsay.

Organic Modification Dependent on Change of Conditions.—
Having thus shown that the physical changes of the earth's surface may have gone on much more rapidly and occupied much
less time than has generally been supposed, we have now to inquire whether there are any considerations which lead to the
conclusion that organic changes may have gone on with corresponding rapidity.

There is no part of the theory of natural selection which is
more clear and satisfactory than that which connects changes of
specific forms with changes of external conditions or environment. If the external world remains for a moderate period unchanged, the organic world soon reaches a state of equilibrium
through the struggle for existence; each species occupies its
place in nature, and there is then no inherent tendency to
change. But almost any change whatever in the external world
disturbs this equilibrium, and may set in motion a whole series
of organic revolutions before it is restored. A change of climate
in any direction will be sure to injure some and benefit other
species. The one will consequently diminish, the other increase,
in number; and the former may even become extinct. But the
extinction of a species will certainly affect other species which
it either preyed upon, or competed with, or served for food;
while the increase of any one animal may soon lead to the extinction of some other to which it was inimical. These changes
will in their turn bring other changes; and before an equilibrium
is again established the proportions, ranges, and numbers of the
species inhabiting the country may be materially altered. The
complex manner in which animals are related to each other is
well exhibited by the importance of insects, which in many
parts of the world limit the numbers or determine the very ex-

istence of some of the higher animals. Mr. Darwin says, "Perhaps Paraguay offers the most curious instance of this; for here neither cattle nor horses nor dogs have ever run wild, though they swarm southward and northward in a wild state ; and Azara and Rengger have shown that this is caused by the greater number in Paraguay of a certain fly which lays its eggs in the navels of these animals when first born. The increase of these flies, numerous as they are, must be habitually checked by some means, probably by other parasitic insects. Hence, if certain insectivorous birds were to decrease in Paraguay, the parasitic insects would probably increase; and this would lessen the number of navel-frequenting flies; then cattle and horses would run wild ; and this would certainly alter (as, indeed, I have observed in parts of South America) the vegetation ; this, again, would largely affect the insects, and this, as we have seen in Staffordshire, the insectivorous birds, and so onward in ever-increasing circles of complexity."

Geographical changes would be still more important, and it is almost impossible to exaggerate the modifications of the organic world that might result from them. A subsidence of land separating a large island from a continent would affect the animals and plants in a variety of ways. It would at once modify the climate, and so produce a series of changes from this cause alone ; but more important would be its effect by isolating small groups of individuals of many species, and thus altering their relations to the rest of the organic world. Many of these would at once be exterminated, while others, being relieved from competition, might flourish and become modified into new species. Even more striking would be the effects when two continents, or any two land areas which had been long separated, were united by an upheaval of the strait which divided them. Numbers of animals would now be brought into competition for the first time. New enemies and new competitors would appear in every part of the country; and a struggle would commence which, after many fluctuations, would certainly result in the extinction of some species, the modification of others, and a considerable alteration in the proportionate numbers and the geographical distribution of almost all.

Any other changes which led to the intermingling of species whose ranges were usually separate would produce corresponding results. Thus, increased severity of winter or summer temperature, causing southward migrations and the crowding together of the productions of distinct regions, must inevitably produce a struggle for existence which would lead to many changes both in the characters and the distribution of animals. Slow elevations of the land would produce another set of changes, by affording an extended area in which the more dominant species might increase their numbers; and, by a greater range and variety of Alpine climates and mountain stations, affording room for the development of new forms of life.

Geographical Mutations as a Motive Power in Bringing about Organic Changes.—Now, if we consider the various geographical changes which, as we have seen, there is good reason to believe have ever been going on in the world, we shall find that the motive power to initiate and urge on organic changes has never been wanting. In the first place, every continent, though permanent in a general sense, has been ever subject to innumerable physical and geographical modifications. At one time the total area has increased, and at another has diminished; great plateaus have gradually risen up, and have been eaten out by denudation into mountain and valley; volcanoes have burst forth, and after accumulating vast masses of eruptive matter have sunk down beneath the ocean, to be covered up with sedimentary rocks, and at a subsequent period again raised above the surface; and the *loci* of all these grand revolutions of the earth's surface have changed their position age after age, so that each portion of every continent has again and again been sunk under the ocean waves, formed the bed of some inland sea, or risen high into plateaus and mountain-ranges. How great must have been the effects of such changes on every form of organic life! And it is to such as these we may perhaps trace those great changes of the animal world which have seemed to revolutionize it, and have led us to class one geological period as the age of reptiles, another as the age of fishes, and a third as the age of mammals.

But such changes as these must necessarily have led to repeated unions and separations of the land masses of the globe, joining together continents which were before divided, and breaking up others into great islands or extensive archipelagoes. Such alterations of the means of transit would probably affect the organic world even more profoundly than the changes of area, of altitude, or of climate, since they afforded the means, at long intervals, of bringing the most diverse forms into competition, and of spreading all the great animal and vegetable types widely over the globe. But the isolation of considerable masses of land for long periods also afforded the means of preservation to many of the lower types, which thus had time to become modified into a variety of distinct forms, some of which became so well adapted to special modes of life that they have continued to exist to the present day, thus affording us examples of the life of early ages which would probably long since have become extinct, had they been always subject to the competition of the more highly organized animals. As examples of such excessively archaic forms, we may mention the mud-fishes and the ganoids, confined to limited fresh-water areas; the frogs and toads, which still maintain themselves vigorously in competition with higher forms; and among mammals the Ornithorhynchus and Echidna of Australia; the whole order of Marsupials—which, out of Australia, where they are quite free from competition, only exist abundantly in South America, which was certainly long isolated from the northern continent; the Insectivora, which, though widely scattered, are generally nocturnal or subterranean in their habits; and the Lemurs, which are most abundant in Madagascar, where they have long been isolated, and almost removed from the competition of higher forms.

Climatal Revolutions as an Agent in Producing Organic Changes.—The geographical and geological changes we have been considering are probably those which have been most effective in bringing about the great features of the distribution of animals, as well as the larger movements in the development of organized beings; but it is to the alternations of warm and cold, or of uniform and excessive climates—of almost perpetual spring in arctic as well as in temperate lands, with occasional phases of cold culminating at remote intervals in glacial epochs —that we must impute some of the more remarkable changes both in the specific characters and in the distribution of organisms.[1] Although the geological evidence is opposed to the belief in early glacial epochs except at very remote and distant intervals, there is nothing which contradicts the occurrence of repeated changes of climate, which, though too small in amount to produce any well-marked physical or organic change, would yet be amply sufficient to keep the organic world in a constant state of movement, and which, by subjecting the whole flora and fauna of a country at comparatively short intervals to decided changes of physical conditions, would supply that stimulus and motive power which, as we have seen, is all that is necessary to keep the processes of "natural selection" in constant operation.

The frequent recurrence of periods of high and of low eccentricity must certainly have produced changes of climate of considerable importance to the life of animals and plants. During periods of high eccentricity with summer in perihelion, that season would be certainly very much hotter, while the winters would be longer and colder than at present; and although geographical conditions might prevent any permanent increase of snow and ice even in the extreme North, yet we cannot doubt that the whole Northern Hemisphere would then have a very different climate than when the changing phase of precession brought a very cool summer and a very mild winter—a perpetual spring, in fact. Now such a change of climate would certainly be calculated to bring about a considerable change of *species*, both by modification and migration, without any such decided change of *type*, either in the vegetation or the animals, that we could say from their fossil remains that any change of climate had taken place. Let us suppose, for instance, that the climate of England and that of Canada were to be mutually exchanged, and that the change took five or six thousand years to bring about; it cannot be doubted that considerable modifications in the fauna and flora of both countries would be the result, although it is impossible to predict what the precise changes would

---

[1] Agassiz appears to have been the first to suggest that the principal epochs of life-extermination were epochs of cold; and Dana thinks that two at least such epochs may be recognized, at the close of the Palæozoic and of the Cretaceous periods, to which we may add the last glacial epoch.

be. We can safely say, however, that some species would stand the change better than others, while it is highly probable that some would be actually benefited by it, and others would be injured. But the benefited would certainly increase and the injured decrease in consequence, and thus a series of changes would be initiated that might lead to most important results. Again, we are sure that some species would become modified in adaptation to the change of climate more readily than others, and these modified species would therefore increase at the expense of others not so readily modified; and hence would arise, on the one hand, extinction of species, and, on the other, the production of new forms.

But this is the very least amount of change of climate that would certainly occur every 10,500 years when there was a high eccentricity, for it is impossible to doubt that a varying distance of the sun in summer from 86 to 99 millions of miles (which is what occurred during — as supposed — the Miocene period, 850,000 years ago) would produce an important difference in the summer temperature and in the actinic influence of sunshine on vegetation. For the intensity of the sun's rays would vary as the square of the distance, or nearly as 74 to 98, so that the earth would be actually receiving one fourth less sun heat during summer at one time than at the other. An equally high eccentricity occurred 2,500,000 years back, and, no doubt, was often reached during still earlier epochs, while a lower but still very high eccentricity has frequently prevailed, and is probably near its average value. Changes of climate, therefore, every 10,500 years, of the character above indicated and of varying intensity, have been the rule rather than the exception in past time; and these changes must have been variously modified by changing geographical conditions so as to produce climatic alterations in different directions, and giving to the ancient lands either dry or wet seasons, storms or calms, equable or excessive temperatures, in a variety of combinations of which the earth perhaps affords no example under the present low phase of eccentricity and consequent slight inequality of sun heat.

Present Condition of the Earth one of Exceptional Stability as regards Climate.—It will be seen by a reference to the dia-

gram at page 163 that during the last 3,000,000 years the eccentricity has been *less* than it is now on eight occasions, for short
periods only, making up a total of about 280,000 years; while
it has been *more* than it is now for many long periods, of from
300,000 to 700,000 years each, making a total of 2,720,000 years,
or nearly as 10 to 1. For nearly half the entire period, or
1,400,000 years, the eccentricity has been nearly double what it
is now, and this is not far from its mean condition. We have
no reason for supposing that this long period of 3,000,000 years
for which we have tables was in any way exceptional as regards
the degree or variation of eccentricity; but, on the contrary, we
may pretty safely assume that its variations during this time
fairly represent its average state of increase and decrease during
all known geological time. But when the glacial epoch ended,
72,000 years ago, the eccentricity was about double its present
amount; it then rapidly decreased till, at 60,000 years back, it
was very little greater than it is now, and since then it has been
uniformly small. It follows that, for about 60,000 years before
our time, the mutations of climate every 10,500 years have been
comparatively unimportant, and that the temperate zones have
enjoyed *an exceptional stability of climate*. During this time
those powerful causes of organic change which depend on considerable changes of climate, and the consequent modifications,
migrations, and extinctions of species, will not have been at
work; the slight changes that did occur would probably be so
slow and so little marked that the various species would be able
to adapt themselves to them without much disturbance; and
the result would be *an epoch of exceptional stability of species*.

But it is from this very period of *exceptional stability* that we
obtain our only *scale* for measuring the rate of organic change.
It includes not only the historical period, but that of the Swiss
Lake dwellings, the Danish shell-mounds, our peat-bogs, our
sunken forests, and many of our superficial alluvial deposits—
the whole, in fact, of the iron, bronze, and neolithic ages. Even
some portion of the palæolithic age and of the more recent
gravels and cave-earths may come into the same general period,
if they were formed when the glacial epoch was passing away.
Now throughout all these ages we find no indication of change

of species, and but little, comparatively, of migration. We thus get an erroneous idea of *the permanence and stability of specific forms*, due to the period immediately antecedent to our own being *a period of exceptional permanence and stability* as regards climatic and geographical conditions.[1]

Date of Last Glacial Epoch, and its Bearing on the Measurement of Geological Time.—Directly we go back from this stable period, we come upon changes both in the forms and in the distribution of species; and when we pass beyond the last glacial epoch into the Pliocene period, we find ourselves in a comparatively new world, surrounded by a considerable number of species altogether different from any which now exist, together with many others which, though still living, now inhabit distant regions. It seems not improbable that what is termed the Pliocene period was really the coming-on of the glacial epoch, and this is the opinion of Professor Jules Marcou.[2] According to our views, a considerable amount of geographical change must have occurred at the change from the Miocene to the Pliocene, favoring the refrigeration of the Northern Hemisphere, and leading, in the way already pointed out, to the glacial epoch whenever a high degree of eccentricity prevailed. As many reasons combine to make us fix the height of the glacial epoch at the period of high eccentricity which occurred 200,000 years back, and as the Pliocene period was probably not of long duration, we must suppose the next great phase of very high eccentricity (850,000 years ago) to fall within the Miocene epoch. Dr. Croll believes that this must have produced a glacial period, but we have shown strong reasons for believing that, in concurrence with favorable geographical conditions, it led to uninterrupted warm climates in the temperate and northern zones.

---

[1] This view was, I believe, first put forth by myself in a paper read before the Geological Section of the British Association in 1869, and subsequently in an article in *Nature*, Vol. I., p. 454. It was also stated by Mr. S. B. K. Skertchley in his "Physical System of the Universe," p. 363 (1878); but we both founded it on what I now consider the erroneous doctrine that actual glacial epochs recurred each 10,500 years during periods of high eccentricity.

[2] "Explication d'une Seconde Édition de la Carte Géologique de la Terre" (1875), p. 64.

This, however, did not prevent the occurrence of local glaciation wherever other conditions led to its initiation, and the
most powerful of such conditions is a great extent of high land.
Now we know that the Alps acquired a considerable part of
their elevation during the latter part of the Miocene period,
since Miocene rocks occur at an elevation of over 6000 feet,
while Eocene beds occur at nearly 10,000 feet. But since that
time there has been a vast amount of denudation, so that these
rocks may first have been raised much higher than we now find
them, and thus a considerable portion of the Alps may once
have been more elevated than now. This would certainly lead
to an enormous accumulation of snow, which would be increased
when the eccentricity reached a maximum, as already fully explained, and may then have caused glaciers to descend into the
adjacent sea, carrying those enormous masses of rock which are
buried in the Upper Miocene of the Superga in Northern Italy.
An earlier epoch of great altitude in the Alps, coinciding with
the very high eccentricity 2,500,000 years ago, may have caused
the local glaciation of the Middle Eocene period when the enormous erratics of the Flysch conglomerate were deposited in the
inland seas of Northern Switzerland, the Carpathians, and the
Apennines. This is quite in harmony with the indications of
an uninterrupted warm climate and rich vegetation during the
very same period in the adjacent low countries, just as we find
at the present day in New Zealand a delightful climate and a
rich vegetation of Metrosideros, fuchsias and tree-ferns on the
very borders of huge glaciers, descending to within seven hundred feet of the sea-level. It is not pretended that these estimates of geological time have any more value than probable
guesses; but it is certainly a curious coincidence that two remarkable periods of high eccentricity should have occurred at
such periods and at such intervals apart as very well accord with
the comparative remoteness of the two deposits in which undoubted signs of ice-action have been found, and that both these
are localized in the vicinity of mountains which are known to
have acquired a considerable elevation at about the same period
of time.

In the tenth edition of the " Principles of Geology," Sir

Charles Lyell, taking the amount of change in the species of mollusca as a guide, estimated the time elapsed since the commencement of the Miocene as one third that of the whole Tertiary epoch, and the latter at one fourth that of geological time since the Cambrian period. Professor Dana, on the other hand, estimates the Tertiary as only one fifteenth of the Mesozoic and Palæozoic combined. On the estimate above given, founded on the dates of phases of high eccentricity, we shall arrive at about four million years for the Tertiary epoch, and sixteen million years for the time elapsed since the Cambrian, according to Lyell, or sixty millions according to Dana. The estimate arrived at from the rate of denudation and deposition (twenty-eight million years) is nearly midway between these, and it is, at all events, satisfactory that the various measures result in figures of the same order of magnitude, which is all one can expect on so difficult and exceedingly speculative a subject.

The only value of such estimates is to define our notions of geological time, and to show that the enormous periods of hundreds of millions of years which have sometimes been indicated by geologists are neither necessary nor warranted by the facts at our command; while the present result places us more in harmony with the calculations of physicists, by leaving a very wide margin between geological time as defined by the fossiliferous rocks and that far more extensive period which includes all possibility of life upon the earth.

Concluding Remarks.—In the present chapter I have endeavored to show that, combining the measured rate of denudation with the estimated thickness and probable extent of the known series of sedimentary rocks, we may arrive at a rude estimate of the time occupied in the formation of those rocks. From another point of departure—that of the probable date of the Miocene period as determined by the epoch of high eccentricity supposed to have aided in the production of the Alpine glaciation during that period, and taking the estimate of geologists as to the proportionate amount of change in the animal world since that epoch—we obtain another estimate of the duration of geological time, which, though founded on far less secure data, agrees pretty nearly with the former estimate. The time thus arrived

at is immensely less than the usual estimates of geologists, and
is so far within the limits of the duration of the earth as calculated by Sir William Thomson as to allow for the development of the lower organisms an amount of time anterior to the
Cambrian period several times greater than has elapsed between
that period and the present day. I have further shown that, in
the continued mutations of climate produced by high eccentricity and opposite phases of precession, even though these did
not lead to glacial epochs, we have a motive power well calculated to produce far more rapid organic changes than have hitherto been thought possible; while in the enormous amount of
specific variation (as demonstrated in an earlier chapter) we have
ample material for that power to act upon, so as to keep the
organic world in a state of rapid change and development pro-
portioned to the comparatively rapid changes in the earth's surface.

We have now finished the series of preliminary studies of the
biological conditions and physical changes which have affected
the modification and dispersal of organisms, and have thus
brought about their actual distribution on the surface of the
earth. These studies will, it is believed, place us in a condition
to solve most of the problems presented by the distribution of
animals and plants, whenever the necessary facts both as to their
distribution and their affinities are sufficiently well known; and
we now proceed to apply the principles we have established to
the interpretation of the phenomena presented by some of the
more important and best known of the islands of our globe,
limiting ourselves to these for reasons which have been already
sufficiently explained in our preface.

INSULAR FAUNAS AND FLORAS

CHAPTER XI.

THE CLASSIFICATION OF ISLANDS.

Importance of Islands in the Study of the Distribution of Organisms.—Classification of Islands with Reference to Distribution.—Continental Islands.—Oceanic Islands.

In the preceding chapters, forming the first part of our work, we have discussed, more or less fully, the general features presented by animal distribution, as well as the various physical and biological changes which have been the most important agents in bringing about the present condition of the organic world.

We now proceed to apply these principles to the solution of the numerous problems presented by the distribution of animals; and in order to limit the field of our inquiry, and at the same time to deal only with such facts as may be rendered intelligible and interesting to those readers who have not much acquaintance with the details of natural history, we propose to consider only such phenomena as are presented by the islands of the globe.

Importance of Islands in the Study of the Distribution of Organisms.—Islands possess many advantages for the study of the laws and phenomena of distribution. As compared with continents they have a restricted area and definite boundaries, and in most cases their geographical and biological limits coincide. The number of species and of genera they contain is always much smaller than in the case of continents, and their peculiar species and groups are usually well defined and strictly limited in range. Again, their relations with other lands are often direct and simple, and even when more complex are far easier to comprehend than those of continents; and they exhibit, besides, certain influences on the forms of life and certain peculiarities of distribution which continents do not present, and whose study offers many points of interest.

In islands we have the facts of distribution often presented to us in their simplest forms, along with others which become gradually more and more complex; and we are therefore able to proceed step by step in the solution of the problems they present. But as in studying these problems we have necessarily to take into account the relations of the insular and continental faunas, we also get some knowledge of the latter, and acquire besides so much command over the general principles which underlie all problems of distribution that it is not too much to say that when we have mastered the difficulties presented by the peculiarities of island life we shall find it comparatively easy to deal with the more complex and less clearly defined problems of continental distribution.

Classification of Islands with Reference to Distribution.—Islands have had two distinct modes of origin : they have either been separated from continents of which they are but detached fragments, or they have originated in the ocean and have never formed part of a continent or any large mass of land. This difference of origin is fundamental, and leads to a most important difference in their animal inhabitants; and we may therefore first distinguish the two classes—oceanic and continental islands.

Mr. Darwin appears to have been the first writer who called attention to the number and importance, both from a geological and biological point of view, of oceanic islands. He showed that with very few exceptions all the remoter islands of the great oceans were of volcanic or coralline formation, and that none of them contained indigenous mammalia or amphibia. He also showed the connection of these two phenomena, and maintained that none of the islands so characterized had ever formed part of a continent. This was quite opposed to the opinions of the scientific men of the day, who almost all held the idea of continental extensions, and of oceanic islands being their fragments, and it was long before Mr. Darwin's views obtained general acceptance. Even now the belief still lingers; and we continually hear of old Atlantic or Pacific continents, of " Atlantis " or " Lemuria," of which hypothetical lands many existing islands, although wholly volcanic, are thought to be the remnants.

We have already seen that Darwin connected the peculiar geo-logical structure of oceanic islands with the permanence of the great oceans which contain them, and we have shown that sev-eral distinct lines of evidence all point to the same conclusion. We may therefore define oceanic islands as follows: Islands of volcanic or coralline formation, usually far from continents, and always separated from them by very deep sea; entirely without indigenous land mammalia or amphibia, but with abundance of birds and insects, and usually with some reptiles. This defini-tion will exclude only two islands which have been sometimes classed as oceanic—New Zealand and the Seychelles. Rodri-guez, which was once thought to be another exception, has been shown by the explorations during the Transit of Venus Expedi-tion to be essentially volcanic, with some upraised coralline lime-stone.

Continental Islands.—Continental islands are always more varied in their geological formation, containing both ancient and recent stratified rocks. They are rarely very remote from a continent, and they always contain some land mammals and amphibia, as well as representatives of the other classes and orders in considerable variety. They may, however, be divided into two well-marked groups—ancient and recent continental islands—the characters of which may be easily defined.

Recent continental islands are always situated on submerged banks connecting them with a continent, and the depth of the intervening sea rarely exceeds 100 fathoms. They resemble the continent in their geological structure, while their animal and vegetable productions are either almost identical with that of the continent, or, if otherwise, the difference consists in the pres-ence of closely allied species of the same types, with occasionally a very few peculiar genera. They possess, in fact, all the charac-teristics of a portion of the continent, separated from it at a recent geological period.

Ancient continental islands differ greatly from the preceding in many respects. They are not united to the adjacent conti-nent by a shallow bank, but are usually separated from it by a depth of sea of a thousand fathoms or upwards. In geological structure they agree generally with the more recent islands;

like them they possess mammalia and amphibia, usually in considerable abundance, as well as all other classes of animals; but these are highly peculiar, almost all being distinct species, and many forming distinct and peculiar genera or families. They are also well characterized by the fragmentary nature of their fauna, many of the most characteristic continental orders or families being quite unrepresented, while some of their animals are allied, not to such forms as inhabit the adjacent continent, but to others found only in remote parts of the world. This very remarkable set of characters mark off the islands which exhibit them as a distinct class, which often present the greatest anomalies and most difficult problems to the student of distribution.

Oceanic Islands.—The total absence of warm-blooded terrestrial animals in an island otherwise well suited to maintain them is held to prove that such island is no mere fragment of any existing or submerged continent, but one that has been actually produced in mid-ocean. It is true that if a continental island were to be completely submerged for a single day and then again elevated, its higher terrestrial animals would be all destroyed; and if it were situated at a considerable distance from land, it would be reduced to the same zoological condition as an oceanic island. But such a complete submergence and re-elevation appears never to have taken place, for there is no single island on the globe which has the physical and geological features of a continental combined with the zoological features of an oceanic island. It is true that some of the coral islands may be formed upon submerged lands of a continental character, but we have no proof of this; and even if it were so, the existing islands are to all intents and purposes oceanic.

We will now pass on to a consideration of some of the more interesting examples of these three classes, beginning with oceanic islands.

All the animals which now inhabit such oceanic islands must either themselves have reached them by crossing the ocean, or be the descendants of ancestors who did so. Let us, then, see what are, in fact, the animal and vegetable inhabitants of these islands, and how far their presence can be accounted for.   We will begin with the Azores, or Western Islands, because they have been thoroughly well explored by naturalists, and in their peculiarities afford us an important clew to some of the most efficient means of distribution among several classes of animals.

CHAPTER XII.

OCEANIC ISLANDS.—THE AZORES AND BERMUDA.

THE AZORES, OR WESTERN ISLANDS.—Position and Physical Features.—Chief Zoological Features of the Azores.—Birds.—Origin of the Azorean Bird Fauna.—Insects of the Azores.—Land Shells of the Azores.—The Flora of the Azores.—The Dispersal of Seeds.—Birds as Seed-carriers.—Facilities for Dispersal of Azorean Plants.—Important Deduction from the Peculiarities of the Azorean Fauna and Flora.

BERMUDA.—Position and Physical Features.—The Red Clay of Bermuda.—Zoology of Bermuda.—Birds of Bermuda.—Comparison of the Bird Faunas of Bermuda and the Azores.—Insects of Bermuda.—Land Mollusca.—Flora of Bermuda.—Concluding Remarks on the Azores and Bermuda.

WE will commence our investigation into the phenomena presented by oceanic islands with two groups of the North Atlantic, in which the facts are of a comparatively simple nature, and such as to afford us a valuable clew to a solution of the more difficult problems we shall have to deal with further on. The Azores and Bermuda offer great contrasts in physical features, but striking similarities in geographical position. The one is volcanic, the other coralline; but both are surrounded by a wide expanse of ocean of enormous depth, the one being about as far from Europe as the other is from America. Both are situated in the temperate zone, and they differ less than six degrees in latitude, yet the vegetation of the one is wholly temperate, while that of the other is almost tropical. The productions of the one are related to Europe, as those of the other are to America, but they present instructive differences; and both afford evidence of the highest value as to the means of dispersal of various groups of organisms across a wide expanse of ocean.

THE AZORES, OR WESTERN ISLANDS.

These islands form a widely scattered group, nine in number, situated between 37° and 39° 40′ N. lat., and stretching in a southeast and northwest direction over a distance of nearly 400 miles. The largest of the islands, San Miguel, is about 40 miles long, and is one of the nearest to Europe, being rather under 900 miles from the coast of Portugal, from which it is separated by an ocean 2500 fathoms deep. The depth between the islands does not seem to be known, but the 1000-fathom line encloses the whole group pretty closely, while a depth of about 1800

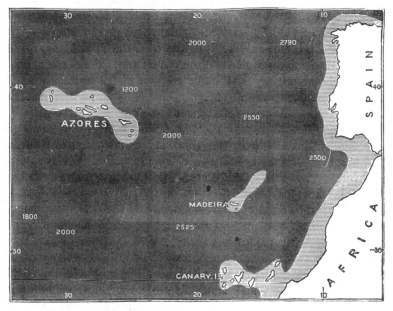

OUTLINE MAP OF THE AZORES.

NOTE.—The light tint shows where the sea is less than 1000 fathoms deep.
　　The dark 　　"　　　　"　　　　"　more　　"　　　　"　　　　"
　　The figures show depths in fathoms.

fathoms is reached within 300 miles in all directions. These great depths render it in the highest degree improbable that the Azores have ever been united with the European continent; while their being wholly volcanic is equally opposed to the view of their having formed part of an extensive Atlantis including Madeira and the Canaries. The only exception to their volcanic structure is the occurrence in one small island only (Santa Ma-

ria) of some marine deposits of Upper Miocene age — a fact
which proves some alterations of level, and perhaps a greater
extension of this island at some former period, but in no way
indicates a former union of the islands, or any greater extension
of the whole group. It proves, however, that the group is of
considerable antiquity, since it must date back to Miocene times;
and this fact may be of importance in considering the origin and
peculiar features of the fauna and flora. It thus appears that in
all physical features the Azores correspond strictly with our
definition of "oceanic islands;" while their great distance from
any other land, and the depth of the ocean around them, make
them typical examples of the class. We should therefore expect
them to be equally typical in their fauna and flora; and this is
the case as regards the most important characteristics, although
in some points of detail they present exceptional phenomena.

Chief Zoological Features of the Azores.[1]—The great feature
of oceanic islands—the absence of all indigenous land mammalia
and amphibia—is well shown in this group; and it is even carried further, so as to include all terrestrial vertebrata, there being
no snake, lizard, frog, or fresh-water fish, although the islands
are sufficiently extensive, possess a mild and equable climate, and
are in every way adapted to support all these groups. On the
other hand, flying creatures, as birds and insects, are abundant;
and there is also one flying mammal—a small European bat. It
is true that rabbits, weasels, rats and mice, and a small lizard
peculiar to Madeira and Teneriffe are now found wild in the
Azores, but there is good reason to believe that these have all
been introduced by human agency. The same may be said of
the gold-fish and eels now found in some of the lakes, there being not a single fresh-water fish which is truly indigenous to the
islands. When we consider that the nearest part of the group
is about 900 miles from Portugal and more than 550 miles from
Madeira, it is not surprising that none of these terrestrial animals can have passed over such a wide expanse of ocean unas-
sisted by man.

---

[1] For most of the facts as to the zoology and botany of these islands I am indebted
to Mr. Godman's valuable work, "Natural History of the Azores, or Western Islands," by Frederick Du Cane Godman, F.L.S., F.Z.S., etc., London, 1870.

Let us now see what animals are believed to have reached the group by natural means, and thus constitute its indigenous fauna. These consist of birds, insects, and land shells, each of which must be considered separately.

Birds.—Fifty-three species of birds have been observed at the Azores, but the larger proportion (thirty-one) are either aquatic or waders—birds of great powers of flight, whose presence in the remotest islands is by no means remarkable. Of these two groups twenty are residents, breeding in the islands, while eleven are stragglers, only visiting the islands occasionally, and all are common European species. The land birds, twenty-two in number, are more interesting, four only being stragglers, while eighteen are permanent residents. The following is a list of these resident land birds:

1. Common Buzzard.............................*Buteo vulgaris.*
2. Long-eared Owl...............................*Asio otus.*
3. Barn Owl.....................................*Strix flammea.*
4. Blackbird....................................*Turdus merula.*
5. Robin ....*Erithacus rubecula.*
6. Blackcap ....*Sylvia atricapilla.*
7. Gold-crest ....*Regulus cristatus.*
8. Wheat-ear ....*Saxicola œnanthe.*
9. Gray Wagtail.................................*Motacilla sulphurea.*
10. Atlantic Chaffinch..........................*Fringilla tintillon.*
11. Azorean Bullfinch...........................*Pyrrhula murina.*
12. Canary ....*Serinus Canarius.*
13. Common Starling ....*Sturnus vulgaris.*
14. Lesser Spotted Woodpecker...................*Dryobates minor.*
14. Wood-pigeon..................................*Columba palumbus.*
16. Rock-dove..................................*Columba livia.*
17. Red-legged Partridge........................*Caccabis rufa.*
18. Common Quail................................*Coturnix communis.*

All the above-named birds are common in Europe and North Africa except three — the Atlantic chaffinch and the canary, which inhabit Madeira and the Canary Islands, and the Azorean bullfinch, which is peculiar to the islands we are considering.

Origin of the Azorean Bird Fauna.—The questions we have now before us are—how did these eighteen species of birds first reach the Azores, and how are we to explain the presence of a single peculiar species while all the rest are identical with European birds? In order to answer them, let us first see what stragglers now actually visit the Azores from the nearest continents. The four species given in Mr. Godman's list are the kestrel, the oriole, the snow-bunting, and the hoopoe; but he also tells us that there are certainly others, and adds, " Scarcely a storm occurs in spring or autumn without bringing one or more species foreign to the islands; and I have frequently been told that swallows, larks, grebes, and other species not referred to here are not uncommonly seen at those seasons of the year."

We have therefore every reason to believe that the birds which are now residents originated as stragglers, which occasionally found a haven in these remote islands when driven out to sea by storms. Some of them, no doubt, still often arrive from the continent, but these cannot easily be distinguished as new arrivals among those which are residents. Many facts mentioned by Mr. Godman show that this is the case. A barn owl, much exhausted, flew on board a whaling-ship when 500 miles southwest of the Azores; and even if it had come from Madeira it must have travelled quite as far as from Portugal to the islands. Mr. Godman also shot a single specimen of the wheat-ear in Flores after a strong gale of wind; and as no one on the island knew the bird, it was almost certainly a recent arrival. Subsequently a few were found breeding in the old crater of Corvo, a small adjacent island; and as the species is not found in any other island of the group, we may infer that this bird is a recent immigrant in process of establishing itself.

Another fact which is almost conclusive in favor of the bird-population having arrived as stragglers is that they are most abundant in the islands nearest to Europe and Africa. The Azores consist of three divisions—an eastern, consisting of two islands, St. Michael's and St. Mary's; a central, of five, Terceira Graciosa, St. George's, Pico, and Fayal; and a western, of two, Flores and Corvo. Now, had the whole group once been united to the continent, or even formed parts of one extensive Atlantic island, we should certainly expect the central group, which is more compact and has a much larger area than all the rest, to have the greatest number and variety of birds. But the fact that birds are most numerous in the eastern group, and diminish as we go westward, is entirely opposed to this theory, while it is strictly in accordance with the view that they are all stragglers from Europe, Africa, or the other Atlantic islands. Omitting oceanic wanderers, and including all birds which have probably arrived involuntarily, the numbers are found to be forty species in the eastern group, thirty-six in the central, and twenty-nine in the western.

To account for the presence of one peculiar species, the bullfinch (which, however, does not differ from the common European bullfinch more than do some of the varieties of North American birds from their type-species), is not difficult, the wonder rather being that there are not more peculiar forms. In our third chapter we have seen how great is the amount of individual variation in birds, and how readily local varieties become established wherever the physical conditions are sufficiently distinct. Now we can hardly have a greater difference of conditions than between the continent of Europe or North Africa and a group of rocky islands in mid-Atlantic, situated in the full course of the Gulf Stream and with an excessively mild though stormy climate. We have every reason to believe that special modifications would soon become established in any animals completely isolated under such conditions. But they are not, as a rule, thus completely isolated, because, as we have seen, stragglers arrive at short intervals; and these, mixing with the residents, keep up the purity of the breed. It follows that only those species which reach the Azores at very remote intervals will be likely to acquire well-marked distinctive characters; and this appears to have happened with the bullfinch alone, a bird which does not migrate, and is therefore less likely to be blown out to sea, more especially as it inhabits woody districts. A few other Azorean birds, however, exhibit slight differences from their European allies.

There is another reason for the very slight amount of peculiarity presented by the fauna of the Azores as compared with many other oceanic islands, dependent on its comparatively recent origin. The islands themselves may be of considerable antiquity, since a few small deposits, believed to be of Miocene age, have been found on them; but there can be little doubt that their present fauna, at all events as concerns the birds, had its origin since the date of the last glacial epoch. Even now icebergs reach the latitude of the Azores only a little to the westward; and when we consider the proofs of extensive ice-action in North America and Europe, we can hardly doubt that these islands were at that time surrounded with pack-ice, while their own mountains, reaching 7600 feet high in Pico, would almost certainly have been covered with perpetual snow and have sent down glaciers to the sea. They might then have had a climate almost as bad as that now endured by the Prince Edward Islands in the Southern Hemisphere, nearly ten degrees farther from the equator, where there are no land birds whatever, although the distance from Africa is not much greater than that of the Azores from Europe, while the vegetation is limited to a few alpine plants and mosses. This recent origin of the birds accounts, in a great measure, for their identity with those of Europe, because, whatever change has occurred must have been effected in the islands themselves, and in a time limited to that which has elapsed since the glacial epoch passed away.

Insects of the Azores.—Having thus found no difficulty in accounting for the peculiarities presented by the birds of these islands, we have only to see how far the same general principles will apply to the insects and land shells. The butterflies, moths, and hymenoptera are few in number, and almost all seem to be common European species, whose presence is explained by the same causes as those which have introduced the birds. Beetles, however, are more numerous, and have been better studied, and these present some features of interest. The total number of species yet known is 212, of which 175 are European; but out of these 101 are believed to have been introduced by human agency, leaving seventy-four really indigenous. Twenty-three of these indigenous species are not found in any of the other Atlantic islands, showing that they have been introduced directly from Europe by causes which have acted more powerfully here than farther south. Besides these, there are thirty-six species not found in Europe, of which nineteen are natives of Madeira or the Canaries, three are American, and fourteen are altogether peculiar to the Azores. These latter are mostly allied to species found in Europe or in the other Atlantic islands, while one is allied to an American species, and two are so. distinct as to constitute new genera. The following list of these peculiar species will be interesting:

CARABIDÆ.

Anchomenus aptinoides............Allied to a species from the Canaries.
Bembidium hesperus...............Allied to the European *B. lætum*.

DYTISCIDÆ.

Agabus Godmanni.................Allied to the European *A. dispar*.

COLYDIIDÆ.

Tarphius Wollastoni...............A genus almost peculiar to the Atlantic Islands.

ELATERIDÆ.

Heteroderes Azoricus.............Allied to a Brazilian species.
Elastrus dolosus..................Belongs to a peculiar Madagascar genus.

MELYRIDÆ.

Attalus miniaticollis..............Allied to a Canarian species.

RHYNCOPHORA.

Phlæophagus variabilis...........Allied to European and Atlantic species.
Acalles Droueti..................A Mediterranean and Atlantic genus.
Laparocerus Azoricus............Allied to Madeiran species.
Asynonychus Godmanni...........A peculiar genus, allied to *Brachyderes* of the South of Europe.
Neocnemis occidentalis...........A peculiar genus, allied to the European genus *Strophosomus*.

HETEROMERA.

Helops Azoricus..................Allied to *H. vulcanus* of Madeira.

STAPHYLINIDÆ.

Xenomma melanocephala..........Allied to *X. filiforme* from the Canaries.

This greater amount of speciality in the beetles than in the birds may be due to two causes. In the first place, many of these small insects have, no doubt, survived the glacial epoch, and may, in that case, represent very ancient forms which have become extinct in their native country; and, in the second place, insects have many more chances of reaching remote islands than birds, for not only may they be carried by gales of wind, but sometimes, in the egg or larva state, or even as perfect insects, they may be drifted safely for weeks over the ocean, buried in the light stems of plants or in the solid wood of trees in which many of them undergo their transformations. Thus we may explain the presence of three common South American species (two elaters and a longicorn), all wood-eaters, and therefore liable to be occasionally brought in floating timber by the Gulf Stream. But insects are also immensely more numerous in species than are land birds, and their transmission would be in most cases quite involuntary, and not dependent on their own powers of flight, as with birds; and thus the chances against the same species being frequently carried to the same island would be considerable. If we add to this the dependence of so many insects on local conditions of climate and vegetation, and their liability to be destroyed by insectivorous birds, we shall see that, although there may be a greater probability of insects as a whole reaching the islands, the chances against any particular insect arriving there, or against the same species arriving frequently, are much greater than in the case of birds. The result is that (as compared with Britain, for example) the birds are proportionately much more numerous than the beetles; while the peculiar species of beetles are much more numerous than among birds, both facts being quite in accordance with what we know of the habits of the two groups. We may also remark that the small size and obscure characters of many of the beetles render it probable that species now supposed to be peculiar really inhabit some parts of Europe or North America.

It is interesting to note that the two families which are pre-eminently wood, root, or seed eaters are those which present the greatest amount of speciality. The two Elateridæ alone exhibit remote affinities, the one with a Brazilian, the other with a Madagascar group; while the only peculiar genera belong to the Rhyncophora, but are allied to European forms. These last almost certainly form a portion of the more ancient fauna of the islands which migrated to them in preglacial times, while the Brazilian elater appears to be the solitary example of a living insect brought by the Gulf Stream to these remote shores. The elater, having its nearest living ally in Madagascar (*Elastrus dolosus*), cannot be held to indicate any independent communication between these distant islands; but is more probably a relic of a once more wide-spread type which has only been able to maintain itself in these localities. Mr. Crotch states that there are some species of beetles common to Madagascar and the Canary Islands, while there are several genera common to Madagascar and South America, and some to Madagascar and Australia. The clew to these apparent anomalies is found in other genera being common to Madagascar, Africa, and South America, while others are Asiatic or Australian. Madagascar, in fact, has insect relations with every part of the globe, and the only rational explanation of such facts is that they are indications of very ancient and wide-spread groups, maintaining themselves only in a few widely separated portions of what was at one time or another the area of their distribution.

Land Shells of the Azores.—Like the insects and birds, the land shells of these islands have a generally European aspect, but with a larger proportion of peculiar species. This was to be expected, because the means by which mollusks are carried over the sea are far less numerous and varied than in the case of insects ;[1] and we may therefore conclude that their introduction is a very rare event, and that a species once arrived remains for long periods undisturbed by new arrivals, and is therefore more likely to become modified by the new conditions, and then fixed as a distinct type. Out of the sixty-nine known species, thirty-seven are common to Europe or the other Atlantic islands; while thirty-two are peculiar, though almost all are distinctly allied to European types. The majority of these shells, especially the peculiar forms, are very small, and many of them may date back to beyond the glacial epoch. The eggs of these would be exceedingly minute, and might occasionally be carried on leaves or other materials during gales of exceptional violence and duration, while others might be conveyed with the earth that often sticks to the feet of birds. There are also, probably, other unknown means of conveyance ; but, however this may be, the general character of the land mollusks is such as to confirm the conclusions we have arrived at from a study of the birds and insects—that these islands have never been connected with a

---

[1] See Chap. V., p. 74.

continent, and have been peopled with living things by such forms only as in some way or other have been able to reach them across many hundred miles of ocean.

The Flora of the Azores.—The flowering plants of the Azores have been studied by one of our first botanists, Mr. H. C. Watson, who has himself visited the islands and made extensive collections; and he has given a complete catalogue of the species in Mr. Godman's volume. As our object in the present work is to trace the past history of the more important islands by means of the forms of life that inhabit them, and as for this purpose plants are sometimes of more value than any class of animals, it will be well to take advantage of the valuable materials here available in order to ascertain how far the evidence derived from the two organic kingdoms agrees in character; and also to obtain some general results which may be of service in our discussion of more difficult and more complex problems.

There are in the Azores 480 known species of flowering plants and ferns, of which no less than 440 are found also in Europe, Madeira, or the Canary Islands; while forty are peculiar to the Azores, but are more or less closely allied to European species. As botanists are no less prone than zoologists to invoke former land-connections and continental extensions to account for the wide dispersal of objects of their study, it will be well to examine somewhat closely what these facts really imply.

The Dispersal of Seeds.—The seeds of plants are liable to be dispersed by a greater variety of agents than any other organisms, while their tenacity of life, under varying conditions of heat and cold, drought and moisture, is also exceptionally great. They have also an advantage, in that the great majority of flowering plants have the sexes united in the same individual, so that a single seed in a state fit to germinate may easily stock a whole island. The dispersal of seeds has been studied by Sir Joseph Hooker, Mr. Darwin, and many other writers, who have made it sufficiently clear that they are in many cases liable to be carried enormous distances. An immense number are specially adapted to be carried by the wind, through the possession of down or hairs or membranous wings or processes; while others are so minute, and produced in such profusion, that it is difficult

to place a limit to the distance they might be carried by gales
of wind or hurricanes. Another class of somewhat heavier
seeds or dry fruits are capable of being exposed for a long time
to sea-water without injury. Mr. Darwin made many experiments on this point, and he found that many seeds, especially
of atriplex, beta, oats, capsicum, and the potato, grew after one
hundred days' immersion, while a large number survived fifty
days. But he also found that most of them sink after a few
days' immersion, and this would certainly prevent them being
floated to very great distances. It is very possible, however,
that dried branches or flower-heads containing seeds would float
longer, while it is quite certain that many tropical seeds do float
for enormous distances, as witness the double cocoa-nuts which
cross the Indian Ocean from the Seychelle Islands to the coast
of Sumatra, and the West Indian beans which frequently reach
the west coast of Scotland. There is therefore ample evidence
of the possibility of seeds being conveyed across the sea for great
distances by winds and surface currents.[1]

Birds as Seed-carriers.—The great variety of fruits that are
eaten by birds afford a means of plant-dispersal in the fact that
seeds often pass through the bodies of birds in a state well fitted
for germination; and such seeds may occasionally be carried
long distances by this means. Of the twenty-two land birds

---

[1] Some of Mr. Darwin's experiments are very interesting and suggestive. Ripe
hazel-nuts sank immediately, but when dried they floated for ninety days, and afterwards germinated. An asparagus-plant with ripe berries, when dried, floated for
eighty-five days, and the seeds afterwards germinated. Out of ninety-four dried
plants experimented with, eighteen floated for more than a month, and some for
three months, and their powers of germination seem never to have been wholly destroyed. Now, as oceanic currents vary from thirty to sixty miles a day, such plants
under the most favorable conditions might be carried $90 \times 60 = 5400$ miles! But even
half of this is ample to enable them to reach any oceanic island, and we must remember that till completely water-logged they might be driven along at a much greater rate
by the wind. Mr. Darwin calculates the distance by the average time of flotation to
be 924 miles; but in such a case as this we are entitled to take the extreme cases,
because such countless thousands of plants and seeds must be carried out to sea annually that the extreme cases in a single experiment with only ninety-four plants
must happen hundreds or thousands of times, and with hundreds or thousands of species, naturally, and thus afford ample opportunities for successful migration. (See
"Origin of Species," 6th ed., p. 325.)

found in the Azores, half are more or less fruit-eaters, and these may have been the means of introducing some plants into the islands.

Birds also frequently have small portions of earth on their feet; and Mr. Darwin has shown by actual experiment that almost all such earth contains seeds. Thus, in nine grains of earth on the leg of a woodcock a seed of the toad-rush was found which germinated; while a wounded red-legged partridge had a ball of earth weighing six and a half ounces adhering to its leg, and from this earth Mr. Darwin raised no less than eighty-two separate plants of about five distinct species. Still more remarkable was the experiment with six and three-quarter ounces of mud from the edge of a little pond, which, carefully treated under glass, produced 537 distinct plants! This is equal to a seed for every six grains of mud; and when we consider how many birds frequent the edges of ponds in search of food, or come there to drink, it is evident that great numbers of seeds may be dispersed by this means.

Many seeds have hispid awns, hooks, or prickles which readily attach them to the feathers of birds, and a great number of aquatic birds nest inland on the ground; and as these are pre-eminently wanderers, they must often aid in the dispersal of such plants.[1]

---

[1] The following remarks, kindly communicated to me by Mr. H. N. Moseley, naturalist to the *Challenger*, throw much light on the agency of birds in the distribution of plants: "Grisebach ("Veg. der Erde," Vol. II., p. 496) lays much stress on the wide ranging of the albatross (Diomedea) across the equator from Cape Horn to the Kurile Islands, and thinks that the presence of the same plants in arctic and antarctic regions may be accounted for, possibly, by this fact. I was much struck at Marion Island, of the Prince Edward group, by observing that the great albatross breeds in the midst of a dense, low herbage, and constructs its nest of a mound of turf and herbage. Some of the indigenous plants, *e. g.* Acæna, have flower-heads which stick like burrs to feathers, etc., and seem specially adapted for transportation by birds. Besides the albatrosses, various species of Procellaria and Puffinus, birds which range over immense distances, may, I think, have played a great part in the distribution of plants, and especially account, in some measure, for the otherwise difficult fact (when occurring in the tropics) that widely distant islands have similar mountain plants. The Procellarias and Puffinus, in nesting, burrow in the ground, as far as I have seen choosing often places where the vegetation is the thickest. The birds in burrowing get their feathers covered with vegetable mould, which must include spores and often seeds. In high latitudes the birds often burrow near the sea-level, as at

Facilities for Dispersal of Azorean Plants.—Now, in the course of very long periods of time, the various causes here enumerated would be sufficient to stock the remotest islands with vegetation, and a considerable part of the Azorean flora appears well adapted to be so conveyed. Of the 439 flowering plants in Mr. Watson's list, I find that about 45 belong to genera that have either pappus or winged seeds; 65 to such as have very minute seeds; 30 have fleshy fruits such as are greedily eaten by birds; several have hispid seeds; and 84 are glumaceous plants, which are all probably well adapted for being carried partly by winds and partly by currents, as well as by some of the other causes mentioned. On the other hand, we have a very suggestive fact in the absence from the Azores of most of the trees and shrubs with large and heavy fruits, however common they may be in Europe. Such are oaks, chestnuts, hazels, apples, beeches, alders, and firs; while the only trees or large shrubs are the Portugal laurel, myrtle, laurestinus, elder, *Laurus Canariensis*, *Myrica Faya*, and a doubtfully peculiar juniper—all small berry-bearers, and therefore likely to have been conveyed by one or other of the modes suggested above.

There can be little doubt that the truly indigenous flora of the islands is far more scanty than the number of plants recorded would imply, because a large but unknown proportion of the species are certainly importations, voluntary or involuntary, by man. As, however, the general character of the whole flora is that of the southwestern peninsula of Europe, and as most of the introduced plants have come from the

Tristan d'Acunha or Kerguelen Land; but in the tropics they choose the mountains for their nesting-place (Finsch and Hartlaub, "Ornith. der Viti- und Tonga-Inseln," 1867, Einleitung, p. xviii.). Thus, *Puffinus Megasi* nests at the top of the Korobasa basaga mountain, Viti Levu, fifty miles from the sea. A Procellaria breeds in like manner in the high mountains of Jamaica, I believe at 7000 feet. Peale describes the same habit of *Procellaria rostrata* at Tahiti, and I saw the burrows myself amidst a dense growth of fern, etc., at 4400 feet elevation in that island. Phaethon has a similar habit. It nests at the crater of Kilauea, Hawaii, at 4000 feet elevation, and also high up in Tahiti. In order to account for the transportation of the plants, it is not, of course, necessary that the same species of Procellaria or Diomedea should now range between the distant points where the plants occur. The ancestor of the now differing species might have carried the seeds. The range of the genus is sufficient."

same country, it is almost impossible now to separate them, and Mr. Watson has not attempted to do so. The whole flora contains representatives of 80 natural orders and 250 genera; and even if we suppose that one half the species only are truly indigenous, there will still remain a wonderfully rich and varied flora to have been carried, by the various natural means above indicated, over 900 miles of ocean, more especially as the large proportion of species identical with those of Europe shows that their introduction has been comparatively recent, and that it is probably (as in the case of the birds) still going on. We may therefore feel sure that we have here by no means reached the limit of distance to which plants can be conveyed by natural means across the ocean; and this conclusion will be of great value to us in investigating other cases where the evidence at our command is less complete and the indications of origin more obscure or conflicting.

Of the forty species which are considered to be peculiar to the islands, all are allied to European plants except six, whose nearest affinities are in the Canaries or Madeira. Two of the Compositæ are considered to be distinct genera, but in this order generic divisions rest on slight technical distinctions; and the *Campanula Vidalii* is very distinct from any other known species. With these exceptions, most of the peculiar Azorean species are closely allied to European plants, and are in several cases little more than varieties of them. While, therefore, we may believe that the larger part of the existing flora reached the islands since the glacial epoch, a portion of it may be more ancient, as there is no doubt that a majority of the species could withstand some lowering of temperature; while in such a warm latitude, and surrounded with sea, there would always be many sunny and sheltered spots in which even tender plants might flourish.

Important Deduction from the Peculiarities of the Azorean Fauna and Flora.—There is one conclusion to be drawn from the almost wholly European character of the Azorean fauna and flora which deserves special attention—namely, that the peopling of remote islands is not due so much to ordinary or normal as to extraordinary and exceptional causes. These islands lie in the course of the southwesterly return trades and also of the

Gulf Stream, and we should therefore naturally expect that American birds, insects, and plants would preponderate if they were conveyed by the regular winds and currents, which are both such as to prevent European species from reaching them. But the violent storms to which the Azores are liable blow from all points of the compass; and it is evidently to these, combined with the greater proximity and more favorable situation of the coasts of Europe and North Africa, that the presence of a fauna and flora so decidedly European is to be traced.

The other North Atlantic islands—Madeira, the Canaries, and the Cape Verds—present analogous phenomena to those of the Azores, but with some peculiarities dependent on their more southern position, their richer vegetation, and, perhaps, their greater antiquity. These have been sufficiently discussed in my " Geographical Distribution of Animals" (Vol. I., pp. 208–215); and as we are now dealing with what may be termed typical examples of oceanic islands, for the purpose of illustrating the laws and solving the problems presented by the dispersal of animals, we will pass on to other cases which have been less fully discussed in that work.

BERMUDA.

The Bermudas are a small group of low islands formed of coral and blown coral-sand consolidated into rock. They are situated in 32° N. lat., about seven hundred miles from North Carolina, and somewhat farther from the Bahama Islands, and are thus rather more favorably placed for receiving immigrants from America and its islands than the Azores are with respect to Europe. There are about one hundred islands and islets in all, but their total area does not exceed fifty square miles. They are surrounded by reefs, some at a distance of thirty miles from the main group; and the discovery of a layer of earth with remains of cedar trees forty-eight feet below the present high-water mark shows that the islands have once been more extensive, and probably included the whole area now occupied by shoals and reefs.[1]

---

[1] *Nature*, Vol. VI., p. 262; "Recent Observations in the Bermudas," by Mr. J. Matthew Jones.

Immediately beyond these reefs, however, extends a very deep ocean, while about four hundred and fifty miles distant in a southeast direction the deepest part of the North Atlantic is reached, where soundings of 3825 and 3875 fathoms have been obtained. It is clear, therefore, that these islands are typically oceanic.

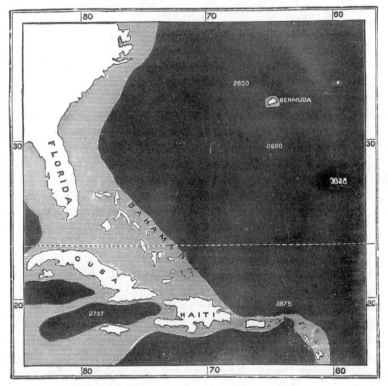

MAP OF BERMUDA AND THE AMERICAN COAST.

NOTE.—The light tint indicates sea less than 1000 fathoms deep.
The dark tint   "   "   more   "   "   "
The figures show the depth in fathoms.

Soundings were taken by the *Challenger* in four different directions around Bermuda, and always showed a rapid deepening of the sea to about 2500 fathoms. This was so remarkable that, in his reports to the Admiralty, Captain Nares spoke of

Bermuda as "a solitary peak rising abruptly from a base only
120 miles in diameter," and in another place as "an isolated
peak rising abruptly from a very small base." These expres-
sions show that Bermuda is looked upon as a typical example of an "oceanic peak;" and on examining the series of official reports of the *Challenger* soundings, I can find no similar
case, although some coasts, both of continents and islands, descend more abruptly. In order to show, therefore, what is the
real character of this peak, I have drawn a section of it on a true
scale from the soundings taken in a north and south direction

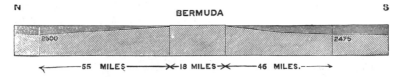

SECTION OF BERMUDA AND ADJACENT SEA-BOTTOM.

The figures show the depth in fathoms at fifty-five miles north and forty-six miles
south of the islands respectively.

where the descent is steepest. It will be seen that the slope is
on both sides very easy, being 1 in 16 on the south, and 1 in
19 on the north. The portion nearest the islands will slope
more rapidly, perhaps reaching in places 1 in 10; but even this
is not steeper than many country roads in hilly countries, while
the remainder would be a hardly perceptible slope. Although
generally very low, some parts of these islands rise to 250
feet above the sea-level, consisting of various kinds of limestone rock, sometimes soft and friable, but often very hard and
even crystalline. It consists of beds which sometimes dip as
much as 30°, and exhibit, besides, great contortions, so that at
first sight the islands appear to exhibit on a small scale the phenomena of a disturbed Palæozoic district. It has, however, long
been known that these rocks are all due to the wind, which
blows up the fine calcareous sand, the product of the disintegration of coral, shells, serpulæ, and other organisms, forming sand-
hills forty and fifty feet high, which move gradually along, overwhelming the lower tracts of land behind them. These are consolidated by the percolation of rain-water which dissolves some

of the lime from the more porous tracts and deposits it lower down, filling every fissure with stalagmite.

The Red Clay of Bermuda.—Besides the calcareous rocks, there is found in many parts of the islands a layer of red earth or clay, containing about thirty per cent. of oxide of iron. This very closely resembles, both in color and chemical composition, the red clay of the ocean-floor, found widely spread in the Atlantic at depths of from 2300 to 3150 fathoms, and occurring abundantly all round Bermuda. It appears, therefore, at first sight, as if the ocean-bed itself has been here raised to the surface, and a portion of its covering of red clay preserved; and this is the view adopted by Mr. Jones in his paper on the "Botany of Bermuda." He says, after giving the analysis, "This analysis tends to convince us that the deep chocolate-colored red clay of the islands found in the lower levels, and from high-water mark some distance into the sea, originally came from the ocean-floor; and that when by volcanic agency the Bermuda column was raised from the depths of the sea, its summit, most probably broken in outline, appeared above the surface covered with this red mud, which in the course of ages has but slightly changed its composition, and yet possesses sufficient evidence to prove its identity with that now lying contiguous to the base of the Bermuda column." But in his "Guide to Bermuda" Mr. Jones tells us that this same red earth has been found, two feet thick, under coral rock, at a depth of forty-two feet below low-water mark, and that it "rested on a bed of compact calcareous sandstone." Now it is quite certain that this "calcareous sandstone" was never formed at the bottom of the deep ocean 700 miles from land; and the occurrence of the red earth at different levels upon coralline sand rock is therefore more probably due to some process of decomposition of the rock itself, or of the minute organisms which abound in the blown sand. The forthcoming volumes on the results of the *Challenger* expedition will probably clear up the difficulty.

Zoology of Bermuda.—As might be expected from their extreme isolation, these islands possess no indigenous land mammalia, frogs, or snakes. There is, however, one lizard, which Professor Cope considers to be distinct from any American spe-

cies, and which he has named *Plestiodon longirostris*. It is said
to be most nearly allied to *P. fasciatus* of the Southeastern
States, from which it differs in having nearly ten more rows of
scales, the tail thicker, and the muzzle longer. In color it is ashy
brown above, greenish blue beneath, with a white line black-
margined on the sides, and it seems to be tolerably abundant in
the islands. This lizard is especially interesting as the only vertebrate animal which exhibits any peculiarity.

Birds.—Notwithstanding its small size, low altitude, and remote position, a great number of birds visit Bermuda annually,
some in large numbers, others only as accidental stragglers. Altogether, over a hundred and eighty species have been recorded,
rather more than half being wading and swimming birds, whose
presence is not so much to be wondered at, as they are great
wanderers; while about eighty-five are land birds, many of
which would hardly be supposed capable of flying so great a
distance. Of the hundred and eighty species, however, about
thirty have only been seen once, and a great many more are
very rare; but about twenty species of land birds are recorded
as tolerably frequent visitors, and nearly half these appear to
come every year.

There are only ten species which are permanent residents on
the island—eight land and two water birds—and of these one has
been almost certainly introduced. These resident birds are as
follows:

1. *Galeoscoptes Carolinensis*. (The Catbird.) Migrates along the east coast of
   the United States.
2. *Sialia sialis*. (The Bluebird.) Migrates along the east coast.
3. *Vireo Novæboracensis*. (The White-eyed Green Tit.) Migrates along the east
   coast.
4. *Passer domesticus*. (The English Sparrow.) ? Introduced.
5. *Corvus Americanus*. (The American Crow.) Common over all North America.
6. *Cardinalis Virginianus*. (The Cardinal-bird.) Migrates from Carolina southward.
7. *Chamæpelia passerina*. (The Ground-dove.) Louisiana, West Indies, and
   Mexico.
8. *Ortyx Virginianus*. (The American Quail.) New England to Florida.
9. *Ardea herodias*. (The Great Blue Heron.) All North America.
10. *Fulica Americana*. (The American Coot.) Temperate and tropical North
    America.

It will be seen that these are all very common North American birds, and most of them are constant visitors from the mainland, so that, however long they may have inhabited the islands, there has been no chance for them to have acquired any distinctive characters through isolation.

Among the most regular visitants which are not resident are the common North American kingfisher (*Ceryle alcyon*), the wood-wagtail (*Siurus Novæboracensis*), the wide-ranging ricebird (*Dolichonyx oryzivora*), and a moor-hen (*Gallinula galeata*); the first three being very common over almost all North America, and the last abundant in the southern portion of it.

Comparison of the Bird Faunas of Bermuda and the Azores. —The bird fauna of Bermuda thus differs from that of the Azores in the much smaller number of resident species and the presence of several regular migrants. This is due, first, to the small area and little-varied surface of these islands, as well as to their limited flora and small supply of insects not affording conditions suitable for the residence of many species all the year round; and, secondly, to the peculiarity of the climate of North America, which causes a much larger number of its birds to be migratory than in Europe. The Northern United States and Canada, with a sunny climate, luxuriant vegetation, and abundant insect life during the summer, supply food and shelter to an immense number of insectivorous and frugivorous birds; so that during the breeding season Canada is actually richer in bird life than Florida. But as the severe winter comes on, all these are obliged to migrate southward—some to Carolina, Georgia, and Florida, others as far as the West Indies, Mexico, or even to Guatemala and South America.

Every spring and autumn, therefore, a vast multitude of birds, belonging to more than a hundred distinct species, migrate northward or southward in Eastern America. A large proportion of these pass along the Atlantic coast; and it has been observed that many of them fly some distance out to sea, passing straight across bays from headland to headland by the shortest route.

Now, as the time of these migrations is the season of storms, especially the autumnal one, which nearly coincides with the

hurricanes of the West Indies and the northerly gales of the
coast of America, the migrating birds are very liable to be carried out to sea. Sometimes they may, as Mr. Jones suggests, be
carried up by local whirlwinds to a great height, where, meeting
with a westerly or northwesterly gale, they are rapidly driven
seaward. The great majority, no doubt, perish, but some reach
the Bermudas and form one of their most striking autumnal features. In October, Mr. Jones tells us, the sportsman enjoys more
shooting than at any other time. The violent revolving gales,
which occur almost weekly, bring numbers of birds of many
species from the American continent, the different members of
the duck tribe forming no inconsiderable portion of the whole;
while the Canada goose and even the ponderous American
swan have been seen amidst the migratory host. With these
come also such delicate birds as the American robin (*Turdus
migratorius*), the yellow-rumped warbler (*Dendrœca coronata*),
the pine-warbler (*Dendrœca pinus*), the wood-wagtail (*Siurus
Novæboracensis*), the summer redbird (*Pyranga œstiva*), the
snow-bunting (*Plectrophanes nivalis*), the redpoll (*Ægiothus
linarius*), the king-bird (*Tyrannus Carolinensis*), and many others. It is, no doubt, in consequence of this repeated immigration
that none of the Bermuda birds have acquired any special peculiarity constituting even a distinct variety; for the few species
that are resident and breed in the islands are continually crossed
by individual immigrants of the same species from the mainland.

Four European birds also have occurred in Bermuda — the
wheat-ear (*Saxicola œnanthe*), which visits Iceland and Lapland
and sometimes the Northern United States; the skylark (*Alauda arvensis*), but this was probably an imported bird or an escape from some ship; the land-rail (*Crex pratensis*), which also
wanders to Greenland and the United States; and the common
snipe (*Scolopax gallinago*), which occurs not unfrequently in
Greenland, but has not yet been noticed in North America. It
is, however, so like the American snipe (*S. Wilsoni*) that a straggler might easily be overlooked.

Two small bats of North American species also occasionally
reach the island, and these are the only wild mammalia except
rats and mice.

Insects of Bermuda.—Insects appear to be very scarce; but it is evident from the lists given by Mr. Jones that only the more conspicuous species have been yet collected. These comprise nineteen beetles, eleven bees and wasps, twenty-six butterflies and moths, nine flies, and the same number of Hemiptera, Orthoptera, and Neuroptera respectively. All appear to be common North American or West Indian species; but until some competent entomological collector visits the islands it is impossible to say whether there are or are not any peculiar species.

Land Mollusca.—The land shells of the Bermudas are somewhat more interesting, as they appear to be the only group of animals except reptiles in which there are any peculiar species. The following list has been kindly furnished me by Mr. Thomas Bland, of New York, who has made a special study of the terrestrial mollusks of the West Indian Islands. The species which are peculiar to the islands are indicated by italics:

LIST OF THE LAND SHELLS OF BERMUDA.

1. Succinea fulgens. (Lea.)...........Also in Cuba.
2. "   Bermudensis. (Pfeiffer.).. " Barbadoes?
3. "   Margarita. (Pfr.)......... " Hayti.
4. *Hyalina Bermudensis.* (Pfr.).......A peculiar form which, according to Mr. Binney, "cannot be placed in any recognized genus." A larger sub-fossil variety also occurs, named *H. Nelsoni* by Mr. Bland, and which appears sufficiently distinct to be classed as another species.
5. "   *circumfirmata.* (Redfield.)
6. "   *discrepans.* (Pfr.)
7. *Patula Reiniana.* (Pfr.)
8. "   hypolepta. (Shuttleworth.)..Probably the same as *P. minuscula* (Binney), a wide-spread American species.
9. Helix vortex. (Pfr.)...............Southern Florida and West Indies.
10. "   microdonta. (Desh.).........Bahama Islands.
11. "   appressa. (Say.).............Virginia and adjacent States; perhaps introduced into Bermuda.
12. "   pulchella. (Müll.)...........Europe; very close to *H. minuta* (Say) of the United States. Introduced into Bermuda?
13. "   ventricosa. (Drap.)..........Azores, Canary Islands, and South Europe.
14. Bulimulus nitidulus. (Pfr.).........Cuba, Hayti, etc.
15. Stenogyra octona. (Ch.)............West Indies and South America.
16. Cionella acicula. (Müll.)...........Florida, New Jersey, and Europe.

17. Pupa pellucida.   (Pfr.)..............West Indies, generally.
18. "   Barbadensis. (Pfr.)..........Barbadoes?
19. "   Jamaicensis. (C.B. Ad.).......Jamaica.
20. Helicina convexa. (Pfr.)...........Barbuda.

Mr. Bland indicates only four species as certainly peculiar to Bermuda, and another sub-fossil species; while one or two of the remainder are indicated as doubtfully identical with those of other countries. We have thus at least one fourth of the land shells peculiar, while almost all the other productions of the islands are identical with those of the adjacent continent and islands. This corresponds, however, with what occurs generally in islands at some distance from continents. In the Azores only one land bird is peculiar out of eighteen resident species; the beetles show about one eighth of the probably non-introduced species as peculiar, the plants about one twentieth; while the land shells have about half the species peculiar. This difference is well explained by the much greater difficulty of transmission over wide seas, in the case of land shells, than of any other terrestrial organisms. It thus happens that when a species has once been conveyed it may remain isolated for unknown ages, and has time to become modified by local conditions unchecked by the introduction of other specimens of the original type.

Flora of Bermuda.—Unfortunately, no good account of the plants of these islands has yet been published. Mr. Jones, in his paper "On the Vegetation of the Bermudas," gives a list of no less than 480 species of flowering plants; but this number includes all the culinary plants, fruit-trees, and garden flowers, as well as all the ornamental trees and shrubs from various parts of the world which have been introduced, mixed up with the European and American weeds that have come with agricultural or garden seeds, and the really indigenous plants, in one undistinguished series. It appears, too, that the late governor, Major-general Lefroy, "has sown and distributed throughout the islands packets of seeds from Kew, representing no less than 600 species, principally of trees and shrubs suited to sandy coast soils"—so that it will be more than ever difficult in future years to distinguish the indigenous from the introduced vegetation.

From the researches of Dr. Rein and Mr. Moseley there appear to be about two hundred and fifty flowering plants in a wild
state, and of these Mr. Moseley thinks less than half are indigenous. The majority are tropical and West Indian, while others are common to the Southern States of North America ; the
former class having been largely brought by means of the Gulf
Stream, the latter by the agency of birds or by winds. Mr.
Jones tells us that the currents bring numberless objects ani-
mate and inanimate from the Caribbean Sea, including the seeds
of trees, shrubs, and other plants, which are continually cast
ashore and sometimes vegetate. The soapberry-tree (*Sapindus*
saponaria) has been actually observed to originate in this way.

Professor Oliver informs me that he knows of no undoubtedly
distinct species of flowering plants peculiar to Bermuda, though
there are some local forms of continental species—instancing
Sisyrinchium, *Bermudianum*, and *Rhus toxicodendron*. There
are, however, two ferns—an Adiantum and a Nephrodium—which
are unknown from any other locality, and this renders it probable that some of the flowering plants are also peculiar. The
juniper, which is so conspicuous a feature of the islands, is said
to be a West Indian species (*Juniperus Barbadensis*) found in
Jamaica and the Bahamas, not the North American red cedar ;
but there seems to be still some doubt about this common plant.

Mr. Moseley, who visited Bermuda in the *Challenger*, has well
explained the probable origin of the vegetation. The large
number of West Indian plants is, no doubt, due to the Gulf
Stream and constant surface drift of warm water in this direction, while others have been brought by the annual cyclones
which sweep over the intervening ocean. The great number of
American migratory birds, including large flocks of the American golden plover, with ducks and other aquatic species, no doubt
occasionally bring seeds, either in the mud attached to their feet
or in their stomachs.[1] As these causes are either constantly in
action or recur annually, it is not surprising that almost all the
species should be unchanged owing to the frequent intercrossing

---

[1] "Notes on the Vegetation of Bermuda," by H. N. Moseley, *Journal of the Linnæan Society*, Vol. XIV., Botany, p. 317.

of freshly arrived specimens. If a competent botanist were thoroughly to explore Bermuda, eliminate the species introduced by human agency, and investigate the source from whence the others were derived and the mode by which they had reached so remote an island, we should obtain important information as to the dispersal of plants, which might afford us a clew to the solution of many difficult problems in their geographical distribution.

Concluding Remarks.—The two groups of islands we have now been considering furnish us with some most instructive facts as to the power of many groups of organisms to pass over from seven hundred to nine hundred miles of open sea. There is no doubt whatever that all the indigenous species have thus reached these islands, and in many cases the process may be seen going on from year to year. We find that, as regards birds, migratory habits and the liability to be caught by violent storms are the conditions which determine the island population. In both islands the land birds are almost exclusively migrants; and in both the non-migratory groups — wrens, tits, creepers, and nuthatches—are absent; while the number of annual visitors is greater in proportion as the migratory habits and prevalence of storms afford more efficient means for their introduction.

We find also that these great distances do not prevent the immigration of some insects of most of the orders, and especially of a considerable number and variety of beetles; while even land shells are fairly represented in both islands, the large proportion of peculiar species clearly indicating that, as we might expect, individuals of this group of organisms arrive only at long and irregular intervals.

Plants are represented by a considerable variety of orders and genera, most of which show some special adaptation for dispersal by wind or water, or through the medium of birds; and there is no reason to doubt that, besides the species that have actually established themselves, many others must have reached the islands, but were not suited to the climate and other physical conditions, or did not find the insects necessary to their fertilization.

If, now, we consider the extreme remoteness and isolation of these islands, their small area, and comparatively recent origin, and that, notwithstanding all these disadvantages, they have acquired a very considerable and varied flora and fauna, we shall, I think, be convinced that, with a larger area and greater antiquity, mere separation from a continent by many hundred miles of sea would not prevent a country from acquiring a very luxuriant and varied flora, and a fauna also rich and peculiar as regards all classes except terrestrial mammals, amphibia, and some groups of reptiles. This conclusion will be of great importance in many cases where the evidence as to the exact origin of the fauna and flora of an island is less clear and satisfactory than in the case of the Azores and Bermuda.

CHAPTER XIII.

THE GALAPAGOS ISLANDS.

Position and Physical Features.—Absence of Indigenous Mammalia and Amphibia.
—Reptiles.—Birds.—Insects and Land Shells.—The Keeling Islands as Illustrating the Manner in which Oceanic Islands are Peopled.—Flora of the Galapagos.
—Origin of the Flora of the Galapagos.—Concluding Remarks.

THE Galapagos differ in many important respects from the
islands we have examined in our last chapter, and the differences are such as to have affected the whole character of their
animal inhabitants. Like the Azores, they are volcanic, but they
are much more extensive, the islands being both larger and more
numerous; while volcanic action has been so recent and extensive that a large portion of their surface consists of barren lava-
fields. They are considerably less distant from a continent than
either the Azores or Bermuda, being about six hundred miles
from the west coast of South America and a little more than
seven hundred from Veragua, with the small Cocos Islands intervening; and they are situated on the equator instead of being
in the north temperate zone. They stand upon a deeply submerged bank, the 1000-fathom line encircling all the more important islands at a few miles' distance, whence there appears to
be a comparatively steep descent all round to the average depth
of that portion of the Pacific, between 2000 and 3000 fathoms.

The whole group occupies a space of about three hundred by
two hundred miles. It consists of five large and twelve small
islands; the largest (Albemarle Island) being about eighty miles
long and of very irregular shape, while the four next in importance—Chatham, Indefatigable, James, and Narborough Islands
—are each about twenty-five or thirty miles long, and of a
rounded or elongate form. The whole are entirely volcanic,
and in the western islands there are numerous active volcanoes.

Unlike the other groups of islands we have been considering, these are situated in a comparatively calm sea, where storms are of rare occurrence and even strong winds almost unknown. They are traversed by ocean currents which are strong and con-

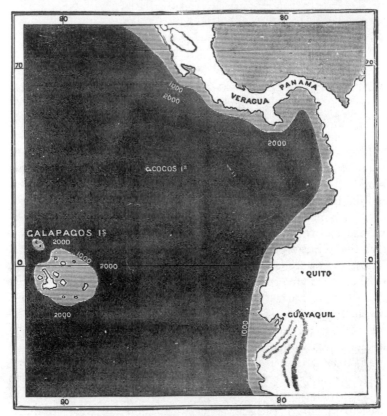

MAP OF THE GALAPAGOS AND ADJACENT COASTS OF SOUTH AMERICA.

The light tint shows where the sea is less than 1000 fathoms deep.
The figures show the depth in fathoms.

stant, flowing towards the northwest from the coast of Peru; and these physical conditions have had a powerful influence on the animal and vegetable forms by which the islands are now inhabited. The Galapagos have also, during three centuries, been frequently visited by Europeans, and were long a favorite resort of buccaneers and traders, who found an ample supply of food in the large tortoises which abound there; and to these visits we may perhaps trace the introduction of some animals whose presence it is otherwise difficult to account for. The vege-

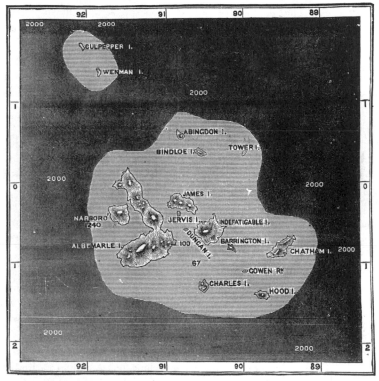

MAP OF THE GALAPAGOS.

The light tint shows a depth of less than 1000 fathoms.
The figures show the depth in fathoms.

tation is generally scanty, but still amply sufficient for the support of a considerable amount of animal life, as shown by the cattle, horses, asses, goats, pigs, dogs, and cats which now run wild in some of the islands.

Absence of Indigenous Mammalia and Amphibia.—As in all other oceanic islands, we find here no truly indigenous mam-

malia, for though there is a mouse of the American genus
Hesperomys, which differs somewhat from any known species,
we can hardly consider this to be indigenous; first, because these
creatures have been little studied in South America, and there
may yet be many undescribed species; and, in the second place,
because, even had it been introduced by some European or native
vessel, there is ample time in two or three hundred years for the
very different conditions to have established a marked diversity
in the characters of the species. This is the more probable because there is also a true rat of the Old World genus Mus, which
is said to differ slightly from any known species; and as this
genus is not a native of the American continents, we are sure
that it must have been recently introduced into the Galapagos.
There can be little doubt, therefore, that the islands are completely destitute of truly indigenous mammalia; and frogs and
toads, the only tropical representatives of the Amphibia, are
equally unknown.

Reptiles.—Reptiles, however, which at first sight appear as
unsuited as mammals to pass over a wide expanse of ocean,
abound in the Galapagos, though the species are not very numerous. They consist of land-tortoises, lizards, and snakes.
The tortoises consist of two peculiar species, *Testudo microphyes*,
found in most of the islands, and *T. Abingdoni*, recently discovered on Abingdon Island, as well as one extinct species, *T.
ephippium*, found on Indefatigable Island. These are all of
very large size, like the gigantic tortoises of the Mascarene
Islands, from which, however, they differ in structural characters; and Dr. Günther believes that they have been originally
derived from the American continent.[1] Considering the well-
known tenacity of life of these animals, and the large number
of allied forms which have aquatic or subaquatic habits, it is
not a very extravagant supposition that some ancestral form,
carried out to sea by a flood, was once or twice safely drifted as
far as the Galapagos, and thus originated the races which now
inhabit them.

---

[1] "Gigantic Land-tortoises Living and Extinct in the Collection of the British
Museum," by A. C. L. G. Günther, F.R.S. 1877.

The lizards are five in number—a peculiar species of gecko, *Phyllodactylus Galapagensis*, and four species of the American family Iguanidæ. Two of these are distinct species of the genus Liocephalus, the other two being large, and so very distinct as to be classed in peculiar genera. One of these is aquatic and found in all the islands, swimming in the sea at some distance from the shore, and feeding on seaweed; the other is terrestrial, and is confined to the four central islands. These were originally described by Mr. Bell as *Amblyrhynchus cristatus* and *A. subcristatus;* they were afterwards placed in two other genera, Trachycephalus and Oreocephalus (see British Museum Catalogue of Lizards); while in a recent paper by Dr. Günther the marine species is again classed as Amblyrhynchus, while the terrestrial form is placed in another genus, Conolophus.

How these lizards reached the islands we cannot tell. The fact that they all belong to American genera or families indicates their derivation from that continent, while their being all distinct species is a proof that their arrival took place at a remote epoch, under conditions perhaps somewhat different from any which now prevail. It is certain that animals of this order have some means of crossing the sea not possessed by any other land vertebrates, since they are found in a considerable number of islands which possess no mammals nor any other land reptiles; but what those means are has not yet been positively ascertained.

It is unusual for oceanic islands to possess snakes, and it is therefore somewhat of an anomaly that two species are found in the Galapagos. Both are closely allied to South American forms, and one is hardly different from a Chilian snake, so that they indicate a more recent origin than in the case of the lizards. Snakes, it is known, can survive a long time at sea, since a living boa-constrictor once reached the Island of St. Vincent from the coast of South America, a distance of two hundred miles by the shortest route. Snakes often frequent trees, and might thus be conveyed long distances if carried out to sea on a tree uprooted by a flood such as often occurs in tropical climates, and especially during earthquakes. To some such accident we may perhaps attribute the presence of these creatures in the Galapagos, and that it is a very rare one is indicated by the fact that only two species have as yet succeeded in obtaining a footing there.

Birds.—We now come to the birds, whose presence here may not seem so remarkable, but which yet present features of interest not exceeded by any other group. Fifty-seven species of birds have now been obtained on these islands, and of these thirty-eight are peculiar to them. But all the species found elsewhere, except one, belong to the aquatic tribes or the waders, which are pre-eminently wanderers, yet even of these eight are peculiar. The true land birds are thirty-one in number, and all but one are entirely confined to the Galapagos; while more than half present such peculiarities that they are classed as distinct genera. All are allied to birds inhabiting tropical America, some very closely; while one—the common American rice-bird, which ranges over the whole northern and part of the southern continent—is the only land bird identical with those of the mainland. The following is a list of these land birds taken from Mr. Salvin's memoir in the *Transactions of the Zoological Society* for the year 1876 :

TURDIDÆ.

1. *Mimus trifasciatus*...............
2. "   *melanotus*................ ⎱ This and the two allied species are related
3. "   *parvulus*.................. ⎰ to a Peruvian bird, *Mimus longicaudus.*

MNIOTILTIDÆ.

4. *Dendrœca aureola*....................Closely allied to the wide-ranging *D. œstiva.*

HIRUNDINIDÆ.

5. *Progne concolor*......................Allied to *P. purpurea* of North and South America.

CŒREBIDÆ.

6. *Certhidea olivacea*.................⎱ A peculiar genus allied to the Andean
7. "   *fusca*....................⎰ genus *Conirostrum.*

FRINGILLIDÆ.

8. *Geospiza magnirostris*..............⎤
9. "   *strenua*............ .....
10. "   *dubia*....................
11. "   *fortis*..................... ⎬ A distinct genus, but allied to the South
12. "   *nebulosa*................... ⎪ American genus *Guiraca.*
13. "   *fuliginosa*.................
14. "   *parvula*.................
15. "   *dentirostris*...............⎦

16. *Cactornis scandens*.................. ⎫
17.   "   *assimilis*.................. ⎪
18.   "   *Abingdoni*.............. ⎬ A genus allied to the last.
19.   "   *pallida*................... ⎪
                                              ⎭
20. *Camarhynchus psittaculus*........... ⎫
21.   "   *crassirostris*........... ⎪ A very peculiar genus allied to *Neorhyn-*
22.   "   *variegatus*............ ⎬ *chus* of the west coast of Peru.
23.   "   *prosthemelas*.......... ⎪
24.   "   *Habeli*.............. ⎭

ICTERIDÆ.

25. *Dolichonyx oryzivora*..................Ranges from Canada to Paraguay.

TYRANNIDÆ.

26. *Pyrocephalus nanus*...................Allied to *P. rubineus* of Ecuador.
27. *Myiarchus magnirostris*...............Allied to West Indian species.

COLUMBIDÆ.

28. *Zenaida Galapagoensis*...............A peculiar species of a South American genus.

FALCONIDÆ.

29. *Buteo Galapagoensis*..................A buzzard of peculiar coloration.

STRIGIDÆ.

30. *Asio Galapagoensis*...................Hardly distinct from the wide-spread *A. brachyotus*.
31. *Strix punctatissima*...................Allied to *S. flammea*, but quite distinct.

We have here every gradation of difference, from perfect identity with the continental species to genera so distinct that it is difficult to determine with what forms they are most nearly allied; and it is interesting to note that this diversity bears a distinct relation to the probabilities of, and facilities for, migration to the islands. The excessively abundant rice-bird, which breeds in Canada and swarms over the whole United States, migrating to the West Indies and South America, visiting the distant Bermudas almost every year, and extending its range as far as Paraguay, is the only species of land bird which remains completely unchanged in the Galapagos; and we may therefore conclude that some stragglers of the migrating host reach the islands sufficiently often to keep up the purity of the breed. Next, we have the almost cosmopolite short-eared owl (*Asio brachyotus*), which ranges from China to Ireland, and from Greenland to the Strait of Magellan, and of this the Galapagos bird is probably only one of the numerous varieties. The little wood-warbler (*Dendrœca aureola*) is closely allied to a species which ranges over the whole of North America and as far south as New Granada. It has also been occasionally met with in Bermuda, an indication that it has considerable powers of flight and endurance. The more distinct species—as the mocking-thrushes (Mimus), the tyrant fly-catchers (Pyrocephalus and Myiarchus), and the ground-dove (Zenaida)—are all allied to non-migratory species peculiar to tropical America, and of a more restricted range; while the distinct genera are allied to South American groups of finches and sugar-birds which have usually restricted ranges, and whose habits are such as not to render them likely to be carried out to sea. The remote ancestral forms of these birds which, owing to some exceptional causes, reached the Galapagos, have thus remained uninfluenced by later migrations, and have in consequence been developed into a variety of distinct types adapted to the peculiar conditions of existence under which they have been placed. Sometimes the different species thus formed are confined to one or two of the islands only, as the two species of Certhidea, which are divided between the islands, but do not appear ever to occur together. *Mimus parvulus* is confined to Albemarle Island, and *M. trifasciatus* to Charles Island; *Cactornis pallida* to Indefatigable Island, and *C. Abingdoni* to Abingdon Island.

Now all these phenomena are strictly consistent with the theory of the peopling of the islands by accidental migrations, if we only allow them to have existed for a sufficiently long period; and the fact that volcanic action has ceased on many of the islands, as well as their great extent, would certainly indicate a considerable antiquity.

The great difference presented by the birds of these islands as compared with those of the equally remote Azores and Bermudas is sufficiently explained by the difference of climatal conditions. At the Galapagos there are none of those periodic storms, gales, and hurricanes which prevail in the North Atlantic, and which every year carry some straggling birds of Europe or North America to the former islands; while, at the same time, the majority of the tropical American birds are non-migratory, and thus afford none of the opportunities presented by the countless hosts of migrants which pass annually northward and southward along the European and especially along the North American coasts. It is strictly in accordance with these different conditions that we find in one case an almost perfect identity with, and in the other an almost equally complete diversity from, the continental species of birds.

Insects and Land Shells.—The other groups of land animals add little of importance to the facts already referred to. The insects are very scanty; the most plentiful group, the beetles, only furnishing about thirty-five species, belonging to twenty-nine genera and eighteen families. The species are almost all peculiar, as are some of the genera. They are mostly small and obscure insects, allied either to American or to world-wide groups. The Carabidæ and the Heteromera are the most abundant groups, the former furnishing six and the latter eight species.[1]

---

[1] The following list of the beetles yet known from the Galapagos shows their scanty proportions and accidental character; the thirty-seven species belonging to thirty-one genera and eighteen families. It is taken from Mr. Waterhouse's enumeration in the *Proceedings of the Zoological Society* for 1877 (p. 81):

CARABIDÆ.
Feronia calathoides.
"   insularis.
"   Galapagoensis.
Amblygnathus obscuricornis.
Solenophorus Galapagoensis.
Notaphus Galapagoensis.

DYTISCIDÆ.
Eunectes occidentalis.

MALACODERMS.
Ablechrus Darwinii.
Corynetes rufipes.
Bostrichus unciniatus.

LAMELLICORNES.
Copris lugubris.
Oryctes Galapagoensis.

ELATERIDÆ.
Physorhinus Galapagoensis.
Acilius incisus.
Copelatus Galapagoensis.

PALPICORNES.
Tropisternus lateralis.
Philhydrus sp.

STAPHYLINIDÆ.
Creophilus villosus.

NECROPHAGA.
Acribis serrativentris.
Phalacrus Darwinii.
Dermestes vulpinus.

CURCULIONIDÆ.
Otiorhynchus cuneiformis.
Anchonus Galapagoensis.

LONGICORNIA.
Eburia amabilis.

HETEROMERA.
Stomion helopoides.
"   lævigatum.
Ammophorus obscurus.
"   Cooksoni.
"   bifoveatus.
Pedonœces Galapagoensis.
"   pubescens.
Phaleria manicata.

ANTHRIBIDÆ.
Ormiscus variegatus.

PHYTOPHAGA.
Diabrotica limbata.
Docema Galapagoensis.
Longitarsus lunatus.

SECURIPALPES.
Scymnus Galapagoensis.

The land shells are not abundant—about twenty in all, most of them peculiar species, but not otherwise remarkable. The observation of Captain Collnet, quoted by Mr. Darwin in his "Journal," that drift-wood, bamboos, canes, and the nuts of a palm are often washed on the southeastern shores of the islands, furnishes an excellent clew to the manner in which many of the insects and land shells may have reached the Galapagos. Whirlwinds also have been known to carry quantities of leaves and other vegetable débris to great heights in the air, and these might be then carried away by strong upper currents and dropped at great distances, and with them small insects and mollusca, or their eggs. We must also remember that volcanic islands are subject to subsidence as well as elevation; and it is quite possible that during the long period the Galapagos have existed some islands may have intervened between them and the coast, and have served as stepping-stones by which the passage to them of various organisms would be greatly facilitated. Sunken banks, the relics of such islands, are known to exist in many parts of the ocean, and countless others, no doubt, remain undiscovered.

The Keeling Islands as Illustrating the Manner in which Oceanic Islands are Peopled.—That such causes as have been here adduced are those by which oceanic islands have been peopled is further shown by the condition of equally remote islands which we know are of comparatively recent origin. Such are the Keeling or Cocos Islands in the Indian Ocean, situated about the same distance from Sumatra as the Galapagos from South America, but mere coral reefs, supporting abundance of cocoa-nut palms as their chief vegetation. These islands were visited by Mr. Darwin, and their natural history carefully examined. The only mammals are rats brought by a wrecked vessel, and said by Mr. Waterhouse to be common English rats, "but smaller and more brightly colored;" so that we have here an illustration of how soon a difference of race is established under a constant and uniform difference of conditions. There are no true land birds, but there are snipes and rails, both apparently common Malayan species. Reptiles are represented by one small lizard, but no account of this is given in the "Zoology

of the Voyage of the Beagle," and we may therefore conclude
that it was an introduced species. Of insects, careful collecting only produced thirteen species belonging to eight distinct
orders. The only beetle was a small Elater, the Orthoptera
were a Gryllus and a Blatta; and there were two flies, two ants,
and two small moths, one a Diopæa which swarms everywhere
in the eastern tropics in grassy places. All these insects were,
no doubt, brought either by winds, by floating timber (which
reaches the islands abundantly), or by clinging to the feathers
of aquatic or wading birds; and we only require more time, to
introduce a greater variety of species, and a better soil and more
varied vegetation, to enable them to live and multiply, in order
to give these islands a fauna and flora equal to those of the Bermudas. Of wild plants there were only twenty species, belonging to nineteen genera and to no less than sixteen natural families, while all were common tropical shore plants. These islands
are thus evidently stocked by waifs and strays brought by the
winds and waves; but their scanty vegetation is mainly due to
unfavorable conditions—the barren coral rock and sand, of which
they are wholly composed, together with exposure to sea-air, being suitable to a very limited number of species which soon
monopolize the surface. With more variety of soil and aspect
a greater variety of plants would establish themselves, and these
would favor the preservation and increase of more insects, birds,
and other animals, as we find to be the case in many small and
remote islands.[1]

---

[1] Juan Fernandez is a good example of a small island which, with time and favorable conditions, has acquired a tolerably rich and highly peculiar flora and fauna. It
is situated in 34° S. lat., 400 miles from the coast of Chili, and, so far as facilities for
the transport of living organisms are concerned, is by no means in a favorable position, for the ocean currents come from the southwest in a direction where there is
no land but the antarctic continent, and the prevalent winds are also westerly. No
doubt, however, there are occasional storms, and there may have been intermediate
islands; but its chief advantages are, no doubt, its antiquity and its varied surface,
offering many chances for the preservation and increase of whatever plants and animals have chanced to reach it. The island consists of basalt, greenstone, and other
ancient rocks, and, though only about twelve miles long, its mountains are three
thousand feet high. Enjoying a moist and temperate climate, it is especially adapted
to the growth of ferns, which are very abundant; and as the spores of these plants
are as fine as dust, and very easily carried for enormous distances by winds, it is not

Flora of the Galapagos.—The plants of these islands are so much more numerous than the known animals, even including the insects, they have been so carefully studied by eminent botanists, and their relations throw so much light on the past history of the group, that no apology is needed for giving a brief outline of the peculiarities and affinities of the flora. The statements we shall make on this subject will be taken from the memoir of Sir Joseph Hooker in the *Linnæan Transactions* for 1851, founded on Mr. Darwin's collections, and a later paper by N. J. Andersson in the *Linnæa* of 1861, embodying more recent discoveries.

The total number of flowering plants known at the latter date was 332, of which 174 were peculiar to the islands, while 158 were common to other countries. Of these latter about 20 have been introduced by man, while the remainder are all natives of some part of America, though about a third part are species of wide range extending into both hemispheres. Of those confined to America, 42 are found in both the northern and southern continents, 21 are confined to South America, while 20 are found only in North America, the West Indies, or Mexico. This equality of North American and South American species in the Galapagos is a fact of great significance in connection with the observation of Sir Joseph Hooker, that the *peculiar* species are allied to the plants of temperate America or to those of the high Andes, while the non-peculiar species are mostly

---

surprising that there are twenty-four species on the island, while the remote period when it first received its vegetation may be indicated by the fact that four of the species are quite peculiar. The same general character pervades the whole flora and fauna. For so small an island it is rich, containing a considerable number of flowering plants, four true land birds, about fifty species of insects, and twenty of land shells. Almost all these belong to South American genera, and a large proportion are South American species; but several of the plants and insects, half the birds, and the whole of the land shells are peculiar. This seems to indicate that the means of transmission were formerly greater than they are now, and that in the case of land shells none have been introduced for so long a period that all have become modified into distinct forms, or have been preserved on the island while they have become extinct on the continent. For a detailed examination of the causes which have led to the modification of the humming-birds of Juan Fernandez, see the author's "Tropical Nature," p. 140; while a general account of the fauna of the island is given in his "Geographical Distribution of Animals," Vol. II., p. 49.

such as inhabit the hotter regions of the tropics near the level
of the sea. He also observes that the seeds of this latter class
of Galapagos plants often have special means of transport, or
belong to groups whose seeds are known to stand long voyages
and to possess great vitality. Mr. Bentham, also, in his elaborate
account of the Compositæ,[1] remarks on the decided Central
American or Mexican affinities of the Galapagos species, so that
we may consider this to be a thoroughly well-established fact.

The most prevalent families of plants in the Galapagos are
the Compositæ (40 species), Gramineæ (32 species), Leguminosæ
(30 species), and Euphorbiaceæ (29 species). Of the Compositæ
most of the species, except such as are common weeds or shore
plants, are peculiar, but there are only two peculiar genera allied
to Mexican forms and not very distinct; while the genus Lipochæta, represented here by a single species, is only found elsewhere in the Sandwich Islands, though it has American affinities.

Origin of the Galapagos Flora.—These facts are explained by
the past history of the American continent, its separation at various epochs by arms of the sea uniting the two oceans across
what is now Central America (the last separation being of recent date, as shown by the identical species of fishes on both
sides of the isthmus), and the influence of the glacial epoch in
driving the temperate American flora southward along the
mountain plateaus.[2] At the time when the two oceans were
united, a portion of the Gulf Stream may have been diverted
into the Pacific, giving rise to a current, some part of which
would almost certainly have reached the Galapagos, and this
may have helped to bring about that singular assemblage of
West Indian and Mexican plants now found there. And as we
now believe that the duration of the last glacial epoch in its successive phases was much longer than the time which has elapsed
since it finally passed away, while throughout the Miocene epoch
the snow-line would often be lowered during periods of high
eccentricity, we are enabled to comprehend the nature of the

---

[1] *Journal of the Linnæan Society*, Vol. XIII., Botany, p. 556.
[2] " Geographical Distribution of Animals," Vol. II., p. 81.

causes which may have led to the islands being stocked with those northern or subalpine types which are so characteristic a feature of that portion of the Galapagos flora which consists of peculiar species.

On the whole, the flora agrees with the fauna in indicating a moderately remote origin, great isolation, and changes of conditions affording facilities for the introduction of organisms from various parts of the American coast, and even from the West Indian Islands and Gulf of Mexico. As in the case of the birds, the several islands differ considerably in their native plants, many species being limited to one or two islands only, while others extend to several. This is, of course, what might be expected on any theory of their origin ; because, even if the whole of the islands had once been united and afterwards separated, long-continued isolation would often lead to the differentiation of species, while the varied conditions to be found upon islands differing in size and altitude as well as in luxuriance of vegetation would often lead to the extinction of a species on one island and its preservation on another. If the several islands had been equally well explored, it might be interesting to see whether, as in the case of the Azores, the number of species diminished in those more remote from the coast; but, unfortunately, our knowledge of the productions of the various islands of the group is exceedingly unequal, and, except in those cases in which representative species inhabit distinct islands, we have no certainty on the subject. All the more interesting problems in geographical distribution, however, arise from the relation of the fauna and flora of the group as a whole to those of the surrounding continents ; and we shall therefore, for the most part, confine ourselves to this aspect of the question in our discussion of the phenomena presented by oceanic or continental islands.

Concluding Remarks.—The Galapagos offer an instructive contrast with the Azores, showing how a difference of conditions that might be thought unimportant may yet produce very striking results in the forms of life. Although the Galapagos are much nearer a continent than the Azores, the number of species of plants common to the continent is much less in the former case than in the latter, and this is still more prominent a characteristic of the insect and the bird fauna. This difference has been shown to depend almost entirely on the one archipelago being situated in a stormy, the other in a calm, portion of the ocean; and it demonstrates the preponderating importance of the atmosphere as an agent in the dispersal of birds, insects, and plants. Yet ocean currents and surface drifts are undoubtedly efficient carriers of plants, and, with plants, of insects and shells, especially in the tropics; and it is probably to this agency that we may impute the recent introduction of a number of common Peruvian and Chilian littoral species, and also at a more remote period of several West Indian types when the Isthmus of Panama was submerged.

In the case of these islands we see the importance of taking past conditions of sea and land and past changes of climate into account, in order to explain the relations of the peculiar or endemic species of their fauna and flora; and we may even see an indication of the effects of climatal changes in the Northern Hemisphere, in the north temperate or Alpine affinities of so many of the plants, and even of some of the birds. The relation between the migratory habits of the birds and the amount of difference from continental types is strikingly accordant with the fact that it is almost exclusively migratory birds that annually reach the Azores and Bermuda; while the corresponding fact that the seeds of those plants which are common to the Galapagos and the adjacent continent have all—as Sir Joseph Hooker states—some special means of dispersal is equally intelligible. The reason why the Galapagos possess four times as many peculiar species of plants as the Azores is clearly a result of the less constant introduction of seeds, owing to the absence of storms; the greater antiquity of the group, allowing more time for specific change; and the influence of cold epochs and of alterations of sea and land in bringing somewhat different sets of plants at different times within the influence of such modified winds and currents as might convey them to the islands.

On the whole, then, we have no difficulty in explaining the probable origin of the flora and fauna of the Galapagos by means of the illustrative facts and general principles already adduced.

CHAPTER XIV.

ST. HELENA.

Position and Physical Features of St. Helena.—Change Effected by European Occupation.—The Insects of St. Helena.—Coleoptera.—Peculiarities and Origin of the Coleoptera of St. Helena.—Land Shells of St. Helena.—Absence of Freshwater Organisms.—Native Vegetation of St. Helena.—The Relations of the St. Helena Compositæ.—Concluding Remarks on St. Helena.

In order to illustrate as completely as possible the peculiar phenomena of oceanic islands, we will next examine the organic productions of St. Helena, and of the Sandwich Islands, since these combine in a higher degree than any other spots upon the globe extreme isolation from all more extensive lands with a tolerably rich fauna and flora whose peculiarities are of surpassing interest. Both, too, have received considerable attention from naturalists; and though much still remains to be done in the latter group, our knowledge is sufficient to enable us to arrive at many interesting results.

Position and Physical Features of St. Helena.—This island is situated nearly in the middle of the South Atlantic Ocean, being more than 1100 miles from the coast of Africa, and 1800 from South America. It is about ten miles long by eight wide, and is wholly volcanic, consisting of ancient basalts, lavas, and other volcanic products. It is very mountainous and rugged, bounded, for the most part, by enormous precipices, and rising to a height of 2700 feet above the sea-level. An ancient crater, about four miles across, is open on the south side, and its northern rim forms the highest and central ridge of the island. Many other hills and peaks, however, are more than two thousand feet high, and a considerable portion of the surface consists of a rugged plateau, having an elevation of about fifteen hundred to two thousand feet. Everything indicates that St. Helena is

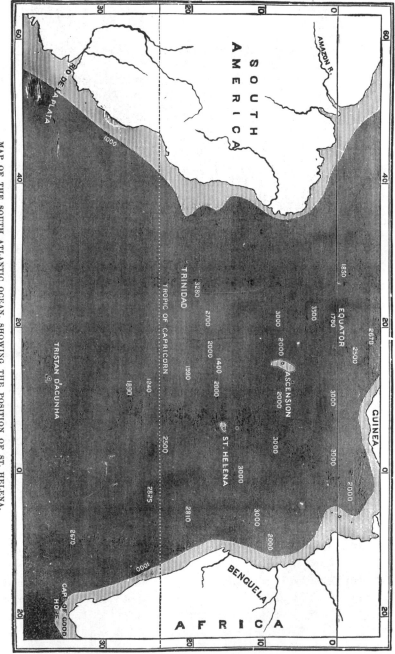

MAP OF THE SOUTH ATLANTIC OCEAN, SHOWING THE POSITION OF ST. HELENA.

The light tint shows depths of less than 1000 fathoms. The figures show depths of the sea in fathoms.

an isolated volcanic mass built up from the depths of the ocean. Mr. Wollaston remarks, "There are the strongest reasons for believing that the area of St. Helena was never *very* much larger than it is at present—the comparatively shallow sea-soundings within about a mile and a half from the shore revealing an abruptly defined ledge, *beyond* which no bottom is reached at a depth of 250 fathoms; so that the original basaltic mass, which was gradually piled up by means of successive eruptions from beneath the ocean, would appear to have its limit definitely marked out by this suddenly terminating submarine cliff—the space between it and the existing coast-line being reasonably referred to that slow process of disintegration by which the island has been reduced, through the eroding action of the elements, to its present dimensions." If we add to this that between the island and the coast of Africa, in a southeasterly direction, is a profound oceanic gulf known to reach a depth of 2860 fathoms, or 17,160 feet, while an equally deep, or perhaps deeper, ocean extends to the west and southwest, we shall be satisfied that St. Helena is a true oceanic island, and that it owes none of its peculiarities to a former union with any continent or other distant land.

Change Effected by European Occupation.—When first discovered, 378 years ago, St. Helena was densely covered with a luxuriant forest vegetation, the trees overhanging the seaward precipices and covering every part of the surface with an evergreen mantle. This indigenous vegetation has been almost wholly destroyed; and although an immense number of foreign plants have been introduced, and have more or less completely established themselves, yet the general aspect of the island is now so barren and forbidding that some persons find it difficult to believe that it was once all green and fertile. The cause of the change is, however, very easily explained. The rich soil formed by decomposed volcanic rock and vegetable deposits could only be retained on the steep slopes so long as it was protected by the vegetation to which it in great part owed its origin. When this was destroyed, the heavy tropical rains soon washed away the soil, and has left a vast expanse of bare rock or sterile clay. This irreparable destruction was caused, in the first place, by goats, which were introduced by the Portuguese in 1513, and increased so rapidly that in 1588 they existed in thousands. These animals are the greatest of all foes to trees, because they eat off the young seedlings, and thus prevent the natural restoration of the forest. They were, however, aided by the reckless waste of man. The East India Company took possession of the island in 1651, and about the year 1700 it began to be seen that the forests were fast diminishing, and required some protection. Two of the native trees, redwood and ebony, were good for tanning, and, to save trouble, the bark was wastefully stripped from the trunks only, the remainder being left to rot; while in 1709 a large quantity of the rapidly disappearing ebony was used to burn lime for building fortifications! By the MS. records quoted in Mr. Melliss's interesting volume on St. Helena,[1] it is evident that the evil consequences of allowing the trees to be destroyed were clearly foreseen, as the following passages show: "We find the place called the Great Wood in a flourishing condition, full of young trees, where the hogs (of which there is a great abundance) do not come to root them up. But the Great Wood is miserably lessened and destroyed within our memories, and is not near the circuit and length it was. But we believe it does not contain now less than fifteen hundred acres of fine woodland and good ground, but no springs of water but what is salt or brackish, which we take to be the reason that that part was not inhabited when the people first chose out their settlements and made plantations; but if wells could be sunk, which the governor says he will attempt when we have more hands, we should then think it the most pleasant and healthiest part of the island. But as to healthiness, we don't think it will hold so if the wood that keeps the land warm were destroyed; for then the rains, which are violent here, would carry away the upper soil, and, it being a clay marl underneath, would produce but little; as it is, we think in case it were enclosed it might be greatly improved. . . . When once this wood is gone, the island will soon be ruined. . . . We viewed the wood's

---

[1] "St. Helena: a Physical, Historical, and Topographical Description of the Island," etc., by John Charles Melliss, F.G.S., etc.   London, 1875.

end which joins the Honorable Company's plantation called the
Hutts, but the wood is so destroyed that the beginning of the
Great Wood is now a whole mile beyond that place; and all the
soil being washed away, that distance is now entirely barren"
(MS. Records, 1716). In 1709 the governor reported to the
Court of Directors of the East India Company that the timber
was rapidly disappearing, and that the goats should be destroyed for the preservation of the ebony wood, and because the isl-
and was suffering from droughts. The reply was, " The goats
are not to be destroyed, being more valuable than ebony." Thus,
through the gross ignorance of those in power, the last opportunity of preserving the peculiar vegetation of St. Helena, and
preventing the island from becoming the comparatively rocky
desert it now is, was allowed to pass away.[1] Even in a mere
pecuniary point of view, the error was a fatal one, for in the next
century (in 1810) another governor reports the total destruction
of the great forests by the goats, and that in consequence the
cost of importing fuel for government use was £2729 7s. 8d. for
a single year! About this time large numbers of European,
American, Australian, and South African plants were imported,
and many of these ran wild and increased so rapidly as to drive
out and exterminate much of the relics of the native flora; so
that now English broom, gorse and brambles, willows and pop-

---

[1] Mr. Marsh, in his interesting work entitled "The Earth as Modified by Human
Action" (p. 51), thus remarks on the effect of browsing quadrupeds in destroying and
checking woody vegetation : "I am convinced that forests would soon cover many parts
of the Arabian and African deserts if man and domestic animals, especially the goat
and the camel, were banished from them. The hard palate and tongue and strong
teeth and jaws of this latter quadruped enable him to break off and masticate tough
and thorny branches as large as the finger. He is particularly fond of the smaller twigs,
leaves, and seed-pods of the *Sont* and other acacias, which, like the American Robinia,
thrive well on dry and sandy soils; and he spares no tree the branches of which are
within his reach, except, if I remember right, the tamarisk that produces manna.
Young trees sprout plentifully around the springs and along the winter water-courses
of the desert, and these are just the halting-stations of the caravans and their routes
of travel. In the shade of these trees, annual grasses and perennial shrubs shoot up,
but are mown down by the hungry cattle of the Bedouin as fast as they grow. A
few years of undisturbed vegetation would suffice to cover such points with groves,
and these would gradually extend themselves over soils where now scarcely any green
thing but the bitter colocynth and the poisonous fox-glove is ever seen."

lars, and some common American, Cape, and Australian weeds, alone meet the eye of the ordinary visitor. These, in Sir Joseph Hooker's opinion, render it absolutely impossible to restore the native flora, which only lingers in a few of the loftiest ridges and most inaccessible precipices, and is rarely seen except by some exploring naturalist.

This almost total extirpation of a luxuriant and highly peculiar vegetation must inevitably have caused the destruction of a considerable portion of the lower animals which once existed on the island, and it is rather singular that so much as has actually been discovered should be left to show us the nature of the aboriginal fauna. Many naturalists have made small collections during short visits, but we owe our present complete knowledge of the two most interesting groups of animals—the insects and the land shells—mainly to the late Mr. T. Vernon Wollaston, who, after having thoroughly explored Madeira and the Canaries, undertook a voyage to St. Helena for the express purpose of studying its terrestrial fauna, and resided for six months (1875–76) in a high central position, whence the loftiest peaks could be explored. The results of his labors are contained in two volumes,[1] which, like all that he wrote, are models of accuracy and research, and it is to these volumes that we are indebted for the interesting and suggestive facts which we here lay before our readers.

Insects—Coleoptera.—The total number of species of beetles hitherto observed at St. Helena is 203; but of these no less than 74 are common and wide-spread insects, which have certainly, in Mr. Wollaston's opinion, been introduced by human agency. There remain 129, which are believed to be truly aborigines, and of these all but one are found nowhere else on the globe. But, in addition to this large amount of specific peculiarity (perhaps unequalled anywhere else in the world), the beetles of this island are equally remarkable for their generic isolation, and for the altogether exceptional proportion in which the great divisions of the order are represented. The species belong to thirty-nine genera, of which no less than twenty-five

---

[1] "Coleoptera Sanctæ Helenæ," 1877; "Testacea Atlantica," 1878.

are peculiar to the island; and many of these are such isolated forms that it is impossible to find their allies in any particular country. Still more remarkable is the fact that more than two thirds of the whole number of indigenous species are Rhyncoph-ora, or weevils, while more than two fifths (fifty-four species) belong to one family, the Cossonidæ. Now, although the Rhyn-cophora are an immensely numerous group and always form a large portion of the insect population, they nowhere else approach such a proportion as this. For example, in Madeira they form one sixth of the whole of the indigenous Coleoptera, in the Azores less than one tenth, and in Britain one seventh. Even more interesting is the fact that the twenty genera to which these insects belong are every one of them peculiar to the island, and in many cases have no near allies elsewhere, so that we cannot but look on this group of beetles as forming the most characteristic portion of the ancient insect fauna. Now, as the great majority of these are wood-borers, and all are closely attached to vegetation, and often to particular species of plants, we might, as Mr. Wollaston well observes, deduce the former luxuriant vegetation of the island from the great preponderance of this group, even had we not positive evidence that it was at no distant epoch densely forest-clad. We will now proceed briefly to indicate the numbers and peculiarities of each of the families of beetles which enter into the St. Helena fauna, taking them, not in systematic order, but according to their importance in the island.

1. RHYNCOPHORA.—This great division includes the weevils and allied groups, and, as above stated, exceeds in number of species all the other beetles of the island. Four families are represented ; the Cossonidæ, with fifteen peculiar genera comprising fifty-four species, and one minute insect (*Stenoscelis hylastoides*) forming a peculiar genus, but which has been found also at the Cape of Good Hope. It is therefore impossible to say of which country it is really a native, or whether it is indigenous to both, and dates back to the remote period when St. Helena received its early immigrants. All the Cossonidæ are found in the highest and wildest parts of the island where the native vegetation still lingers, and many of them are only found

in the decaying stems of tree-ferns, box-wood, arborescent Compositæ, and other indigenous plants. They are all pre-eminently peculiar and isolated, having no direct affinity to species found in any other country. The next family, the Tanyrhynchidæ, has one peculiar genus in St. Helena, with ten species. This genus (Nesiotes) is remotely allied to European, Australian, and Madeiran insects of the same family: the habits of the species are similar to those of the Cossonidæ. The Trachyphlœidæ are represented by a single species belonging to a peculiar genus not very remote from a European form. The Anthribidæ, again, are highly peculiar. There are twenty-six species, belonging to three genera, all endemic, and so extremely peculiar that they form two new sub-families. One of the genera, Acarodes, is said to be allied to a Madeiran genus.

2. GEODEPHAGA.—These are the terrestrial carnivorous beetles, very abundant in all parts of the world, especially in the temperate regions of the Northern Hemisphere. In St. Helena there are fourteen species, belonging to three genera, one of which is peculiar. This is the *Haplothorax Burchellii*, the largest beetle on the island, and now very rare. It resembles a large black Carabus. There is also a peculiar Calosoma, very distinct, though resembling in some respects certain African species. The rest of the Geodephaga, twelve in number, belong to the wide-spread genus Bembidium; but they are altogether peculiar and isolated, except one, which is of European type, and alone has wings, all the rest being wingless.

3. HETEROMERA.—This group is represented by three peculiar genera containing four species, with two species belonging to European genera. They belong to the families Opatridæ, Mordellidæ, and Anthicidæ.

4. BRACHYELYTRA.—Of this group there are six peculiar species, belonging to four European genera—Homalota, Philonthus, Xantholinus, and Oxytelus.

5. PRIOCERATA. — The families Elateridæ and Anobiidæ are each represented by a peculiar species of a European genus.

6. PHYTOPHAGA.—There are only three species of this tribe, belonging to the European genus Longitarsus.

7. LAMELLICORNIS.—Here are three species, belonging to two

genera. One is a peculiar species of Trox, allied to South African forms; the other two belong to the peculiar genus Melissius, which Mr. Wollaston considers to be remotely allied to Australian insects.

8. Pseudo-trimera.—Here we have the fine lady-bird *Chilomenus lunata*, also found in Africa, but apparently indigenous in St. Helena; and a peculiar species of Euxestes, a genus only found elsewhere in Madeira.

9. Trichopterygidæ.—These, the minutest of beetles, are represented by one species of the European and Madeiran genus Ptinella.

10. Necrophaga. — One indigenous species of Cryptophaga inhabits St. Helena, and this is said to be very closely allied to a Cape species.

Peculiarities and Origin of the Coleoptera of St. Helena.—We see that the great mass of the indigenous species are not only peculiar to the island, but so isolated in their characters as to show no close affinity with any existing insects; while a small number (about one third of the whole) have some relations, though often very remote, with species now inhabiting Europe, Madeira, or South Africa. These facts clearly point to the very great antiquity of the insect fauna of St. Helena, which has allowed time for the modification of the originally introduced species, and their special adaptation to the conditions prevailing in this remote island. This antiquity is also shown by the remarkable specific modification of a few types. Thus the whole of the Cossonidæ may be referred to three types, one species only (*Hexacoptus ferrugineus*) being allied to the European Cossonidæ, though forming a distinct genus; a group of three genera and seven species remotely allied to the *Stenoscelis hylastoides*, which occurs also at the Cape; while a group of twelve genera with forty-six species have their only (remote) allies in a few insects widely scattered in South Africa, New Zealand, Europe, and the Atlantic islands. In like manner, eleven species of Bembidium form a group by themselves; and the Heteromera form two groups—one consisting of three genera and species of Opatridæ allied to a type found in Madeira; the other, Anthicodes, altogether peculiar.

Now each of these types may well be descended from a single species which originally reached the island from some other land; and the great variety of generic and specific forms into which some of them have diverged is an indication, and to some extent a measure, of the remoteness of their origin. The rich insect fauna of Miocene age found in Switzerland consists mostly of genera which still inhabit Europe, with others which now inhabit the Cape of Good Hope or the tropics of Africa and South America; and it is not at all improbable that the origin of the St. Helena fauna dates back to at least as remote, and not improbably to a still earlier epoch. But if so, many difficulties in accounting for its origin will disappear. We know that at that time many of the animals and plants of the tropics of North America, and even of Australia, inhabited Europe; while during the changes of climate which, as we have seen, there is good reason to believe periodically occurred there would be much migration from the temperate zones towards the equator, and the reverse. If, therefore, the nearest ally of any insular group now inhabits a particular country, we are not obliged to suppose that it reached the island from that country, since we know that most groups have ranged in past times over wider areas than they now inhabit. Neither are we limited to the means of transmission across the ocean that now exist, because we know that those means have varied greatly. During such extreme changes of conditions as are implied by glacial periods and by warm polar climates, great alterations of winds and of ocean currents are inevitable; and these are, as we have already proved, the two great agencies by which the transmission of living things to oceanic islands has been brought about. At the present time the southeast trade-winds blow almost constantly at St. Helena, and the ocean currents flow in the same direction, so that any transmission of insects by their means must almost certainly be from South Africa. Now there is undoubtedly a South African element in the insect fauna, but there is no less clearly a European, or at least a north temperate element, and this is very difficult to account for by causes now in action. But when we consider that this northern element is chiefly represented by remote generic affinity, and has therefore all the signs of great antiquity, we find a possible means of accounting for it. We have seen that during early Tertiary times an almost tropical climate extended far into the Northern Hemisphere, and a temperate climate to the arctic regions. But if at this time (as is not improbable) the antarctic regions were as much ice-clad as they are now, it is certain that an enormous change must have been produced in the winds. Instead of a great difference of temperature between each pole and the equator, the difference would be mainly between one hemisphere and the other, and this might so disturb the trade-winds as to bring St. Helena within the south temperate region of storms—a position corresponding to that of the Azores and Madeira in the North Atlantic, and thus subject it to violent gales from all points of the compass. At this remote epoch, the mountains of equatorial Africa may have been more extensive than they are now, and may have served as intermediate stations by which some northern insects may have migrated to the Southern Hemisphere.

We must remember, also, that these peculiar forms are said to be northern only because their nearest allies are now found in the North Atlantic islands and Southern Europe; but it is not at all improbable that they are really wide-spread Miocene types which have been preserved mainly in favorable insular stations. They may, therefore, have originally reached St. Helena from Southern Africa, or from some of the Atlantic islands, and may have been conveyed by oceanic currents as well as by winds.[1]

---

[1] On Petermann's map of Africa in the new edition of Stieler's "Hand-Atlas" (1879), the Island of Ascension is shown as seated on a much larger and shallower submarine bank than St. Helena. The 1000-fathom line round Ascension encloses an oval space 170 miles long by 70 wide, and even the 300-fathom line one over 60 miles long; and it is therefore probable that a much larger island once occupied this site. Now Ascension is nearly equidistant between St. Helena and Liberia, and such an island might have served as an intermediate station through which many of the immigrants to St. Helena passed. As the distances are hardly greater than in the case of the Azores, this removes whatever difficulty may have been felt of the possibility of *any* organisms reaching so remote an island. The present island of Ascension is probably only the summit of a huge volcanic mass, and any remnant of the original fauna and flora it might have preserved may have been destroyed by great volcanic eruptions. Mr. Darwin collected some masses of tufa which were found to be mainly organic, containing, besides remains of fresh-water infusoria, the siliceous tissue of plants! In the light of the great extent of the submarine bank on which

This is the more probable, as a large proportion of the St. Helena beetles live even in the perfect state within the stems of plants or trunks of trees, while the eggs and larvæ of a still larger number are likely to inhabit similar stations. Drift-wood might therefore be one of the most important agencies by which these insects reached the island.

Let us now see how far the distribution of other groups supports the conclusions derived from a consideration of the beetles. The Hemiptera have been studied by Dr. F. Buchanan White; and though far less known than the beetles, indicate somewhat similar relations. Eight out of twenty-one genera are peculiar, and the thirteen other genera are, for the most part, widely distributed, while one of the peculiar genera is of African type. The other orders of insects have not been collected or studied with sufficient care to make it worth while to refer to them in detail; but the land shells have been carefully collected and minutely described by Mr. Wollaston himself, and it is interesting to see how far they agree with the insects in their peculiarities and affinities.

Land Shells of St. Helena.—The total number of species is only twenty-nine, of which seven are common in Europe or the other Atlantic islands, and are, no doubt, recent introductions. Two others, though described as distinct, are so closely allied to European forms that Mr. Wollaston thinks they have probably been introduced and have become slightly modified by new conditions of life; so that there remain exactly twenty species which may be considered truly indigenous. No less than thirteen of these, however, appear to be extinct, being now only found on the surface of the ground or in the surface soil in places where the native forests have been destroyed and the land not cultivated. These twenty peculiar species belong to the following genera: Hyalina (3 sp.), Patula (4 sp.), Bulimus (7 sp.), Subulina (3 sp.), Succinea (3 sp.); of which one species of Hyalina, three of Patula, all the Bulimi, and two of Subulina are extinct. The

the island stands, Mr. Darwin's remark, that "we may feel sure that at some former epoch the climate and productions of Ascension were very different from what they are now," has received a striking confirmation. (See " Naturalist's Voyage round the World," p. 495.)

three Hyalinas are allied to European species, but all the rest appear to be highly peculiar, and to have no near allies with the species of any other country. Two of the Bulimi (*B. auris vulpinæ* and *B. Darwinianus*) are said to somewhat resemble Brazilian, New Zealand, and Solomon Island forms, while neither Bulimus nor Succinea occurs at all in the Madeira group.

Omitting the species that have probably been introduced by human agency, we have here indications of a somewhat recent immigration of European types which may perhaps be referred to the glacial period; and a much more ancient immigration from unknown lands, which must certainly date back to Miocene, if not to Eocene, times.

Absence of Fresh-water Organisms.—A singular phenomenon is the total absence of indigenous aquatic forms of life in St. Helena. Not a single water-beetle or fresh-water shell has been discovered; neither do there seem to be any water-plants in the streams except the common water-cress, one or two species of Cyperus, and the Australian *Isapis prolifera*. The same absence of fresh-water shells characterizes the Azores, where, however, there is one indigenous water-beetle. In the Sandwich Islands also recent observations refer to the absence of water-beetles, though here there are a few fresh-water shells. It would appear, therefore, that the wide distribution of the same generic and specific forms which so generally characterizes fresh-water organisms, and which has been so well illustrated by Mr. Darwin, has its limits in the *very remote* oceanic islands, owing to causes of which we are at present ignorant.

The other classes of animals in St. Helena need occupy us little. There are no indigenous mammals, reptiles, fresh-water fishes, or true land birds; but there is one species of wader—a small plover (*Ægialitis Sanctæ Helenæ*) very closely allied to a species found in South Africa, but presenting certain differences which entitle it to the rank of a peculiar species. The plants, however, are of especial interest from a geographical point of view, and we must devote a few pages to their consideration as supplementing the scanty materials afforded by the animal life, thus enabling us better to understand the biological relations and probable history of the island.

Native Vegetation of St. Helena.—Plants have certainly more varied and more effectual means of passing over wide tracts of ocean than any kinds of animals. Their seeds are often so minute, of such small specific gravity, or so furnished with downy or winged appendages, as to be carried by the wind for enormous distances. The bristles or hooked spines of many small fruits cause them to become easily attached to the feathers of aquatic birds, and they may thus be conveyed for thousands of miles by these pre-eminent wanderers; while many seeds are so protected by hard outer coats and dense inner albumen that months of exposure to salt water does not prevent them from germinating, as proved by the West Indian seeds that reach the Azores or even the west coast of Scotland, and, what is more to the point, by the fact stated by Mr. Melliss, that large seeds which have floated from Madagascar or Mauritius round the Cape of Good Hope have been thrown on the shores of St. Helena and have then sometimes germinated!

We have therefore little difficulty in understanding *how* the island was first stocked with vegetable forms. *When* it was so stocked (generally speaking) is equally clear. For, as the peculiar Coleopterous fauna, of which an important fragment remains, is mainly composed of species which are specially attached to certain groups of plants, we may be sure that the plants were there long before the insects could establish themselves. However ancient, then, is the insect fauna, the flora must be more ancient still. It must also be remembered that plants, when once established in a suitable climate and soil, soon take possession of a country, and occupy it almost to the complete exclusion of later immigrants. The fact of so many European weeds having overrun New Zealand and temperate North America may seem opposed to this statement, but it really is not so. For in both these cases the native vegetation has first been artificially removed by man and the ground cultivated; and there is no reason to believe that any similar effect would be produced by the scattering of any amount of foreign seed on ground already completely clothed with an indigenous vegetation. We might therefore conclude, *a priori*, that the flora of such an island as St. Helena would be of an excessively ancient type, preserving

for us in a slightly modified form examples of the vegetation
of the globe at the time when the island first rose above the
ocean.   Let us see, then, what botanists tell us of its character
and affinities.

The truly indigenous flowering plants are about fifty in number, besides twenty-six ferns.   Forty of the former and ten of
the latter are absolutely peculiar to the island, and, as Sir Joseph
Hooker tells us, " with scarcely an exception, cannot be regarded
as very close specific allies of any other plants at all.   Seventeen of them belong to peculiar genera, and of the others all
differ so markedly as species from their congeners that not one
comes under the category of being an insular form of a continental species."   The affinities of this flora are, Sir Joseph
Hooker thinks, mainly African and especially South African, as
indicated by the presence of the genera Phylica, Pelargonium,
Mesembryanthemum, Oteospermum, and Wahlenbergia, which
are eminently characteristic of southern extratropical Africa.
The sixteen ferns which are not peculiar are common either to
Africa, India, or America, a wide range sufficiently explained by
the dust-like spores of ferns, capable of being carried to unknown distances by the wind, and the great stability of their
generic and specific forms, many of those found in the Miocene
deposits of Switzerland being hardly distinguishable from living species.   This shows that identity of species of ferns between St. Helena and distant countries does not necessarily imply a recent origin.

The Relation of the St. Helena Compositæ.—In an elaborate
paper on the Compositæ,[1] Mr. Bentham gives us some valuable
remarks on the affinities of the seven endemic species belonging
to the genera Commidendron, Melanodendron, Petrobium, and
Pisiadia, which form so important a portion of the existing
flora of St. Helena.   He says, " Although nearer to Africa than
to any other continent, those composite denizens which bear evidence of the greatest antiquity have their affinities, for the most
part, in South America, while the colonists of a more recent
character are South African. . . . Commidendron and Melano-

---

[1] " Notes on the Classification, History, and Geographical Distribution of Compositæ," *Journal of the Linnæan Society*, Vol. XIII., p. 563 (1873).

dendron are among the woody Asteroid forms exemplified in the Andine Diplostephium, and in the Australian Olearia. Petrobium is one of three genera, remains of a group probably of great antiquity, of which the two others are Podanthus in Chili and Astemma in the Andes. The Pisiadia is an endemic species of a genus otherwise Mascarene or of Eastern Africa, presenting a geographical connection analogous to that of the St. Helena Melhaniæ[1] with the Mascarene Trochetia."

Whenever such remote and singular cases of geographical affinity as the above are pointed out, the first impression is to imagine some mode by which a communication between the distant countries implicated might be effected; and this way of viewing the problem is almost universally adopted, even by naturalists. But if the principles laid down in this work and in my "Geographical Distribution of Animals" are sound, such a course is very unphilosophical. For, on the theory of evolution, nothing can be more certain than that groups now broken up and detached were once continuous, and that fragmentary groups and isolated forms are but the relics of once wide-spread types, which have been preserved in a few localities where the physical conditions were especially favorable, or where organic competition was less severe. The true explanation of all such remote geographical affinities is that they date back to a time when the ancestral group of which they are the common descendants had a wider or a different distribution; and they no more imply any closer connection between the distant countries the allied forms now inhabit than does the existence of living Equidæ in South Africa and extinct Equidæ in the Pliocene deposits of the Pampas imply a continent bridging the South Atlantic to allow of their easy communication.

Concluding Remarks on St. Helena.—The sketch we have now given of the chief members of the indigenous fauna and flora of St. Helena shows that by means of the knowledge we have obtained of past changes in the physical history of the earth, and of the various modes by which organisms are conveyed across

---

[1] The Melhaniæ comprise the two finest timber trees of St. Helena, now almost extinct—the red-wood and native ebony.

the ocean, all the more important facts become readily intelligible. We have here an island of small size and great antiquity, very distant from every other land, and probably at no time very much less distant from surrounding continents, which became stocked by chance immigrants from other countries at some remote epoch, and which has preserved many of their more or less modified descendants to the present time. When first visited by civilized man, it was in all probability far more richly stocked with plants and animals, forming a kind of natural museum or vivarium in which ancient types, perhaps dating back to the Miocene period, or even earlier, had been saved from the destruction which has overtaken their allies on the great continents. Unfortunately, many—we do not know how many—of these forms have been exterminated by the carelessness and improvidence of its civilized but ignorant rulers; and it is only by the extreme ruggedness and inaccessibility of its peaks and crater-ridges that the scanty fragments have escaped by which alone we are able to obtain a glimpse of this interesting chapter in the life-history of our earth.

CHAPTER XV.

THE SANDWICH ISLANDS.

Position and Physical Features.—Zoology of the Sandwich Islands.—Birds.—Reptiles.—Land Shells.—Insects.—Vegetation of the Sandwich Islands.—Peculiar Features of the Hawaiian Flora.—Antiquity of the Hawaiian Fauna and Flora.—Concluding Observations on the Fauna and Flora of the Sandwich Islands.—General Remarks on Oceanic Islands.

THE Sandwich Islands are an extensive group of large islands situated in the centre of the North Pacific, being 2350 miles from the nearest part of the American coast—the Bay of San Francisco—and about the same distance from the Marquesas and the Samoa Islands to the south, and the Aleutian Islands a little west of north. They are therefore wonderfully isolated in mid-ocean, and are only connected with the other Pacific islands by widely scattered coral reefs and atolls, the nearest of which, however, are six or seven hundred miles distant, and are all nearly destitute of animal or vegetable life. The group consists of seven large inhabited islands besides four rocky islets; the largest, Hawaii, being seventy miles across, and having an area of 3800 square miles—being somewhat larger than all the other islands together. A better conception of this large island will be formed by comparing it with Devonshire, with which it closely agrees both in size and shape, though its enormous volcanic mountains rise to nearly 14,000 feet. Three of the smaller islands are each about the size of Hertfordshire or Bedfordshire, and the whole group stretches from northwest to southeast for a distance of about 350 miles. Though so extensive, the entire archipelago is volcanic, and the largest island is rendered sterile and comparatively uninhabitable by its three active volcanoes and their wide-spread deposits of lava.

The ocean depths by which these islands are separated from the nearest continents are enormous. North, east, and south, soundings have been obtained a little over or under 3000 fathoms, and these profound deeps extend over a large part of the North Pacific. We may be quite sure, therefore, that the Sandwich Islands have during their whole existence been as completely severed from the great continents as they are now; but on the west and south there is a possibility of more extensive

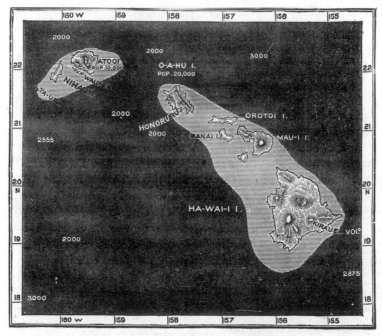

MAP OF THE SANDWICH ISLANDS.

The light tint shows where the sea is less than 1000 fathoms deep.
The figures show the depth in fathoms.

islands having existed, serving as stepping-stones to the island groups of the mid-Pacific. This is indicated by a few widely scattered coral islets, around which extend considerable areas of less depth, varying from 200 to 1000 fathoms, and which *may* therefore indicate the sites of submerged islands of considerable extent. When we consider that east of New Zealand and New Caledonia all the larger and loftier islands are of volcanic origin, with no trace of any ancient stratified rocks (except, perhaps, in the Marquesas, where, according to Jules Marcou, granite and gneiss are said to occur), it seems probable that the innumerable coral reefs and atolls, which occur in groups on deeply submerged banks, mark the sites of bygone volcanic islands similar

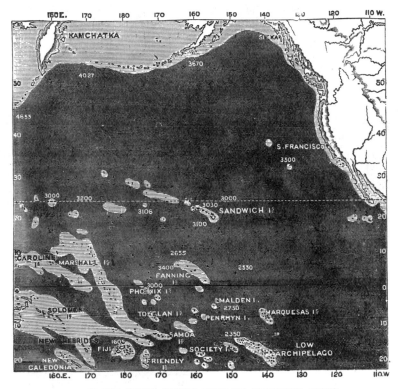

MAP OF THE NORTH PACIFIC WITH ITS SUBMERGED BANKS.

The light tint shows where the sea is less than 1000 fathoms deep.
The dark   "   "   "   more   "   "   "
The figures show the depth in fathoms.

to those which now exist, but which, after becoming extinct, have been lowered or destroyed by denudation, and finally, by subsidence of the earth's crust, have altogether disappeared, except where their sites are indicated by the upward-growing coral reefs. If this view is correct, we should give up all idea of there ever having been a Pacific continent, but should look upon that vast ocean as having from the remotest geological epochs been the seat of volcanic forces, which from its profound depths have gradually built up the islands which now dot its surface, as well as many others which have sunk beneath its waves. The number of islands, as well as the total quantity of land surface, may sometimes have been greater than it is now, and may thus have facilitated the transfer of organisms from one group to another, and, more rarely, even from the American, Asiatic, or Australian continent. Keeping these various facts and considerations in view, we may now proceed to examine the fauna and flora of the Sandwich Islands, and discuss the special phenomena they present.

Zoology of the Sandwich Islands: Birds.—It need hardly be said that indigenous mammalia are quite unknown in the Sandwich Islands, the most interesting of the higher animals being the birds, which are tolerably numerous and highly peculiar. Many aquatic and wading birds which range over the whole Pacific visit these islands, twenty-four species having been observed; but even of these five are peculiar—a coot, *Fulica alai;* a moor-hen, *Gallinula Sandvichensis;* a rail with rudimentary wings, *Pennula millei;* and two ducks, *Anas Wyvilliana* and *Bernicla Sandvichensis.* The birds of prey are also great wanderers. Four have been found in the islands—the short-eared owl, *Otus brachyotus,* which ranges over the greater part of the globe, but is here said to resemble the variety found in Chili and the Galapagos; the barn-owl, *Strix flammea,* of a variety common in the Pacific; a peculiar sparrow-hawk, *Accipiter Hawaii;* and *Buteo solitarius,* a buzzard of a peculiar species, and colored so as to resemble a hawk of the American sub-family Polyborinæ. It is to be noted that the genus Buteo abounds in America, but is not found in the Pacific; and this fact, combined with the remarkable coloration, renders it almost certain that this peculiar species is of American origin.

Coming now to the Passeres, or true perching-birds, we find sixteen species, all peculiar, belonging to ten genera, all but one of which are also peculiar. The following is a list of these extremely interesting birds:

I. Muscicapidæ (Flycatchers).
1. *Chasiempis Sandvichensis.*
2. *Phœornis obscura.*

II. Meliphagidæ (Honeysuckers).
3. *Mohoa nobilis.*
4.   "   *braccata.*
5.   "   *apicalis.*
6. *Chætoptila angustipluma.*

III. Drepanididæ.
7. *Drepanis coccinea.*
8.   "   *rosea.*
9.   "   *flava.*
10.   "   *sanguinea.*

Drepanididæ—*Continued.*
11. *Hemignathus olivaceus.*
12.   "   *obscurus.*
13.   "   *lucidus.*
14. *Loxops coccinea.*
15.   "   *aurea.*
16. *Loxioides bailloni.*
17. *Psittirostra psittacea.*
18. *Fringilla anna* (recently described, perhaps belongs also to this group).

IV. Corvidæ (Crows).
19. *Corvus Hawaiensis.*

Taking the above in the order here given, we have, first, two peculiar genera of flycatchers, a family confined to the Old World, but extending over the Pacific as far as the Marquesas Islands. Next we have two peculiar genera (with four species) of honeysuckers, a family confined to the Australian region, and also ranging over all the Pacific islands to the Marquesas. We now come to the most important group of birds in the Sandwich Islands, comprising five peculiar genera, and eleven or twelve species, which are believed to form a peculiar family allied to the Oriental flower-peckers (Diceidæ), and perhaps remotely to the American greenlets (Vireonidæ) or tanagers (Tanagridæ). They possess singularly varied beaks, some having this organ much thickened like those of finches, to which family some of them have been supposed to belong. In any case, they form a most peculiar group, and cannot be associated with any other known birds. The last species, and the only one not belonging to a peculiar genus, is the Hawaiian crow, belonging to the almost universally distributed genus Corvus.

On the whole, the affinities of these birds are, as might be expected, chiefly with Australia and the Pacific Islands; but they exhibit in the buzzard, one of the owls, and perhaps in some of the Drepanididæ, slight indications of very rare or very remote communication with America. The amount of speciality is, however, wonderful, far exceeding that of any other islands; the only approach to it being made by New Zealand and Madagascar, which have a much more varied bird fauna and a smaller

proportionate number of peculiar genera. These facts undoubtedly indicate an immense antiquity for this group of islands, or
the vicinity of some very ancient land (now submerged), from
which some portion of their peculiar fauna might be derived.

Reptiles.—The only other vertebrate animals are two lizards.
One of these is a very wide-spread species, *Ablepharus pacilo-
pleurus*, said by Dr. Günther to be found in Timor, Australia,
the Samoa Islands, and the Sandwich Islands. It seems hardly
likely that such a range can be due to natural causes. The
other is said to form a peculiar genus of geckoes, but both its
locality and affinities appear to be somewhat doubtful.

Land Shells.—The only other group of animals which has
been carefully studied, and which presents features of especial interest, are the land shells. These are very numerous, about thirty
genera and between three and four hundred species having been
described; and it is remarkable that this single group contains
as many species of land shells as all the other Polynesian islands
from the Pelew Islands and Samoa to the Marquesas. All the
species are peculiar, and about three fourths of the whole belong
to peculiar genera, fourteen of which constitute the sub-family
Achatinellinæ, entirely confined to this group of islands and constituting its most distinguishing feature. Thirteen genera (comprising sixty-four species) are found also in the other Polynesian
islands, but three genera of Auriculidæ (Plecotrema, Pedipes,
and Blauneria) are not found in the Pacific, but inhabit—the
former genus Australia, China, Bourbon, and Cuba, the two latter the West Indian Islands. Another remarkable peculiarity
of these islands is the small number of Operculata, which are
represented by only one genus and five species, while the other
Pacific islands have twenty genera and 115 species, or more
than half the number of the Inoperculata. This difference is
so remarkable that it is worth stating in a comparative form:

	Inoperculata.	Operculata.	Auriculidæ.
Sandwich Islands	332	5	9
Rest of Pacific islands	200	115	16

When we remember that in the West Indian Islands the
Operculata abound in a greater proportion than even in the
Pacific islands generally, we are led to the conclusion that lime-

stone, which is plentiful in both these areas, is especially favorable to them, while the purely volcanic rocks are especially unfavorable. The other peculiarities of the Sandwich Islands, however, such as the enormous preponderance of the strictly endemic Achatinellinæ, and the presence of genera which occur elsewhere only beyond the Pacific area in various parts of the great continents, undoubtedly point to a very remote origin, at a time when the distribution of many of the groups of Mollusca was very different from that which now prevails.

A very interesting feature of the Sandwich group is the extent to which the species and even the genera are confined to separate islands. Thus the genera Carelia and Catinella, with eight species, are peculiar to the island of Kaui; Bulimella, Apex, Frickella, and Blauneria to Oahu; Perdicella to Maui; and Eburnella to Lanai. The Rev. John T. Gulick, who has made a special study of the Achatinellinæ, informs us that the average range of the species in this sub-family is five or six miles, while some are restricted to but one or two square miles, and only very few have the range of a whole island. Each valley, and often each side of a valley, and sometimes even every ridge and peak, possesses its peculiar species.[1] The island of Oahu, in which the capital is situated, has furnished about half the species already known. This is partly due to its being more forest-clad, but also, no doubt, in part to its being better explored; so that, notwithstanding the exceptional riches of the group, we have no reason to suppose that there are not many more species to be found in the less explored islands. Mr. Gulick tells us that the forest region that covers one of the mountain-ranges of Oahu is about forty miles in length, and five or six miles in width, yet this small territory furnishes about 175 species of Achatinellinæ, represented by 700 or 800 varieties. The most important peculiar genus, not belonging to the Achatinella group, is Carelia, with six species and several named varieties, all peculiar to Kaui, the most westerly of the large islands. This would seem to show that the small islets stretching westward,

---

[1] *Journal of the Linnæan Society*, 1873, p. 496, "On Diversity of Evolution under one Set of External Conditions." *Proceedings of the Zoological Society of London*, 1873, p. 80, "On the Classification of the Achatinellinæ."

and situated on an extensive bank with less than a thousand fathoms of water over it, may indicate the position of a large submerged island whence some portion of the Sandwich Island fauna was derived.

Insects.—Unfortunately we have as yet no such knowledge of the insects of these islands as we possess in the case of the Azores and St. Helena, but some considerable collections have been sent over by Mr. T. Blackburn, now resident there, and we may therefore soon possess fuller and more accurate information. Although insects are said to be very scarce, yet all the chief tribes of Coleoptera appear to be represented, though as yet by very few species. These appear to be, for the most part, peculiar, but to have wide-spread affinities. The majority, as might be expected, are allied to Polynesian, Australian, or Malayan forms; some few are South American (perhaps introduced), while others show north temperate affinities. There are several new genera, and one peculiar group of six species is said to form a new family allied to the Anthribidæ. A new genus of Lucanidæ is said to be allied to a Chilian genus. If we consider the greater facilities of insects for dispersal when compared with birds or land shells, the characteristics of the insect fauna, so far as yet known, are sufficiently in harmony with the amount of speciality and isolation presented by the latter groups.

Vegetation of the Sandwich Islands.—The flora of these islands is in many respects so peculiar and remarkable, and so well supplements the information derived from its interesting but scanty fauna, that a brief account of its more striking features will not be out of place; and we fortunately have a pretty full knowledge of it, owing to the researches of the American botanist Horace Mann, and of Dr. Pickering, who accompanied the United States Exploring Expedition.

Considering their extreme isolation, their uniform volcanic soil, and the large proportion of the chief island which consists of barren lava-fields, the flora of the Sandwich Islands is extremely rich, consisting, so far as at present known, of 554 species of flowering plants and 135 ferns. This is considerably richer than the Azores (439 Phanerogams and 39 ferns), which, though less extensive, are far better known, or than the Galapagos (332 Phanerogams), which are more strictly comparable, being equally volcanic, while their somewhat smaller area may perhaps be compensated by their proximity to the American continent. Even New Zealand, with more than twenty times the area of the Sandwich group, whose soil and climate are much more varied, and whose botany has been thoroughly explored, has not double the number of flowering plants (935 species), while in ferns it is barely equal.

Peculiar Features of the Flora.—This rich insular flora is, wonderfully peculiar, for if we deduct 69 species, which are believed to have been introduced by man, there remain 620 species, of which 377, or more than three fifths, are quite peculiar to the islands. There are no less than 39 peculiar genera out of a total of 253, and these 39 genera comprise 153 species, so that the most isolated forms are those which most abound and thus give a special character to the flora. Besides these peculiar types, several genera of wide range are here represented by highly peculiar species. Such are Lobelia, the Hawaiian species of which are woody shrubs from six to twenty feet high, one even being a tree reaching a height of forty feet. Shrubby geraniums fifteen feet high grow as epiphytes on forest trees, as do some Vacciniums and Epacrids. Violets and plantains also form tall shrubby plants, and there are many strange arborescent Compositæ, as in other oceanic islands.

The affinities of the flora generally are very wide. Although there are many Polynesian groups, yet Australian, New Zealand, and American forms are equally represented. Dr. Pickering notes the total absence of a large number of families found in Southern Polynesia, such as Dilleniaceæ, Anonaceæ, Olacaceæ, Aurantiaceæ, Guttiferæ, Malpighiaceæ, Meliaceæ, Combretaceæ, Rhizophoraceæ, Melastomaceæ, Passifloraceæ, Cunoniaceæ, Jasminaceæ, Acanthaceæ, Myristicaceæ, Casuaraceæ, Scitamineæ, and Aracæ, as well as the genera Clerodendrum, Ficus, and epidendric orchids. Australian affinities are shown by the genera Exocarpus, Cyathodes, Melicope, Pittosporum, and by a phyllodinous Acacia. New Zealand is represented by Ascarina, Coprosma, Acæna, and several Cyperaceæ; while America is represented by the genera Nama, Gunnera, Phyllostegia, Sisyrin-

chium, and by a red-flowered Rubus and a yellow-flowered Sanic-
ula allied to Oregon species.

There is no true alpine flora on the higher summits, but several of the temperate forms extend to a great elevation. Thus
Mr. Pickering records Vaccinium, Ranunculus, Silene, Gnaphalium, and Geranium as occurring above ten thousand feet elevation; while Viola, Drosera, Acæna, Lobelia, Edwardsia, Do-
donæa, Lycopodium, and many Compositæ range above six
thousand feet. Vaccinium and Silene are very interesting, as
they are peculiar to the north temperate zone, except one Silene
in South Africa.

The proportionate abundance of the different families in this
interesting flora is as follows:

1.	Compositæ	47 species.	11.	Piperaceæ	12 species.
2.	Cyperaceæ	39 "	12.	Convolvulaceæ	12 "
3.	Lobeliaceæ	35 "	13.	Malvaceæ	12 "
4.	Rubiaceæ	33 "	14.	Amarantaceæ	9 "
5.	Labiatæ	27 "	15.	Araliaceæ	8 "
6.	Leguminosæ	20 "	16.	Violaceæ	6 "
7.	Rutaceæ	17 "	17.	Pittosporaceæ	6 "
8.	Caryophyllaceæ	14 "	18.	Myrtaceæ	6 "
9.	Gesneriaceæ	14 "	19.	Goodeniaceæ	6 "
10.	Urticaceæ	13 "	20.	Thymelaceæ	6 "

Four other orders—Geraniaceæ, Rhamnaceæ, Rosaceæ, and
Cucurbitaceæ—have five species each; and among the more important orders which have less than five species each are Ranunculaceæ, Ericaceæ, Primulaceæ, Polygonaceæ, Orchidaceæ, and
Juncaceæ. In the above enumeration the grasses (Graminaceæ)
are omitted, as they were not described at the time Mr. Mann's
article was written. The most remarkable feature here is the
great abundance of Lobeliaceæ, a character of the flora which is
probably unique; while the superiority of Labiatæ to Leguminosæ and the scarcity of Rosaceæ and Orchidaceæ are also very
unusual. Composites, as in most temperate floras, stand at the
head of the list, and as these have been carefully studied by Mr.
Bentham, it will be interesting to note the affinities which they
indicate. Omitting four genera and species which are cosmopolitan, and have no doubt entered with civilized man, there re-

main twelve genera and forty-four species of Compositæ in the islands. All the species are peculiar, as are six of the genera; and in another genus, Coreopsis, the six species form a peculiar named section or sub-genus, Campylotheca; while the genus Lipochæta, with ten species, is only known elsewhere in the Galapagos, where a single species occurs. We may therefore consider that eight out of the twelve genera of Hawaiian Compositæ are really confined to the archipelago. The relations of the genera are thus given by Mr. Bentham :

	No. of Species.	External Relations of the Species.
Lagenophora	1	With the Old World and Extratropical America.
Aster	1	American and Extratropical Old World.
Tetramolobium	6	South Extratropical American.
Vittadinia	1	South Extratropical American and Australian.
Campylotheca (s.g.)	6	With the Tropical American and very few Old World species of Coreopsis and Bidens.
Bidens	1	The Tropical American species.
Lipochæta	10	American Wedelioidæ and Helianthioidæ.
Argyroxiphium	2	With Madieæ of the Mexican region.
Wilkesia	1	With Madieæ of the Mexican region.
Dubantia	3	Distantly with Madieæ and Galinsogeæ of the Mexican region.
Raillardia	11	With Raillardella of the Mexican region.
Hesperomannia	1	With Stifftia and Wunderlichia of the Brazilian region.

The great preponderance of American relations of the Compositæ, as above indicated, is very interesting and suggestive. It is here that we meet with some of the most isolated and remarkable forms, implying great antiquity; and when we consider the enormous extent and world-wide distribution of this order (comprising ten thousand species), its distinctness from all others, the great specialization of its flowers to attract insects, and of its seeds for dispersal by wind and other means, we can hardly doubt that its origin dates back to a very remote epoch. We may therefore look upon the Compositæ as representing the most ancient portion of the existing flora of the Sandwich Islands, carrying us back to a very remote period when the facilities for communication with America were greater than they are now. This may be indicated by the two deep submarine banks in the North Pacific, between the Sandwich Islands and

San Francisco, which, from an ocean-floor nearly 3000 fathoms deep, rise up to within a few hundred fathoms of the surface, and seem to indicate the subsidence of two islands, each about as large as Hawaii. The plants of north temperate affinity may be nearly as old, but these may have been derived from Northern Asia by way of Japan and the extensive line of shoals which run northwestward from the Sandwich Islands as shown on our map. Those which exhibit Polynesian or Australian affinities, consisting, for the most part, of less highly modified species usually of the same genera, may have had their origin at a later, though still somewhat remote, period, when large islands, indicated by the extensive shoals to the south and southwest, offered facilities for the transmission of plants from the tropical portions of the Pacific Ocean.

Antiquity of the Hawaiian Fauna and Flora.—The great antiquity implied by the peculiarities of the fauna and flora, no less than by the geographical conditions and surroundings, of this group will enable us to account for another peculiarity of its flora—the absence of so many families found in other Pacific islands. For the earliest immigrants would soon occupy much of the surface, and become specially modified in accordance with the conditions of the locality, and these would serve as a barrier against the intrusion of many forms which at a later period spread over Polynesia. The extreme remoteness of the islands, and the probability that they have always been more isolated than those of the Central Pacific, would also necessarily result in an imperfect and fragmentary representation of the flora of surrounding lands.

Concluding Observations on the Fauna and Flora of the Sandwich Islands.—The indications thus afforded by a study of the flora seem to accord well with what we know of the fauna of the islands. Plants, having so much greater facilities for dispersal than animals, and also having greater specific longevity and greater powers of endurance under adverse conditions, exhibit in a considerable degree the influence of the primitive state of the islands and their surroundings; while members of the animal world, passing across the sea with greater difficulty, and subject to extermination by a variety of adverse conditions, retain much more of the impress of a recent state of things, with, perhaps, here and there an indication of that ancient approach to America so clearly shown in the Compositæ and some other portions of the flora.

General Remarks on Oceanic Islands.—We have now reviewed the main features presented by the assemblages of organic forms which characterize the more important and best-known of the oceanic islands. They all agree in the total absence of indigenous mammalia and amphibia ; while their reptiles, when they possess any, do not exhibit indications of extreme isolation and antiquity. Their birds and insects present just that amount of specialization and diversity from continental forms which may be best explained by the known means of dispersal acting through long periods ; their land shells indicate greater isolation, owing to their admittedly less effective means of conveyance across the ocean ; while their plants show most clearly the effects of those changes of conditions which we have reason to believe have occurred during the Tertiary epoch, and preserve to us in highly specialized and archaic forms some record of the primeval immigration by which the islands were originally clothed with vegetation. But in every case the series of forms of life in these islands is scanty and imperfect as compared with far less favorable continental areas, and no one of them presents such an assemblage of animals or plants as we always find in an island which we know has once formed part of a continent.

It is still more important to note that none of these oceanic archipelagoes present us with a single type which we may suppose to have been preserved from Mesozoic times ; and this fact, taken in connection with the volcanic or coralline origin of all of them, powerfully enforces the conclusion at which we have arrived in the earlier portion of this volume, that during the whole period of geologic time, as indicated by the fossiliferous rocks, our continents and oceans have, speaking broadly, been permanent features of our earth's surface. For had it been otherwise—had sea and land changed place repeatedly, as was once supposed ; had our deepest oceans been the seat of great continents, while the site of our present continents was occupied by an oceanic abyss—is it possible to imagine that no fragments of such continents would remain in the present oceans, bringing down to us some of their ancient forms of life preserved with but little change? The correlative facts that the islands of our great oceans are all volcanic (or coralline built, probably, upon degraded and submerged volcanic islands), and that their productions are all more or less clearly related to the existing inhabitants of the nearest continents, are hardly consistent with any other theory than the permanence of oceanic and continental areas.

We may here refer to the one apparent exception, which, however, lends additional force to the argument. New Zealand is sometimes classed as an oceanic island, but it is not so really; and we shall discuss its peculiarities and probable origin further on.

CHAPTER XVI.

CONTINENTAL ISLANDS OF RECENT ORIGIN: GREAT BRITAIN.

Characteristic Features of Recent Continental Islands.—Recent Physical Changes of the British Isles.—Proofs of Former Elevation.—Submerged Forests.—Buried River Channels.—Time of Last Union with the Continent.—Why Britain is Poor in Species.—Peculiar British Birds.—Fresh-water Fishes.—Cause of Great Speciality in Fishes.—Peculiar British Insects.—Lepidoptera Confined to the British Isles.—Peculiarities of the Isle of Man Lepidoptera.—Coleoptera Confined to the British Isles.—Trichoptera Peculiar to the British Isles.—Land and Fresh-water Shells.—Peculiarities of the British Flora.—Peculiarities of the Irish Flora.—Peculiar British Mosses and Hepaticæ.—Concluding Remarks on the Peculiarities of the British Fauna and Flora.

WE now proceed to examine those islands which are the very reverse of the "oceanic" class, being fragments of continents or of larger islands from which they have been separated by subsidence of the intervening land at a period which, geologically, must be considered recent. Such islands are always still connected with their parent land by a shallow sea, usually, indeed, not exceeding a hundred fathoms deep; they always possess mammalia and reptiles either wholly or in large proportion identical with those of the mainland; while their entire flora and fauna are characterized either by the total absence or comparative scarcity of those endemic or peculiar species and genera which are so striking a feature of all oceanic islands. Such islands will, of course, differ from each other in size, in antiquity, and in the richness of their respected faunas, as well as in their distance from the parent land and the facilities for intercommunication with it; and these diversities of conditions will manifest themselves in the greater or less amount of speciality of their animal productions.

This speciality, when it exists, may have been brought about in two ways. A species or even a genus may on a continent have a very limited area of distribution, and this area may be wholly, or almost wholly, contained in the separated portion or island, to which it will henceforth be peculiar. Even when the area occupied by a species is pretty equally divided at the time of separation between the island and the continent, it may happen that it will become extinct on the latter, while it may survive on the former, because the limited number of individuals after division may be unable to maintain themselves against the severer competition or more contrasted climate of the continent, while they may flourish under the more favorable insular conditions. On the other hand, when a species continues to exist in both areas, it may on the island be subjected to some modifications ·by the altered conditions, and may thus come to present characters which differentiate it from its continental allies and constitute it a new species. We shall in the course of our survey meet with cases illustrative of both these processes.

The best examples of recent continental islands are Great Britain and Ireland, Japan, Formosa, and the larger Malay islands, especially Borneo, Java, and Celebes; and as each of these presents special features of interest, we will give a short outline of their zoology and past history in relation to that of the continents from which they have recently been separated, commencing with our own islands, to which the present chapter will be devoted.

Recent Physical Changes in the British Isles.—Great Britain is perhaps the most typical example of a large and recent continental island now to be found upon the globe. It is joined to the continent by a shallow bank which extends from Denmark to the Bay of Biscay, the 100-fathom line from these extreme points receding from the coasts so as to include the whole of the British Isles and about fifty miles beyond them to the westward. (See map.) Beyond this line the sea deepens rapidly to the 500 and 1000 fathom lines, the distance between 100 and 1000 fathoms being from twenty to fifty miles, except where there is a great outward curve to include the Porcupine Bank, 170 miles west of Galway, and to the northwest of Caithness, where a narrow ridge less than 500 fathoms below the surface joins the extensive bank under 300 fathoms, on which are situated the Faroe Islands and Iceland, and which stretches across to Greenland. In the North Channel between Ireland and Scotland, and in the Minch between the outer Hebrides and Skye, are a series of hollows in the sea-bottom from 100 to 150 fathoms

MAP SHOWING THE SHALLOW BANK CONNECTING THE BRITISH ISLES WITH THE CONTINENT.

The light tint indicates a depth of less than 100 fathoms.
The figures show the depth in fathoms.
The narrow channel between Norway and Denmark is 2580 feet deep.

deep. These correspond exactly to the points between the opposing highlands where the greatest accumulations of ice would necessarily occur during the glacial epoch, and they may well be termed submarine lakes, of exactly the same nature as those which occur in similar positions on land.

Proofs of Former Elevation—Submerged Forests.—What renders Britain particularly instructive as an example of a recent continental island is the amount of direct evidence that exists, of several distinct kinds, showing that the land has been sufficiently elevated (or the sea depressed) to unite it with the continent—and this at a very recent period. The first class of evidence is the existence, all round our coasts, of the remains of submarine forests often extending far below the present low-water mark. Such are the submerged forests near Torquay in Devonshire, and near Falmouth in Cornwall, both containing stumps of trees in their natural position rooted in the soil, with deposits of peat, branches, and nuts, and often with remains of insects and other land animals. These occur in very different conditions and situations, and some have been explained by changes in the height of the tide, or by pebble banks shutting out the tidal waters from estuaries; but there are numerous examples to which such hypotheses cannot apply, and which can only be explained by an actual subsidence of the land (or rise of the sea-level) since the trees grew.

We cannot give a better idea of these forests than by quoting the following account by Mr. Pengelly of a visit to one which had been exposed by a violent storm on the coast of Devonshire, at Blackpool, near Dartmouth:

" We were so fortunate as to reach the beach at spring-tide low water, and to find, admirably exposed, by far the finest example of a submerged forest which I have ever seen. It occupied a rectangular area, extending from the small river or stream at the western end of the inlet about one furlong eastward, and from the low-water line thirty yards up the strand. The lower or seaward portion of the forest area, occupying about two thirds of its entire breadth, consisted of a brownish drab-colored clay, which was crowded with vegetable débris, such as small twigs, leaves, and nuts. There were also numerous prostrate trunks and branches of trees, lying partly embedded in the clay, without anything like a prevalent direction. The trunks varied from six inches to upwards of two feet in diameter.

Much of the wood was found to have a reddish or bright-pink hue when fresh surfaces were exposed. Some of it, as well as many of the twigs, had almost become a sort of ligneous pulp; while other examples were firm, and gave a sharp crackling sound on being broken. Several large stumps projected above the clay in a vertical direction, and sent roots and rootlets into the soil in all directions and to considerable distances. It was obvious that the movement by which the submergence was effected had been so uniform as not to destroy the approximate horizontality of the old forest ground. One fine example was noted of a large prostrate trunk having its roots still attached, some of them sticking up above the clay, while others were buried in it. Hazel-nuts were extremely abundant—some entire, others broken, and some obviously gnawed. . . . It has been stated that the forest area reached the spring-tide low-water line; hence as the greatest tidal range on this coast amounts to eighteen feet, we are warranted in inferring that the subsidence amounted to eighteen feet as a minimum, even if we suppose that some of the trees grew in a soil the surface of which was not above the level of high water. There is satisfactory evidence that in Torbay it was not less than forty feet, and that in Falmouth harbor it amounted to at least sixty-seven feet."[1]

On the coast of the Bristol Channel similar deposits occur, as well as along much of the coast of Wales and in Holyhead harbor. It is believed by geologists that the whole Bristol Channel was, at a comparatively recent period, an extensive plain, through which flowed the river Severn; for, in addition to the evidence of submerged forests, there are on the coast of Glamorganshire numerous caves and fissures in the face of high sea-cliffs, in one of which no less than a thousand antlers of the reindeer were found, the remains of animals which had been devoured there by bears and hyenas—facts which can only be explained by the existence of some extent of dry land stretching seaward from the present cliffs, but since submerged and washed away. This plain may have continued down to very recent times, since the whole of the Bristol Channel to beyond Lundy Island is under

---

[1] *Geological Magazine*, 1870, p. 165.

twenty-five fathoms deep. In the East of England we have a similar forest-bed at Cromer in Norfolk; and in the North of Holland an old land surface has been found fifty-six feet below high-water mark.

Buried River Channels.—Still more remarkable are the buried river channels which have been traced on many parts of our coasts. In order to facilitate the study of the glacial deposits of Scotland, Dr. James Croll obtained the details of about two hundred and fifty bores put down in all parts of the mining districts of Scotland for the purpose of discovering minerals.[1] These revealed the interesting fact that there are ancient valleys and river channels at depths of from 100 to 260 feet below the present sea-level. These old rivers sometimes run in quite different directions from the present lines of drainage, connecting what are now distinct valleys; and they are so completely filled up and hidden by boulder clay, drift, and sands that there is no indication of their presence on the surface, which often consists of mounds or low hills more than 100 feet high. One of these old valleys connects the Clyde near Dumbarton with the Forth at Grangemouth, and appears to have contained two streams flowing in opposite directions from a water-shed about midway at Kilsith. At Grangemouth the old channel is 260 feet below the sea-level. The water-shed at Kilsith is now 160 feet above the sea—the old valley-bottom being 120 feet deep, or 40 feet above the sea. In some places the old valley was a ravine with precipitous rocky walls, which have been found in mining excavations. Dr. Geikie, who has himself discovered many similar buried valleys, is of opinion that "they unquestionably belong to the period of the boulder clay."

We have here a clear proof that when these rivers were formed the land must have stood in relation to the sea *at least* 260 feet higher than it does now, and probably much more; and this is sufficient to join England to the continent. Supporting this evidence, we have fresh-water or littoral shells found at great depths off our coasts. Mr. Godwin Austen records the dredging-up of a fresh-water shell (*Unio pictorum*) off the mouth of the

---

[1] *Transactions of the Edinburgh Geological Society*, Vol. I., p. 330.

English Channel between the 50-fathom and 100-fathom lines, while in the same locality gravel banks with littoral shells now lie under sixty or seventy fathoms water.[1] More recently Mr. Gwyn Jeffreys has recorded the discovery of eight species of fossil arctic shells off the Shetland Isles in about ninety fathoms water, all being characteristic shallow-water species, so that their association at this great depth is a distinct indication of considerable subsidence.[2]

Time of Last Union with the Continent.—The period when this last union with the continent took place was comparatively recent, as shown by the identity of the shells with living species, and the fact that the buried river channels are all covered with clays and gravels of the glacial period, of such a character as to indicate that most of them were deposited above the sea-level. From these and various other indications geologists are all agreed that the last continental period, as it is called, was subsequent to the greatest development of the ice, but probably before the cold epoch had wholly passed away. But if so recent, we should naturally expect our land still to show an almost perfect community with the adjacent parts of the continent in its natural productions; and such is found to be the case. All the higher and more perfectly organized animals are, with but few exceptions, identical with those of France and Germany; while the few species still considered to be peculiar may be accounted for either by an original local distribution, by preservation here owing to favorable insular conditions, or by slight modifications having been caused by these conditions resulting in a local race, sub-species, or species.

Why Britain is Poor in Species.—The former union of our islands with the continent is not, however, the only recent change they have undergone. There is equally good evidence that a considerable portion, if not the entire area, had been submerged to a depth of nearly 2000 feet (see Chap. IX., p. 166), at which time only what are now the highest mountains would remain as groups of rocky islets. This submersion must have destroyed

---

[1] *Quarterly Journal of Geological Society,* 1850, p. 96.
[2] "British Association Report," Dundee, 1867, p. 431.

the greater part of the life of our country; and as it certainly
occurred during the latter part of the glacial epoch, the subsequent elevation and union with the continent cannot have been
of very long duration, and this fact must have had an important
bearing on the character of the existing fauna and flora of Britain. We know that just before and during the glacial period
we possessed a fauna almost or quite identical with that of adjacent parts of the continent, and equally rich in species. The
submergence destroyed this fauna; and the permanent change
of climate on the passing-away of the glacial conditions appears
to have led to the extinction or migration of many species in
the adjacent continental areas, where they were succeeded by the
assemblage of animals now occupying Central Europe. When
England became continental, these entered our country; but
sufficient time does not seem to have elapsed for the migration to
have been completed before subsidence again occurred, cutting
off the further influx of purely terrestrial animals, and leaving
us without the number of species which our favorable climate
and varied surface entitle us to.

To this cause we must impute our comparative poverty in
mammalia and reptiles—more marked in the latter than the former, owing to their lower vital activity and smaller powers of
dispersal. Germany, for example, possesses nearly ninety species
of land mammalia, and even Scandinavia about sixty, while Britain has only forty, and Ireland only twenty-two. The depth of
the Irish Sea being somewhat greater than that of the German
Ocean, the connecting land would thus probably be of small extent and of less duration, thus offering an additional barrier to
migration, whence has arisen the comparative zoological poverty
of Ireland. This poverty attains its maximum in the reptiles, as
shown by the following figures:

> Belgium has 22 species of reptiles and amphibia.
> Britain   " 13   "   "   "
> Ireland   " 4   "   "   "

Where the power of flight existed, and thus the period of migration was prolonged, the difference is less marked; so that
Ireland has seven bats to twelve in Britain, and about 110 as
against 130 land birds.

Plants, which have considerable facilities for passing over the sea, are somewhat intermediate in proportionate numbers, there being about 970 flowering plants and ferns in Ireland to 1425 in Great Britain—or almost exactly two thirds, a proportion intermediate between that presented by the birds and the mammalia.

Peculiar British Birds.—Among our native mammalia, reptiles, and amphibia, it is the opinion of the best authorities that we possess neither a distinct species nor distinguishable variety. In birds, however, the case is different, since some of our species, in particular our coal-tit (*Parus ater*) and long-tailed tit (*Parus caudatus*) present well-marked differences of color as compared with continental specimens; and in Mr. Dresser's work on the "Birds of Europe" they are considered to be distinct species; while Professor Newton, in his new edition of Yarrell's "British Birds," does not consider the difference to be sufficiently great or sufficiently constant to warrant this, and therefore classes them as insular races of the continental species. We have, however, one undoubted case of a bird peculiar to the British Isles in the red grouse (*Lagopus Scoticus*), which abounds in Scotland, Ireland, the North of England, and Wales, and is very distinct from any continental species, though closely allied to the willow-grouse of Scandinavia. This latter species resembles it considerably in its summer plumage, but becomes pure white in winter; whereas our species retains its dark plumage throughout the year, becoming even darker in winter than in summer. We have here, therefore, a most interesting example of an insular form in our own country; but it is difficult to determine how it originated. On the one hand, it may be an old continental species which during the glacial epoch found a refuge here when driven from its native haunts by the advancing ice; or, on the other hand, it may be a descendant of the Northern willow-grouse, which has lost its power of turning white in winter owing to its long residence in the lowlands of an island where there is little permanent snow, and where assimilation in color to the heather among which it lurks is at all times its best protection. In either case it is equally interesting, as the one large and handsome bird which is peculiar to our islands, notwithstanding their recent separation from the continent.

The following is a list of birds now held to be peculiar to the British Isles:

1. *Parus Britannicus*..Closely allied to *P. ater* of the continent; a local race or subspecies.
2. " *rosea* ....Allied to *P. caudatus* of the continent.
3. *Lagopsus Scoticus*...Allied to *L. albus* of Scandinavia, but very distinct.

Fresh-water Fishes.—Although the productions of fresh waters have generally, as Mr. Darwin has shown, a wide range, fishes appear to form an exception, many of them being extremely limited in distribution. Some are confined to particular river valleys or even to single rivers, others inhabit the lakes of a limited district only, while some are confined to single lakes —often of small area—and these latter offer examples of the most restricted distribution of any organisms whatever. Cases of this kind are found in our own islands, and deserve our especial attention. It has long been known that some of our lakes possessed peculiar species of trout and char; but how far these were unknown on the continent, and how many of these in different parts of our islands were really distinct, had not been ascertained till Dr. Günther, so well known for his extensive knowledge of the species of fishes, obtained numerous specimens from every part of the country, and by comparison with all known continental species determined their specific differences. The striking and unexpected result has thus been attained that no less than fifteen well-marked species of fresh-water fishes are altogether peculiar to the British Islands. The following is the list, with their English names and localities: [1]

FRESH-WATER FISHES PECULIAR TO THE BRITISH ISLES.

Latin Name.	English Name.	Locality.
1. *Salmo brachypoma*...Short-headed salmon....		Firth of Forth, Tweed, Ouse.
2. " *Gallivensis*...Galway sea-trout........		Galway, West of Ireland.
3. " *Orcadensis*...Loch Stennis trout......		Lakes of Orkney.

[1] The list of names was furnished to me by Dr. Günther, and I have added the localities from the papers containing the original descriptions, and from Dr. Haughton's " British Fresh-water Fishes."

Latin Name.	English Name.	Locality.
4. *Salmo ferox*.........Great lake-trout.........		Larger lakes of Scotland, the North of England, and Wales.
5. " *stomachicus*...Gillaroo trout...........		Lakes of Ireland.
6. " *nigripennis*....Black-finned trout.......		Mountain lochs of Wales and Scotland.
7. " *Levenensis*....Loch Leven trout.......		Loch Leven, Loch Lomond, Windermere.
8. " *Perisii*.......Welsh char.............		Llanberris lakes, North Wales.
9. " *Willughbii*....Windermere char.......		Lake Windermere, and others in North of England, and Lake Bruiach in Scotland.
10. " *Killinensis*....Lough Killin char.......		Killin Lake, in Mayo, Ireland.
11. " *Colii*.........Cole's char.............		Lough Eske and Lough Dan, Ireland.
12. " *Grayi*........Gray's char.............		Lough Melvin, Leitrim, Northwest of Ireland.
13. *Coregonus clupeoides*.The gwyniad, or schelly..		Loch Lomond, Ulleswater, Haweswater, and Bala Lake.
14. " *vandesius*..The vendace............		Lochmaben, Dumfriesshire.
15. " *pollan*.....The pollan.............		Lough Neagh and Lough Erne, North of Ireland.

These fifteen peculiar fishes differ from each other and from all British and continental species, not in color only, but in such important structural characters as the form and size of the fins, the number of the fin-rays, and the form or proportions of the head, body, or tail. They are, in fact, as Dr. Günther assures me, just as good and distinct species as any other recognized species of fish. It may, indeed, be objected that, until all the small lakes of Scandinavia are explored and their fishes compared with ours, we cannot be sure that we have any peculiar species. But this objection has very little weight if we consider how our own species vary from lake to lake and from island to island, so that the Orkney species is not found in Scotland, and not one of the peculiar British species extends to Ireland, which has no less than six species altogether peculiar to it. If the species of our own two islands are thus distinct, what reason have we for believing that they will be otherwise than distinct from those of Scandinavia? At all events, with the amount of evidence we already possess of the very restricted ranges of many of our species, we must certainly hold them to be peculiar till they have been proved to be otherwise.

The great speciality of the Irish fishes is very interesting, because it is just what we should expect on the theory of evolution.   In Ireland the two main causes of specific change—isolation and altered conditions—are each more powerful than in
Britain. Whatever difficulty continental fishes may have in passing over to Britain, that difficulty will certainly be increased by
the second sea passage to Ireland; and the latter country has
been longer isolated, for tho Irish Sea with its northern and
southern channels is considerably deeper than the German Ocean
and the eastern half of the English Channel; so that, when the
last subsidence occurred, Ireland would have been an island for
some length of time when England and Scotland still formed
part of the continent.   Again, whatever differences have been
produced by the exceptional climate of our islands will have
been greater in Ireland, where insular conditions are at a maximum, the abundance of moisture and the equability of temperature being far more pronounced than in any other part of
Europe.

Among the remarkable instances of limited distribution afforded by these fishes, we have the Loch Stennis trout confined
to the little group of lakes in the mainland of Orkney, occupying altogether an area of about ten miles by three; the Welsh
char confined to the Llanberris lakes, about three miles in
length; Gray's char confined to Lough Melvin, about seven miles
long; while the Lough Killin char, known only from a small
mountain lake in Ireland, and the vendace, from the equally
small lakes at Lochmaben in Scotland, are two examples of restricted distribution which can hardly be surpassed.

Cause of Great Speciality in Fishes.—The reason why fishes
alone should exhibit such remarkable local modifications in lakes
and islands is sufficiently obvious.   It is due to the extreme
rarity of their transmission from one lake to another.   Just as
we found to be the case in oceanic islands, where the means
of transmission were ample hardly any modification of species
occurred; while, where these means were deficient, and individuals once transported remained isolated during a long succession
of ages, their forms and characters became so much changed as
to bring about what we term distinct species or even distinct

genera—so these lake fishes have become modified because the
means by which they are enabled to migrate so rarely occur. It
is quite in accordance with this view that some of the smaller
lakes contain no fishes, because none have ever been conveyed
to them. Others contain several; and some fishes which have
peculiarities of constitution or habits which render their transmission somewhat less difficult occur in several lakes over a wide
area of country, though none appear to be common to the British and Irish lakes.

The manner in which fishes are enabled to migrate from lake
to lake is unknown, but many suggestions have been made. It
is a fact that whirlwinds and waterspouts sometimes carry living
fish in considerable numbers and drop them on the land. Here
is one mode which might certainly have acted now and then in
the course of thousands of years, and the eggs of fishes may have
been carried with even greater ease. Again, we may well suppose that some of these fish have once inhabited the streams that
enter or flow out of the lakes, as well as the lakes themselves;
and this opens a wide field for conjecture as to modes of migration, because we know that rivers have sometimes changed their
courses to such an extent as to form a union with distinct river
basins. This has been effected either by floods connecting low
water-sheds, by elevations of the land changing lines of drainage, or by ice blocking up valleys and compelling the streams to
flow over water-sheds to find an outlet. This is known to have
occurred during the glacial epoch, and is especially manifest in
the case of the Parallel Roads of Glenroy, and it probably affords the true solution of many of the cases in which existing
species of fish inhabit distinct river basins whether in streams or
lakes. If a fish thus wandered out of one river basin into another, it might then retire up the streams to some of the lakes,
where alone it might find conditions favorable to it. By a combination of the modes of migration here indicated, it is not difficult to understand how so many species are now common to the
lakes of Wales, Cumberland, and Scotland, while others less able
to adapt themselves to different conditions have survived only
in one or two lakes in a single district; or these last may have
been originally identical with other forms, but have become

modified by the particular conditions of the lake in which they
have found themselves isolated.

Peculiar British Insects.—We now come to the class of insects, and here we have much more difficulty in determining
what are the actual facts, because new species are still being
yearly discovered, and considerable portions of Europe are but
imperfectly explored. It often happens that an insect is discovered in our islands, and for some years Britain is its only recorded locality ; but at length it is found on some part of the continent, and not unfrequently has been all the time known there,
but disguised by another name, or by being classed as a variety
of some other species. This has occurred so often that our best
entomologists have come to take it for granted that *all* our supposed peculiar British species are really natives of the continent,
and will one day be found there ; and, owing to this feeling, little
trouble has been taken to bring together the names of such as
from time to time remain known from this country only. The
view of the probable identity of our entire insect-fauna with that
of the continent is held by such well-known authorities as Mr.
E. C. Rye and Dr. D. Sharp for the beetles, and by Mr. H. T.
Stainton for butterflies and moths ; but as we have already seen
that among two orders of vertebrates—birds and fishes—there
are undoubtedly peculiar British species, it seems to me that
all the probabilities are in favor of there being a much larger
number of peculiar species of insects. In every other island
where some of the vertebrates are peculiar—as in the Azores,
the Canaries, the Andaman Islands, and Ceylon—the insects
show an equal, if not a higher, proportion of speciality, and there
seems no reason whatever why the same law should not apply
to us. Our climate is undoubtedly very distinct from that of
any part of the continent, and in Scotland, Ireland, and Wales
we possess extensive tracts of wild mountainous country where
a moist uniform climate, an alpine or northern vegetation, and a
considerable amount of isolation offer all the conditions requisite for the preservation of some species which may have become extinct elsewhere, and for the slight modification of others
since our last separation from the continent. I think, therefore,
that it will be very interesting to take stock, as it were, of our

recorded peculiarities in the insect world, for it is only by so doing that we can hope to arrive at any correct solution of the question on which there is at present so much difference of opinion. For the list of Coleoptera with the accompanying notes I am indebted to Mr. E. C. Rye; and Dr. Sharp has also given me valuable information as to the recent occurrence of some of the supposed peculiar species on the continent. For the Lepidoptera I first noted all the species and varieties marked as British only in Staudinger's "Catalogue of European Lepidoptera." This list was carefully corrected by Mr. Stainton, who weeded out all the species known by him to have been since discovered, and furnished me with valuable information on the distribution and habits of the species. This information often has a direct bearing on the probability of the insect being peculiar to Britain, and in some cases may be said to explain why it should be so. For example, the larvæ of some of our peculiar species of Tineina feed during the winter, which they are enabled to do owing to our mild and insular climate, but which the severer continental winters render impossible. A curious example of the effect this habit may have on distribution is afforded by one of our commonest British species, *Elachista rufocinerea*, the larva of which mines in the leaves of *Holcus mollis* and other grasses from December to March. This species, though common everywhere with us, extending to Scotland and Ireland, is quite unknown in similar latitudes on the continent, but appears again in Italy, the South of France, and Dalmatia, where the mild winters enable it to live in its accustomed manner.

Such cases as this afford an excellent illustration of those changes of distribution, dependent probably on recent changes of climate, which may have led to the restriction of certain species to our islands. For should any change of climate lead to the extinction of the species in South Europe, where it is far less abundant than with us, we should have a common and wide-spread species entirely restricted to our islands. Other species feed in the larva state on our common gorse, a plant found only in limited portions of Western and Southern Europe; and the presence of this plant in a mild and insular climate such as ours may well be supposed to have led to the preservation of some of the numerous species which are or have been dependent on it.

Mr. McLachlan has kindly furnished me with some valuable information on certain species of Trichoptera, or caddis-flies, which seem to be peculiar to our islands; and this completes the list of orders which have been studied with sufficient care to afford materials for such a comparison. We will now give the list of peculiar British insects, beginning with the Lepidoptera, and adding such notes as have been kindly supplied by the gentlemen already referred to.

List of the Species or Varieties of Lepidoptera which, so far as at present Known, are Confined to the British Islands.

(*The Figures Show the Dates when the Species were First Described.*)

DIURNI.

1. *Polyommatus dispar.* "The larger copper." This fine insect, once common in the fens, but now extinct owing to extensive drainage, is generally admitted to be peculiar to our island, at all events as a variety or local form. Its continental ally differs constantly in being smaller and in having smaller spots; but the difference, though constant, is so slight that it is now classed as a variety under the name of *rutilus.* Our insect may therefore be stated to be a well-marked local form of a continental species.

2. Lycæna astrarche, *var. Artaxerxes.* This very distinct form is confined to Scotland and the North of England. The species of which it is considered a variety (more generally known to English entomologists as *P. agestis*) is found in the southern half of England, and almost everywhere on the continent.

BOMBYCES.

3. *Lithosia sericea.* North of England (1861).

4. Hepialus humuli, *var. Hethlandica.* Shetland Islands (1865). A remarkable form, in which the male is usually yellow and buff instead of pure white, as in the common form, but exceedingly variable in tint and markings.

5. *Epichnopteryx reticella.* Sheerness, Gravesend, and other localities along the Thames (1847).

6. E. pulla, *var. radiella.* Near London (1830?). Rare; the species in Central and Southern Europe. (Doubtfully peculiar, in Mr. Stainton's opinion.)

NOCTUÆ.

7. *Acronycta myricæ.* Scotland only (1852). A distinct species.

8. *Agrotis subrosea.* Cambridgeshire and Huntingdonshire fens, perhaps extinct (1835.) The *var. subcærulea* is found in Finland and Livonia.

9. *A. Ashworthii.* South and West (1855). Distinct and not uncommon.

10. *Dianthecia Barretti.* Ireland (1864). Perhaps a form of the continental *D. luteago*, Mr. McLachlan thinks.

11. Aporophyla australis, *var. Pascuea.* South of England (1830 ?). This is a variety of a species otherwise confined to the *South* of Europe, and is thus especially interesting.

GEOMETRÆ.

12. Boarmia gemmaria, *var. perfumaria.* Near London (1866). A large dark variety of a common species, distinctly marked; perhaps a good species, as the larva feeds on ivy, while the larva of *B. gemmaria* is said to refuse this plant, and to die if it has nothing else to eat; but Mr. McLachlan thinks this wants confirmation.

13. Cidaria albulata, *var. griseata.* East of England (1835). A variety of a species otherwise confined to Central and Southern Europe.

14. *Eupithecia constrictata.* Widely spread, but local (1857). Larva on thyme.

PYRALIDINA.

15. Aglossa pinguinalis, *var. Streatfieldi.* Mendip Hills; unique (1830 ?). A remarkable and distinct variety of the common "tabby."

16. *Asopia pictalis.* Unique (1830?). Perhaps an imported species.

17. *Scoparia alpina.* Scotland (1859).

TORTRICINA.

18. *Teras Shepherdana.* Fens of Cambridgeshire (1852).

19. *Cochylis dilucidana.* South of England (1829). Scarce; larva in stems of the wild parsnip.

20. *Aphelia nigrovittana.* Scotland (1852). A local form of the generally distributed *A. lanceolana.*

21. *Eudemis fuligana.* Southeast of England (1828). Rare; on fleabane.

22. *Grapholitha nævana.* Generally distributed (1845). Doubtfully distinct from continental species, in Mr. Stainton's opinion.

23. *G. parvulana.* Isle of Wight (1858 ?). Rare; a distinct species.

24. *G. Weirana.* South of England (1850). A distinct species.

TINEINA.

25. *Tinea cochylidella.* Sanderstead, near Croydon (1854). Unique.

26. *T. pallescentella.* Near Liverpool (1854). Abundant; probably imported in wool, Mr. Stainton thinks.

27. *T. flavescentella.* Near London (1829). Scarce; perhaps imported.

28. *Acrolepia betuletella.* Yorkshire and county of Durham (1840). Rare.

29. *Argyresthia semifusca.* North and West of England (1829). Scarce; a distinct species.

30. *Gelechia divisella.* A fen insect (1854). Rare.

31. *G. celerella.* West of England (1854).

32. *Bryophila politella.* Moors of North of England (1854).

33. *Lita fraternella.* Widely scattered (1834). Larva feeds in shoots of *Stellaria uliginosa* in spring. Mr. Stainton thinks it has been overlooked abroad.

34. *Anacampsis sircomella.* North and West of England (1854). Perhaps a melanic variety of the more widely spread *A. tæniolella.*

35. *A. immaculatella.* West Wickham (1834). Unique; a distinct species.

36. *Glyphipteryx cladiella.* Eastern counties (1859). Abundant.

37. *Glyphipteryx schœnicolella.* In several localities (1859).
38. *Gracillaria straminella.* North Britain (1850). Perhaps a local form of the more southern *G. elongella.*
39. *Ornix Loganella.* Scotland (1848). Abundant, and a distinct species.
40. *O. Devoniella.* In Devonshire (1854). Unique.
41. *Coleophora albicosta.* Widely spread (1829). Common on furze (*Ulex Europœus*). May probably be found in the Northwest of France, where the food-plant abounds.
42. *C. saturatella.* South of England (1850). Abundant on broom.
43. *C. inflata.* South of England (1857). On *Silene inflata.*
44. *C. squamosella.* Surrey (1856). Very rare, but an obscure species.
45. *C. salinella.* On sea-coast (1859). Abundant.
46. *Perittia obscurepunctella.* Widely scattered (1848). Larva feeds on common honeysuckle in July. Mr. Stainton thinks it must have been overlooked on the continent.
47. *Elachista flavicomella.* Dublin (1856). Excessively rare, two specimens only known.
48. *E. consortella.* Scotland (1854). A doubtful species.
49. *E. megeriella.* Widely distributed (1854). Common. Larva feeds in grass during winter and early spring.
50. *E. obliquella.* Near London (1854). Unique.
51. *E. eleochariella.* North and East of England (1854). An obscure form.
52. *E. subocellea.* Widely distributed (1835). An obscure form; perhaps mixed on the continent with other species.
53. *E. triatomea.* In chalk and limestone districts (1812). Abundant and distinct.
54. *E. triseriatella.* South of England (1854). Very local; an obscure species.
55. *Lithocolletis nigrescentella.* Northumberland (1850). Rare; a dark form of *L. Bremiella*, which is widely distributed.
56. *L. irradiella.* North Britain (1854). A northern form of the more southern and wide-spread *L. lautella.*
57. *L. triguttella.* Sanderstead, near Croydon (1848). Unique; very peculiar.
58. *L. uliciolella.* In a few wide-spread localities (1854). A peculiar form.
59. *L. Caledoniella.* North Britain (1854). A local variety of the more wide-spread *L. corylifoliella.*
60. *L. Dunningiella.* North of England (1852). A somewhat doubtful species.
61. *Bucculatrix demaryella.* Widely distributed (1848). Rather common.
62. *Trifurcula squamatella.* South of England (1854). A doubtful species.
63. *T. atrifrontella.* South of England, also in Lancashire (1854). Very rare and peculiar.
64. *Nepticula ignobiliella.* Widely scattered (1854). On hawthorn; not common.
65. *N. poterii.* South of England (1858). Bred from larvæ in *Poterium sanguisorba.*
66. *N. quinquella.* South of England (1848). On oak-leaves, very local.
67. *N. apicella.* Local (1854). Probably confused with allied species on the continent.
68. *N. Headleyella.* Local (1854). A rare species.

PTEROPHORINA.

69. *Agdistis Bennettii.* East coast (1840). Common on *Statice limonium.*

We have here a list of sixty-nine species, which, according to the best authority, are, in the present state of our knowledge, peculiar to Britain. It is a curious fact that only five of these have been described less than twenty years ago; and as during all that time they have not been recognized on the continent, notwithstanding that good colored figures exist of almost all of them, it seems highly probable that many of them are really confined to our island. At the same time, we must not apply this argument too rigidly; for the very day before my visit to Mr. Stainton, he had received a letter from Professor Zeller announcing the discovery on the continent of a species of our last family, Pterophorina, which for more than forty years had been considered to be exclusively British. This insect, *Platyptilia similidactyla* (*Pterophorus isodactylus*, Stainton's "Manual"), had been taken rarely in the extreme north and south of our islands—Teignmouth and Orkney—a fact which seemed somewhat indicative of its being a straggler. Again, seven of the species are unique—that is, have only been captured once; and it may be supposed that as they are so rare as to have been found only once in England, they may be all equally rare and not yet found on the continent. But this is hardly in accordance with the laws of distribution. Widely scattered species are generally abundant in some localities; while, when a species is on the point of extinction, it must for a time be very rare in the single locality where it last maintains itself. It is then more probable that some of these unique species represent such as are almost extinct than that they have a wide range and are equally rare everywhere; and the peculiarity of our insular climate, combined with our varied soil and vegetation, offers conditions which may favor the survival of some species with us after they have become extinct on the continent.

In the list here given nine are recorded as varieties, while ten more, in Mr. Stainton's opinion, ought probably to be classed as varieties or local forms of other species, making nineteen in all. This leaves no less than fifty undoubted species not yet found beyond our islands; and though Mr. Stainton thinks that most of these will ultimately be found on the continent, we can hardly doubt, both from general considerations dependent on the laws of distribution, and from the peculiar habits, conspicuous appearance, and restricted range of many of our species, that a very considerable number will remain permanently as peculiar British insects.

Peculiarities of the Isle of Man Lepidoptera.—Before quitting the Lepidoptera, it will be well to notice some very interesting examples of local modification, apparently brought about by extreme conditions of exposure and insulation, and which throw some light on the way in which local forms, varieties, or species may be produced. This interesting phenomenon occurs in the Isle of Man, where Mr. Edwin Birchall has collected Lepidoptera assiduously, and has discovered a number of varieties, apparently peculiar to the island, of which he has been so good as to send me specimens accompanied by some valuable notes.

The Isle of Man has no woods, bogs, or heaths, the mountains being mostly covered with grass and rocks, so that a very abundant insect fauna cannot be expected. Sixteen species of butterflies have been observed, and of these only one—the common tortoise-shell (*Vanessa urticæ*)—presents any peculiarity. This, however, is always remarkably small, a specimen rarely being found to equal the smallest English specimens; so that we must look upon it as a dwarf race developed in the island and confined to it.

The following moths also present definite peculiarities:

1. *Agrotis lucernea, var.* This is of a grayish-black color, with hardly any markings. All are alike, and are very distinct from the common type of the species, which is abundant in Wales.
2. *Cirrhædia xerampelina, var.* This is much darker and more richly colored than the English form, the yellow band being reduced to a narrow line, sometimes a mere thread. This would doubtless be regarded as a distinct species if it occurred with equal constancy in some more remote island.
3. *Dianthæcia capsophila, var.* This is an exceedingly dark and richly marked form of the Irish *D. capsophila*, itself a local variety, Mr. Birchall thinks, 'of *D. carpophaga.*
4. *Dianthæcia cæsia, var.* This is another dark form of a rare Irish and continental species.
5. *Tephrosia biundularia, var.* This is an exceedingly dark form, and differs so much from North of England specimens as to have all the appearance of another species. Mr. Birchall has bred it from captured parents, and finds that the produce is this dark form only.

We will now pass on to the Coleoptera, or beetles, an order which has been of late years energetically collected and carefully studied by British entomologists.

LIST OF THE SPECIES OF BEETLES WHICH, SO FAR AS AT PRESENT KNOWN, ARE CONFINED TO THE BRITISH ISLANDS.

CARABIDÆ.

1. *Dromius vectensis* (Rye). Common in the Isle of Wight, not known elsewhere.
2. *Harpalus latus, var. metallescens* (Rye). Unique, but very marked. South coast.
3. *Stenolophus derelictus* (Dawson). Unique. North Kent.

HELOPHORIDÆ.

4. *Ochthebius Poweri* (Rye). Very marked. South coast. A few specimens only.

BRACHYELYTRA.

5. *Aleochara Hibernica* (Rye). Ireland. Mountain-tops.
6. *Oxypoda rupicola* (Rye). Scotland. Mountain-tops. Several specimens.
7. * " *Edinensis* (Sharp). Scotland.
8. " *verecunda* (Sharp). Scotland.
9. " *Waterhousei* (Rye). London district.
10. *Homalota eximia* (Sharp).
11. " *clavipes* (Sharp). Scotland; on mountains. Not rare.
12. " *oblongiuscula* (Sharp). Scotland, perhaps also Swiss.
13. " *princeps* (Sharp). A coast insect.
14. " *curtipennis* (Sharp).
15. " *exarata* (Sharp).
16. " *puberula* (Sharp).
17. " *indiscreta* (Sharp).⎫
18. " *atricolor* (Sharp). ⎪Some Continental authors deny that there are
19. " *germana* (Sharp). ⎬ good species (Sharp).
20. " *setigera* (Sharp). ⎭
21. * " *Sharpi* (Rye). Very marked; unique.
22. *Bryoporus castaneus* (Hardy and Bold). Very marked; unique. Northumberland hills.
23. *Stenus oscillator* (Rye). Unique. South coast.
24. *Scopæus Ryei* (Wollaston). Very distinct. Dorset coast. Several specimens.
25. *Trogophlœus spinicollis* (Rye). Mersey estuary. Unique. Most distinguishable; nothing like it in Europe.
26. *Lesteva Sharpi* (Rye). Scotch hills.
27. *Eudectus Whitei* (Sharp). Scotch hills. Probably a variety of *E. Giraudi* of Austria (the only European species) *fide* Kraatz (Sharp).
28. *Homalium rugulipenne* (Rye). Exceedingly marked form. Northern, western, and southern coasts. Rare.

PSELAPHIDÆ.

29. *Bryaxis cotus* (Sharp). Coast.
30. " *Waterhousei* (Rye). Coast.

31. *Bythinus glabratus (Rye). Sussex coast. A few specimens; very distinguishable; myrmecophilous (lives in ants' nests).

TRICHOPTERYGIDÆ.

32. Ptinella maria   (Matthews).
33. Trichopteryx Saræ   "
34.   "   Poweri   "
35.   "   Edithia   "
36.   "   cantiana   "
37.   "   fuscula   "
38.   "   Kirbii   "
39.   "   fratercula   "
40.   "   Waterhousei   "
41.   "   championis   "
42.   "   Jansoni   "
43.   "   seminitens   "
44.   "   suffocata (Haliday). Ireland.
45.   "   dispar (Matthews).
46.   "   carbonaria (Matthews).
47. Ptilium Halidayi (Matthews).
48.   "   Caledonicum (Sharp). Scotland. Very marked form.
49.   "   insigne (Matthews).
50. Actidium concolor (Sharp). Scotland. Very marked. *
51. Ptenidium Kraatzii (Matthews).

ANISOTOMIDÆ.

52. *Agathidium rhinoceros (Sharp). Old fir-woods in Perthshire. Local; many
    specimens. A very marked species.
53. Anisotoma similata (Rye). Unique. South of England.
54.   "   lunicollis (Rye). Northeast and South of England. A very marked
    form. Several specimens.
55. *Anisotoma clavicornis (Rye). Unique. Scotland.

PHALACRIDÆ.

56. *Phalacrus Brisouti (Rye). A few specimens. South of England.

CRYPTOPHAGIDÆ.

57. Atomaria Wollastoni (Sharp). Unique. Scotland.
58.   "   divisa (Rye). Unique. South of England.

LATHRIDIDÆ.

59. Corticaria Wollastoni (Waterhouse). South coast.

BYRRHIDÆ.

60. Syncalypta hirsuta (Sharp).

ELATERIDÆ.

61. Elater coccinatus (Rye). Very marked, but possibly a variety of the European
    E. præustus. South of England.

TELEPHORIDÆ.

62. *Telephorus Darwinianus (Sharp). Scotland, sea-coast. A stunted form, of abnormal habits.

CYPHONIDÆ.

63. Cyphon punctipennis (Sharp). Scotland.

ANTHICIDÆ.

64. Anthicus salinus (Crotch). South coast.
65. *   "   Scoticus (Rye). Loch Leven. Very distinct. Many specimens.

CURCULIONIDÆ.

66. *Cathormiocerus maritimus (Rye). A few specimens on our south coast. A curious genus, only found elsewhere on the coasts of the Mediterranean.
67. *Ceuthorhynchus contractus, var. pallipes (Crotch). Lundy Island. Several specimens. A curious variety only known from this island.
68. *Liosomus troglodytes (Rye). A very queer form. Two or three specimens. South of England.
69. *Apion Ryei (Blackburn). Shetland Islands. Several specimens.

HALTICIDÆ.

70. Thyamis agilis (Rye). South of England. Many specimens.
     "   distinguenda (Rye). South of England. Many specimens.
71. *Psylliodes luridipennis (Kutschera). Lundy Island. A very curious form, not uncommon in this small island, to which it appears to be confined.

COCCINELLIDÆ.

72. Scymnus lividus (Bald). Northumberland. A doubtful species.

Of the seventy-two species of beetles in the preceding list, a considerable number, no doubt, owe their presence there to the fact that they have not yet been discovered or recognized on the continent. This is almost certainly the case with many of those which have been separated from other species by very minute and obscure characters, and especially with the excessively minute Trichopterygidæ described by Mr. Matthews. There are others, however, to which this mode of getting rid of them will not apply, as they are so marked as to be at once recognized by any competent entomologist, and often so plentiful that they can be easily obtained when searched for. Of this class are the twenty-three species whose names are marked with an asterisk (*), being those which, in Mr. Rye's opinion, are most likely to be peculiar to the localities where they are found, if any are; but of this he is still somewhat sceptical. Six of these are unique, leaving seventeen which have occurred either rarely or in some abundance. Dividing the probably peculiar species according to locality, we find that the South of England has produced 9, North of England 2, Scotland 6, Ireland 1, Shetland Islands 1, and Lundy Island 2. These numbers are, generally speaking, proportionate to the richness of the district and the amount of work bestowed upon it; Scotland, however, giving more than its due proportion in this respect, which must be imputed to its really possessing a greater amount of speciality. The single peculiar Irish species stands as a monument of our comparative ignorance of the entomology of the sister isle. The peculiar species of Apion in the Shetland Islands is interesting, and may be connected with the very peculiar climatal conditions there prevailing, which have led in some cases to a change of habits, so that a species of weevil (*Otiorhynchus maurus*) always found on mountain-sides in Scotland, here occurs on the sea-shore. Still more curious is the occurrence of two distinct forms (a species and a well-marked variety) on the small granitic Lundy Island in the Bristol Channel. This island is about three miles long, and twelve from the coast of Devonshire, consisting mainly of granite with a little of the Devonian formation, and the presence here of peculiar insects can only be due to isolation with special conditions, and immunity from enemies or competing forms. When we consider the similar islands off the coasts of Scotland and Ireland, with the Isle of Man and the Scilly Islands, none of which have been yet thoroughly explored for beetles, it is probable that many similar examples of peculiar isolated forms remain to be discovered.

Mr. Rye hardly thinks it possible that the *Dromius vectensis* can really be peculiar to the Isle of Wight, although it is abundant there, and has never been found elsewhere; but the case of Lundy Island renders it less improbable; and when we consider that the *Arum Italicum, Calamintha sylvatica*, and perhaps one or two other plants are found nowhere else in the British Isles, we must admit that the same causes which have acted to restrict the range of a plant may have had a similar effect with a beetle.

I must also notice the *Cathormiocerus maritimus*, because its only near ally inhabits the coasts of the Mediterranean; and it thus offers an analogous case to the small moth, *Elachista rufo-cinerea*, which is found only in Britain and the extreme South of Europe. Looking then at what seems to me the probabilities of the case from the standpoint of evolution and natural selection, and giving due weight to the facts of local distribution as they are actually presented to us, I am forced to differ from the opinion held by our best entomological authorities, and to believe that some considerable proportion of the species which in the present state of our knowledge appear to be peculiar to our islands are, not only apparently, but really, so peculiar.

I am indebted to Mr. Robert McLachlan for the following information on certain Trichopterous Neuroptera, or caddis-flies, which appear to be confined to our islands. The peculiar aquatic habits of the larvæ of these insects—some living in ponds or rivers, others in lakes, and others again only in clear mountain streams—render it not improbable that some of them should have become isolated and preserved in the mountain districts of our western coasts, or that they should be modified owing to such isolation. In these insects the characters depended on to separate the species are wholly structural, and the care with which Mr. McLachlan has studied them renders it certain that the species here referred to are not mere varieties of known continental forms, however closely they may resemble them in form and coloration.

TRICHOPTERA PECULIAR TO THE BRITISH ISLES.

1. *Setodes argentipunctella.*—This species is known only from the lakes Windermere and Killarney. It has recently been described by Mr. McLachlan, and is quite distinct from any known species, though allied to *S. punctata* and *S. viridis*, which inhabit France and Western Europe.

2. *Rhyacophila munda.*—Described by Mr. McLachlan in 1863. A very distinct species, found only in mountain streams in Wales and Devonshire.

3. *Philopotamus insularis* (? a variety of *P. montanus*). — This can hardly be termed a British species or variety, because, so far as at present known, it is peculiar to the island of Guernsey. It agrees structurally with *P. montanus*, a species found both in Britain and on the continent, but it differs in its strikingly yellow color, and less pronounced markings. All the specimens from Guernsey are alike, and resident entomologists assured Mr. McLachlan that no other kind is known. Strange to say, some examples from Jersey differ considerably, resembling the common European and British form. Even should this peculiar variety be at some future time found on the continent, it would still be a remarkable fact that the form of insect inhabit-

ing two small islands only twenty miles apart should constantly differ; but as Jersey is between Guernsey and the coast, it seems just possible that the more insular conditions, and perhaps some peculiarity of the soil and water in the former island, have really led to the production or preservation of a well-marked variety of insect.

Land and Fresh-water Shells.—As regards the land and freshwater mollusca, it seems difficult to obtain accurate information. Several species have been recorded as British only; but I am informed by Mr. Gwyn Jeffries that most of these are decidedly continental, while a few may be classed as varieties of continental species. According to the late Mr. Lovell Reeve, the following species are peculiar to our islands; and although the first two seem exceedingly doubtful, yet the last two, to which alone we accord the dignity of capital type, may not improbably be peculiar to Ireland, being only found in the remote southwestern mountain region, where the climate possesses in the highest degree the insular characteristics of a mild and uniform temperature with almost perpetual moisture, and where several of the peculiar Irish plants alone occur.

1. *Cyclas pisidioides.*—A small bivalve shell found in canals. Perhaps a variety of *C. corneum* or *C. rivicola*, according to Mr. Gwyn Jeffries.

2. *Assiminia Grayana.*—A small univalve shell allied to the periwinkles, found on the banks of the Thames between Greenwich and Gravesend, on the mud at the roots of aquatic plants.

3. Geomalacus maculosus.—A beautiful slug, black, spotted with yellow or white. It is found on rocks on the shores of Lake Carogh, south of Castlemain Bay, in Kerry. It was discovered in 1842, and has never been found in any other locality. An allied species is found in Portugal and France, which Mr. Gwyn Jeffries thinks may be identical.

4. Limnæa involuta.—A beautiful pond-snail with a small polished amber-colored shell, found only in a small alpine lake and its inflowing stream on Cromaghaun Mountain, near the Lakes of Killarney. It appears to be a very distinct species, most nearly allied to *L. glutinosa*, which is not found in Ireland. It was discovered in 1832, and has frequently been obtained since in the same locality.

The facts—that these two last-named species have been known for about forty or fifty years respectively; that they have never been found in any other locality than the above-named very restricted stations; and that they have not yet been clearly identified with any continental species—all point to the conclusion that they are the last remains of peculiar forms which have everywhere else become extinct.

Peculiarities of the British Flora.—Thinking it probable that there must also be some peculiar British plants, but not finding any enumeration of such in the British Floras of Babington, Hooker, or Bentham, I applied to the greatest living authority on the distribution of British plants, Mr. H. C. Watson, who has very kindly given me all the information I required, and I cannot do better than quote his words. He says, "It may be stated pretty confidently that there is no 'species' (generally accepted among botanists as a good species) peculiar to the British Isles. True, during the past hundred years, nominally new species have been named and described on British specimens only, from time to time. But these have gradually come to be identified with species described elsewhere under other names; or they have been reduced in rank by succeeding botanists, and placed or replaced as varieties of more widely distributed species. In his 'British Rubi,' Professor Babington includes as good species some half-dozen which he has, apparently, not identified with any foreign species or variety. None of these are accepted as 'true species,' nor even as 'sub-species,' in the 'Students' Flora,' where the brambles are described by Baker, a botanist well acquainted with the plants of Britain. And as all these nominal species of Rubi are of late creation, they have truly never been subjected to real or critical tests as 'species.'"

But besides these obscure forms, about which there is so much difference of opinion among botanists, there are a few flowering plants which, as *varieties* or *sub-species*, are apparently peculiar to our islands. These are: (1) *Helianthemum Breweri*, an annual rock-rose found only in Anglesea and Holyhead Island (classed as a sub-species of *H. guttatum* by Hooker and Babington); (2) *Rosa Hibernica*, found only in North Britain and Ireland (a species long considered peculiar to the British Isles, but said to have been recently discovered in France); (3) *Œnanthe fluviatilis*, a water-dropwort, found only in the South of England and in one locality in Ireland (classed as a sub-species of *Œ. phellandrium* by Hooker); (4) *Hieracium iricum*, a hawkweed found in North Britain and Ireland (classed by Hooker as a sub-species of *H. Lawsoni*, and said to be "confined to Great Britain").

Two other species are, so far as the European flora is concerned, peculiar to Britain, being natives of North America, and they are very interesting because they are certainly both truly indigenous; that is, not introduced by human agency. These are: (1) *Spiranthes Romanzoviana*, an orchid allied to our lady's-tresses, widely distributed in North America, but only found elsewhere in the extreme southwest corner of Ireland; and (2) *Eriocaulon septangulare*, the pipewort—a curious North American water-plant, found in lakes in the Hebrides and the West of Ireland. Along with these we may perhaps class the beautiful Irish filmy fern *Trichomanes radicans*, which inhabits the Azores, Madeira, and Canary Islands, the Southwest of Ireland, Wales, and formerly Yorkshire, but is not certainly known to occur in any part of continental Europe (except, perhaps, the Southwest of Spain), though found in many tropical countries.

We may here notice the interesting fact that Ireland possesses no less than twenty species or sub-species of flowering plants not found in Britain, and some of these *may* be altogether peculiar. As a whole they show the effect of the pre-eminently mild and insular climate of Ireland in extending the range of some South European species. The following lists of these plants, with a few remarks on their distribution, will be found interesting:

LIST OF IRISH FLOWERING PLANTS WHICH ARE NOT FOUND IN BRITAIN.

1. *Helianthemum guttatum.* Ireland, near Cork, and on an island off the coast of Galway (also Channel Islands, France, Italy).
2. *Arenaria ciliata.* Southwest of Ireland (also Auvergne, Pyrenees, Crete). A variety of this species has been recently found in Pembrokeshire.
3. *Saxifraga umbrosa.* West of Ireland (also Northern Spain, Portugal).
4. "   *geum.* Southwest of Ireland (also Pyrenees).
5. "   *hirsuta.* Southwest of Ireland (also Pyrenees).
6. *Saxifraga hirta* (*hypnoides* sub-sp.). South of Ireland, apparently unknown on the continent.
7. *Inula salicina.* West of Ireland (Middle and South Europe).
8. *Erica Mediterranea.* West of Ireland (West of France, Spain, Mediterranean).
9. "   *Mackiana* (*tetralix* sub-sp.) West of Ireland (Spain).
10. *Arbutus unedo.* Southwest of Ireland (South of France and Spain).
11. *Dabeocia polifolia.* West of Ireland (West of France and Spain).
12. *Pinguicula grandiflora.* Southwest of Ireland (West of France, Spain, Alps, etc.).

13. *Neotinea intacta.*   West of Ireland (France, South of Europe).
14. *Spiranthes Romanzoviana.*   Southwest of Ireland (North America).
15. *Sisyrinchium Bermudianum.*   West of Ireland (? introduced ; North America).
16. *Potamogeton longifolius (lucens* var.).   West of Ireland, unique specimen.
17.   "   *Kirkii (natans* sub-sp.).   West of Ireland (Arctic Europe).
18. *Eriocaulon septangulare.*   West of Ireland, Skye, Hebrides (North America).
19. *Carex Buxbaumii.*   Northeast of Ireland, on island in Lough Neagh (Arctic and
      Alpine Europe, North America).
20. Calamagrostis stricta, *var. Hookeri.*   On the shores and islands of Lough Neagh.
      The species occurs at one locality in Cheshire (Germany, Arctic Europe, and
      North America).

We find here nine Southwest European species which probably
had a wider range in mild preglacial times, and have been preserved in the South and West of Ireland owing to its milder
climate.   It must be remembered that during the height of the
glacial epoch Ireland was continental, so that these plants may
have followed the retreating ice to their present stations and
survived the subsequent depression.   This seems more probable
than that so many species should have reached Ireland for the
first time during the last union with the continent subsequent
to the glacial epoch.   The arctic, alpine, and American plants
may all be examples of species which once had a wider range,
and which, owing to the more favorable conditions, have continued to exist in Ireland while becoming extinct in the adjacent
parts of Britain and Western Europe.

As contrasted with the extreme scarcity of peculiar species
among the flowering plants, it is the more interesting and unexpected to find a considerable number of peculiar mosses and
Hepaticæ, some of which present us with phenomena of distribution of a very remarkable character.   For the following lists
and the information as to the distribution of the genera and
species I am indebted to Mr. William Mitten, one of the first
authorities on these beautiful little plants.

List of the Species of Mosses and Hepaticæ which are Peculiar to the
British Isles or not found in Europe.
(*Those belonging to non-European genera in italics.*)

Mosses.

1. Systegium multicapsulare............Central and South England.
2.   "   Mittenii.................South of England.

3. Campylopus Shawii..................North Britain.
4.   "   setifolius...............Ireland.
5. Seligeria calcicola...................South of England.
6. Pottia viridifolia.....................South of England.
7. Leptodontium recurvifolium..........Ireland and Scotland.
8. Tortula Woodii.....................Ireland.
9.   "   Hibernica..................Ireland.
10. *Streptopogon gemmascens*............Sussex.
11. Grimmia subsquarrosa..............North Britain.
12.   "   Stirtoni.....................North Britain.
13. Glyphomitrium Daviesii.............On basalt generally.
14. Zygodon Nowellii...................North Britain.
15. Bryum Barnesii....................North Britain.
16. *Hookeria lætevirens*................Ireland and Cornwall (also Madeira).
17. *Daltonia splachnoides*...............Ireland.

HEPATICÆ.

1. Gymnomitrium crenulatum..........West of England, Ireland.
2. Radula voluta. ....Ireland and Wales.
3. *Acrobolbus Wilsoni*.................Ireland.
4. *Lejeunia calyptrifolia*...............Cornwall, Lake district, Ireland.
5.   "   *microscopica*...............Ireland.
6. Lophocolea spicata..................Ireland.
7. Jungermannia cuneifolia............Ireland.
8.   "   doniana...............Scotland.
9. *Petalophyllum Ralfsii*..............West Britain, Ireland.

Many of the above are minute or obscure plants, and are closely allied to other European species with which they may have been confounded. We cannot, therefore, lay any stress on these individually as being absent from the continent of Europe, so much of which is imperfectly explored, though it is probable that some of them are really confined to Britain. But there are a few—indicated by italics—which are in a very different category; for they belong to genera which are altogether unknown in any other part of Europe, and their nearest allies are to be found in the tropics or in the Southern Hemisphere. The three non-European genera of mosses to which we refer all have their maximum of development in the Andes, while the three non-European Hepaticæ appear to have their maximum in the temperate regions of the Southern Hemisphere. Mr. Mitten has kindly furnished me with the following particulars of the distribution of these genera:

STREPTOPOGON is a comparatively small genus, with seven species in the Andes, one in the Himalayas, and three in the south temperate zone, besides our English species.

DALTONIA is a large genus of inconspicuous mosses, having seventeen species in the Andes, two in Brazil, two in Mexico, one in the Galapagos, six in India and Ceylon, five in Java, two in Africa, and three in the antarctic islands, and one in Ireland.

HOOKERIA (restricting that term to the species referable to Cyclodictyon) is still a large genus of handsome and remarkable mosses, having twenty-six species in the Andes, eleven in Brazil, eight in the Antilles, one in Mexico, two in the Pacific islands, one in New Zealand, one in Java, one in India, and five in Africa—besides our British species, which is found also in Madeira and the Azores, but in no part of Europe proper.

These last two are very remarkable cases of distribution, since Mr. Mitten assures me that the plants are so markedly different from all other mosses that they would scarcely be overlooked in Europe.

The distribution of the non-European genera of Hepaticæ is as follows:

ACROBOLBUS.   A small genus found only in New Zealand and the adjacent islands, besides Ireland.

LEJEUNIA.   A very extensive genus abounding in the tropical regions of America, Africa, the Indian Archipelago, and the Pacific islands, reaching to New Zealand and antarctic America, sparingly represented in the British and Atlantic islands and in North America.

PETALOPHYLLUM.   A small genus confined to Australia and New Zealand in the Southern Hemisphere, and Ireland in the Northern.

We have also a moss—*Myurium Hebridarum*—found only in Scotland and the Atlantic islands ; and one of the Hepaticæ—*Mastigophara Woodsii*—found in Ireland and the Himalayas, the genus being most developed in New Zealand and unknown in any part of continental Europe.

These are certainly very interesting facts, but they are by no means so exceptional in this group of plants as to throw any doubt upon their accuracy. The Atlantic islands present very similar phenomena in the *Rhamphidium purpuratum*, whose nearest allies are in the West Indies and South America; and in three species of Sciaromium, whose only allies are in New Zealand, Tasmania, and the Andes of Bogotá. An analogous and equally curious fact is the occurrence in the Drontheim mountains, in Central Norway, of a little group of four or five peculiar species of mosses of the genus Mnium, which are found nowhere else; although the genus extends over Europe, India, and the Southern Hemisphere, but always represented by a very few wide-ranging species except in this one mountain group.[1]

Such facts show us the wonderful delicacy of the balance of conditions which determine the existence of particular species in any locality. The spores of mosses and Hepaticæ are so minute that they must be continually carried through the air to great distances, and we can hardly doubt that, so far as its powers of diffusion are concerned, any species which fruits freely might soon spread itself over the whole world. That they do not do so must depend on peculiarities of habit and constitution, which fit the different species for restricted stations and special climatic conditions; and according as the adaptation is more general, or the degree of specialization extreme, species will have wide or restricted ranges. Although their fossil remains have been rarely detected, we can hardly doubt that mosses have as high an antiquity as ferns or Lycopods; and, coupling this antiquity with their great powers of dispersal, we may understand how many of the genera have come to occupy a number of detached areas scattered over the whole earth, but always such as afford the peculiar conditions of climate and soil best suited to them. The repeated changes of temperature and other climatic conditions, which, as we have seen, occurred through all the later geological epochs, combined with those slower changes caused by geographical mutations, must have greatly affected the distribution of such ubiquitous yet delicately organized plants as mosses. Throughout countless ages they must have been in a constant state of comparatively rapid migration, driven to and fro by every physical and organic change, often subject to modification of structure or habit, but always seizing upon every available spot in which they could even temporarily maintain themselves.

Here, then, we have a group in which there is no question of the means of dispersal, and where the difficulties that present themselves are not how the species reached the remote localities in which they are now found, but rather why they have not es-

---

[1] I am indebted to Mr. Mitten for this curious fact.

tablished themselves in many other stations which, so far as we can judge, seem equally suitable to them. Yet it is a curious fact that the phenomena of distribution actually presented by this group do not essentially differ from those presented by the higher flowering plants which have apparently far less diffusive power, as we shall find when we come to treat of the floras of oceanic islands ; and we believe that the explanation of this is, that the life of species, and especially of genera, is often so prolonged as to extend over whole cycles of such terrestrial mutations as we have just referred to; and that thus the majority of plants are afforded means of dispersal which are usually sufficient to carry them into all suitable localities on the globe. Hence it follows that their actual existence in such localities depends mainly upon vigor of constitution and adaptation to conditions, just as it does in the case of the lower and more rapidly diffused groups, and only partially on superior facilities for diffusion. This important principle will be used further on to afford a solution of some of the most difficult problems in the distribution of plant-life.

Concluding Remarks on the Peculiarities of the British Fauna and Flora.—The facts, now, I believe, for the first time brought together, respecting the peculiarities of the British fauna and flora are sufficient to show that there is considerable scope for the study of geographical distribution, even in so apparently unpromising a field as one of the most recent of continental islands. Looking at the general bearing of these facts, they prove that the idea so generally entertained as to the biological identity of the British Isles with the adjacent continent is not altogether correct. Among birds we have undoubted peculiarities in at least three instances ; peculiar fishes are much more numerous, and in this case the fact that the Irish species are all different from the British, and those of the Orkneys distinct from those of Scotland, renders it almost certain that the great majority of the fifteen peculiar British fishes are really peculiar, and will never be found on the European continent. The mosses and Hepaticæ also have been sufficiently collected in Europe to render it pretty certain that the more remarkable of the peculiar British forms are not found there. Why, therefore, it may well be asked, should there not be a proportionate number of peculiar British insects? It is true that numerous species have been first discovered in Britain, and subsequently on the continent; but we have many species which have been known for twenty, thirty, or forty years, some of which are not rare with us, and yet have never been found on the continent. We have also the curious fact of our outlying islands, such as the Shetland Isles, the Isle of Man, and the little Lundy Island, possessing each some peculiar forms which *certainly* do not exist on our principal island, which has been so very thoroughly worked. Analogy, therefore, would lead us to conclude that many other species would exist on our islands and not on the continent; and when we find that a very large number (150), in three orders only, are so recorded, we may, I think, be sure that a considerable portion of these (though how many we cannot say) are really endemic British species.

The general laws of distribution also lead us to expect such phenomena. Very rare and very local species are such as are becoming extinct; and it is among insects, which are so excessively varied and abundant, which present so many isolated forms, and which, even on continents, afford numerous examples of very rare species confined to restricted areas, that we should have the best chance of meeting with every degree of rarity down to the point of almost complete extinction. But we know that in all parts of the world islands are the refuge of species or groups which have become extinct elsewhere; and it is therefore in the highest degree probable that some species which have ceased to exist on the continent should be preserved in some part or other of our islands, especially as these present favorable climatic conditions such as do not exist elsewhere.

There is therefore a considerable amount of harmony in the various facts adduced in this chapter, as well as a complete accordance with what the laws of distribution in islands would lead us to expect. In proportion to the species of birds and fresh-water fishes, the number of insect-forms is enormously great, so that the numerous species here recorded as not yet known on the continent are not to be wondered at; while it would, I think, be almost an anomaly if, with peculiar birds and fishes there were *not* a fair proportion of peculiar insects. Our entomologists should therefore give up the assumption that all our insects do exist on the continent, and will some time or other be found there, as not in accordance with the evidence; and when this is done, and the interesting peculiarities of some of our smaller islands are remembered, the study of our native animals and plants in relation to those of other countries will acquire a new interest. The British Isles are said to consist of more than a thousand islands and islets. How many of these have ever been searched for insects? With the case of Lundy Island before us, who shall say that there is not yet scope for extensive and interesting investigations into the British fauna and flora?

CHAPTER XVII.

BORNEO AND JAVA.

Position and Physical Features of Borneo.—Zoological Features of Borneo: Mammalia.—Birds.—The Affinities of the Bornean Fauna.—Java, its Position and Physical Features.—General Character of the Fauna of Java.—Differences between the Fauna of Java and that of the other Malay Islands.—Special Relations of the Javan Fauna to that of the Asiatic Continent.—Past Geographical Changes of Java and Borneo.—The Philippine Islands.—Concluding Remarks on the Malay Islands.

As a representative of recent continental islands situated in the tropics, we will take Borneo, since, although perhaps not much more ancient than Great Britain, it presents a considerable amount of speciality, and, in its relations to the surrounding islands and the Asiatic continent, offers us some problems of great interest and considerable difficulty.

The accompanying map shows that Borneo is situated on the eastern side of a submarine bank of enormous extent, being about 1200 miles from north to south, and 1500 from east to west, and embracing Java, Sumatra, and the Malay Peninsula. This vast area is all included within the 100-fathom line; but by far the larger part of it—from the Gulf of Siam to the Java Sea—is under fifty fathoms, or about the same depth as the sea that separates our own island from the continent. The distance from Borneo to the southern extremity of the Malay Peninsula is about 350 miles, and it is nearly as far from Sumatra and Java, while it is more than 600 miles from the Siamese Peninsula, opposite to which its long northern coast extends. There is, I believe, nowhere else upon the globe an island so far from a continent, yet separated from it by so shallow a sea. Recent changes of sea and land must have occurred here on a grand scale, and this adds to the interest attaching to the study of this large island.

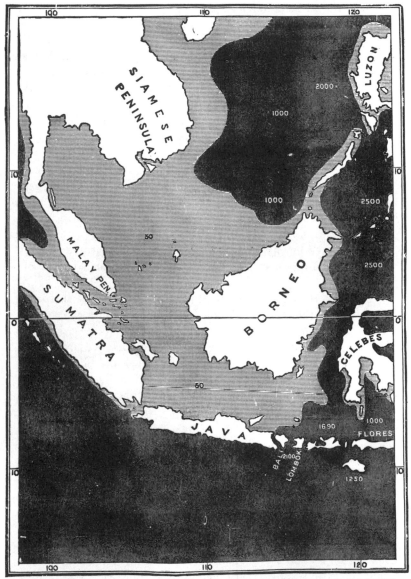

MAP OF BORNEO AND JAVA, SHOWING THE GREAT SUBMARINE BANK OF SOUTH-
EASTERN ASIA.

The light tint shows a depth less than 100 fathoms.
The figures show the depth of the sea in fathoms.

The internal geography of Borneo is somewhat peculiar. A large portion of its surface is lowland, consisting of great alluvial valleys which penetrate far into the interior; while the mountains, except in the north, are of no great elevation, and there are no extensive plateaus. A subsidence of 500 feet would allow the sea to fill the great valleys of the Pontianak, Banjermassin, and Coti rivers, almost to the centre of the island, greatly reducing its extent, and causing it to resemble in form the island of Celebes to the east of it.

In geological structure Borneo is thoroughly continental, possessing formations of all ages, with basalt and crystalline rocks, but no recent volcanoes. It possesses vast beds of coal of Tertiary age; and these, no less than the great extent of alluvial deposits in its valleys, indicate great changes of level in recent geological times.

Having thus briefly indicated those physical features of Borneo which are necessary for our inquiry, let us turn to the organic world.

Neither as regards this great island nor those which surround it have we the amount of detailed information in a convenient form that is required for a full elucidation of its past history. We have, however, a tolerable acquaintance with the two higher groups—mammalia and birds—both of Borneo and of all the surrounding countries, and to these alone will it be necessary to refer in any detail. The most convenient course, and that which will make the subject easiest for the reader, will be to give, first, a connected sketch of what is known of the zoology of Borneo itself, with the main conclusions to which they point; and then to discuss the mutual relations of some of the adjacent islands, and the series of geographical changes that seem required to explain them.

Zoological Features of Borneo.

Mammalia. — About ninety-six species of mammalia have been discovered in Borneo, and of these nearly two thirds are identical with those of the surrounding countries, and nearly one half with those of the continent. Among these are two lemurs, three civets, three cats, three deer, the tapir, the elephant, and several squirrels—an assemblage which could certainly only have reached the country by land. The following species of mammalia are supposed to be peculiar to Borneo :

QUADRUMANA.

1. Simia morio. A small orang-outang with large incisor teeth.
2. Hylobates concolor.
3. Nasalis larvatus.
4. Semnopithecus rubicundus.
5.   "   chrysomelas.
6.   "   frontatus.
7. Macacus melanotus.

CARNIVORA.

8. Cynogale Bennettii.
9. Paradoxurus stigmaticus.
10. Herpestes semitorquatus.
11.   "   brachyurus.
12. Felis badia.
13. Lutra Lovii (Günther, *P. Z. S.*, 1876, p. 736).

UNGULATA.

14. Sus barbatus.

RODENTIA.

15. Pteromys phæomelas.

16. Sciurus ephippium.
17.   "   pluto.
18.   "   macrotis.
19.   "   Sarawakensis.
20.   "   Borneonensis.
21.   "   rufogularis.
22.   "   atricapillus.
23.   "   rufogaster.
24. Acanthion crassispinis.
25. Trichys lipura.

INSECTIVORA.

26. Tupaia splendidula.
27.   "   minor (Günther, *P. Z. S.*, 1876, p. 426).
28. Dendrogale murina.
29. Ptilocerus Lowii.

CHIROPTERA.

30. Phyllorina doriæ.
31. Vesperugo stenopterus.
32.   "   doriæ.
33.   "   tylopus.
34. Taphozous affinis.

Of the thirty-four peculiar species here enumerated, it is probable that when they are more carefully studied some will be found to be identical with those of Malacca or Sumatra ; but there are also four peculiar genera which are less likely to be discovered elsewhere. These are Nasalis, the remarkable long-nosed monkey ; Cynogale, a semi-aquatic civet ; Trichys, a tailless porcupine ; and Ptilocerus, a feather-tailed arboreal insectivore. These peculiar forms do not, however, imply that the separation of the island from the continent is of very ancient date, for the country is so vast, and so much of the connecting land is covered with water, that the amount of speciality is hardly, if at all, greater than occurs in many continental areas of equal extent. This will be more evident if we consider that Borneo is as large as the Indo-Chinese Peninsula, or as the Indian Peninsula south of Bombay ; and if either of these countries were separated from the continent by the submergence of the whole area north of it as far as the Himalayas, it would be found to contain about as many peculiar genera and species as Borneo actually does now. A more decisive test of the lapse of time since the separation took place is to be found in the presence of a number of representative species closely allied to those of the surrounding countries, such as the tailed monkeys and the numerous squirrels. These, however, are best seen among the birds, which have been more thoroughly collected and more carefully studied than the mammalia.

Birds. — About 400 species of birds are known to inhabit Borneo, of which 340 are land birds. There are about 70 peculiar species; and, according to Count Salvadori, 34 of these (39 with later additions) are very distinct forms, while no less than 31 are slight modifications of species found in Sumatra or the Malay Peninsula. The following are the species of birds considered by Count Salvadori to be peculiar to Borneo, with the addition of a few species since added:

First Series.
Very Distinct Species.

Second Series.
Representative Species.

STRIGIDÆ (Owls).

   1. Ninox Borneonensis.
   2. Ciccaba leptogrammica.

MEGALÆMIDÆ (Barbets).

   3. Chotorea chrysopsis.
   4. Calorhamphus fuliginosus.

PICIDÆ (Woodpeckers).

   5. Hemilophus Fischeri.
   6. Jungipicus aurantiiventris.
   7. Micropternus badiosus.

CUCULIDÆ (Cuckoos).

1. Indicator archipelagus.
2. Heterococcyx neglectus.

   8. Rhopodytes Borneonensis.

ALCEDINIDÆ (Kingfishers).

3. Ceyx Sharpei.
4.   " Dillwynni.

   9. Pelargopsis leucocephala.
   10. Dacelo melanops.

PODARGIDÆ.

   11. Batrachostomus adspersus.

CAPRIMULGIDÆ (Goatsuckers).

5. Caprimulgus arundinaceus.
6.   " concretus.
7.   " Salvadorii.

   12. Caprimulgus Bulweri.

FIRST SERIES.   SECOND SERIES.
Very Distinct Species.   *Representative Species.*

HIRUNDINIDÆ (Swallows).

8. Delichon dasypus. |

MUSCICAPIDÆ (Flycatchers).

9. Cyornis rufifrons. |
10.   "   turcosa. |
11.   "   beccariana. |
12. Schwaneria cærulata. |

ARTAMIDÆ (Swallow-shrikes).

13. Artamus clemenciæ. |

LANIIDÆ (Shrikes).

14. Lanius Schwaneri. | 13. Volvocivora Schierbrandi.
15. Pityriasis gymnocephala. |

NECTARINIIDÆ (Sunbirds).

16. Arachnothera crassirostris. |

DICEIDÆ (Flower-peckers).

                                        | 14. Prionochilus xanthopygius.
                                        | 15. Diceum nigrimentum.
17. Zosterops melanura. | 16. Zosterops parvula.

PYCNONOTIDÆ (Bulbuls).

18. Pycnonotus Gourdinii. |
19. Criniger Diardi. |
20.   "   Finschii. |

TIMALIIDÆ (Babblers).

21. Turdinus leucogrammicus. | 17. Pomatorhinus Borneonensis.
22. Setaria pectoralis. | 18. Mixornis Borneonensis.
23.   "   cinereicapilla. | 19. Drymocataphus capistratioides.
                                        | 20. Brachypteryx umbratilis.
                                        | 21. Malacocincla rufiventris.

PITTIDÆ (Pittas).

24. Pitta Bertæ. | 22. Pitta granatina.
25.   "   arcuata. | 23.   "   Schwaneri.
26.   "   Baudii. | 24.   "   Usheri.

SYLVIIDÆ (Warblers).

27. Abrornis Schwaneri. |
28. Prinia superciliaris. | 25. Orthotomus Borneonensis.
29. Calamodyta doriæ. |
30. Kittacincla Stricklandi. | 26. Kittacincla suavis.

CORVIDÆ (Crows and Jays).

                                        | 27. Dendrocitta cinerascens.
                                        | 28. Platysmurus aterrimus.

Representative forms of the same character as these are, no doubt, found in all extensive continental areas, but they are rarely so numerous. Thus, in Mr. Elwes's paper on the "Distribution of Asiatic Birds," he states that 12.5 per cent. of the land birds of Burmah and Tenasserim are peculiar species, whereas we find that in Borneo they are about 20 per cent., and the difference may fairly be imputed to the greater proportion of slightly modified representative species due to a period of complete isolation. Of peculiar genera, the Indo-Chinese Peninsula has one—Ampeliceps, a remarkable yellow-crowned starling, with bare pink-colored orbits; while two others, Temnurus and Crypsirhina—singular birds allied to the jays—are found in no other part of the Asiatic continent, though they occur in some of the Malay Islands. Borneo has three peculiar genera—Schwaneria, a flycatcher; Hematortyx, a crested partridge; and Lobiophasis, a pheasant hardly distinct from Euplocamus; while two others, Pityriasis, an extraordinary bare-headed bird between a jay and a shrike, and Carpococcyx, a pheasant-like ground-cuckoo, formerly thought to be peculiar, are said to have been discovered also in Sumatra.

The insects and land shells of Borneo and of the surrounding countries are too imperfectly known to enable us to arrive at any accurate results with regard to their distribution. They agree, however, with the birds and mammals in their general approximation to Malayan forms, but the number of peculiar species is perhaps larger.

The proportion here shown of one third peculiar species of mammalia to about one fifth peculiar species of land birds teaches us that the possession of the power of flight only affects the distribution of animals in a limited degree, and gives us confidence in the results we may arrive at in those cases where we have, from whatever cause, to depend on a knowledge of the birds alone. And the difference we here find to exist is almost wholly due to the wide range of certain groups of powerful flight—as the birds of prey, the swallows and swifts, the king-crows, and some others; while the majority of forest birds appear to remain confined, by even narrow watery barriers, to almost as great an extent as do the mammalia.

The Affinities of the Bornean Fauna.—The animals of Borneo exhibit an almost perfect identity in general character, and a close similarity in species, with those of Sumatra and the Malay Peninsula. So great is this resemblance that it is a question whether it might not be quite as great were the whole united; for the extreme points of Borneo and Sumatra are 1500 miles apart—as far as from Madrid to Constantinople, or from Bombay to Rangoon. In this distance we should expect to meet with many local species, and even representative forms, so that we hardly require a lapse of time sufficient to have produced specific change. So far as the forms of life are concerned, Borneo, as an island, may be no older than Great Britain; for the time that has elapsed since the glacial epoch would be amply sufficient to produce such a redistribution of the species, consequent on their mutual relations being disturbed, as would bring the islands into their present zoological condition. There are, however, other facts to be considered, which seem to imply much greater and more complex revolutions than the recent separation of Borneo from Sumatra and the Malay Peninsula, and that these changes must have been spread over a considerable lapse of time. In order to understand what these changes probably were, we must give a brief sketch of the fauna of Java, the peculiarities of which introduce a new element into the question we have to discuss.

JAVA.

The rich and beautiful island of Java, interesting alike to the politician, the geographer, and the naturalist, is more especially attractive to the student of geographical distribution, because it furnishes him with some of the most curious anomalies and difficult problems in a place where such would be least expected. As Java forms with Sumatra one almost unbroken line of volcanoes and volcanic mountains, interrupted only by the narrow Strait of Sunda, we should naturally expect a close resemblance between the productions of the two islands. But in point of fact there is a much greater difference between them than between Sumatra and Borneo, so much farther apart, and so very unlike in physical features. Java differs from the three great land masses Borneo, Sumatra, and the Malay Peninsula far more than either of these does from each other; and this is the first anomaly we encounter. But a more serious difficulty than this remains to be stated. Java has certain close resemblances to the Siamese Peninsula, and also to the Himalayas, which Borneo and Sumatra do not exhibit; and, looking at the relative position of these lands respectively, this seems most incomprehensible. In order fully to appreciate the singularity and difficulty of the problem, it will be necessary to point out the exact nature and amount of these peculiarities in the fauna of Java.

General Character of the Fauna of Java.—If we were only to take account of the number of peculiar species in Java, and the relations of its fauna generally to that of the surrounding lands, we might pass it over as a less interesting island than Borneo or Sumatra. Its mammalia (ninety species) are nearly as numerous as those of Borneo, but are apparently less peculiar, none of the genera and only five or six of the species being confined to the island. In land birds it is decidedly less rich, having only 270 species, of which 40 are peculiar, and only one or two belong to peculiar genera; so that here again the amount of speciality is less than in Borneo. It is only when we proceed to analyze the species of the Javan fauna, and trace their distribution and affinities, that we discover its interesting nature.

Difference between the Fauna of Java and that of the other Great Malay Islands.—Comparing the fauna of Java with that which may be called the typical Malayan fauna as exhibited in Borneo, Sumatra, and the Malay Peninsula, we find the following differences. No less than thirteen genera of mammalia, each of which is known to inhabit at least two, and generally all three, of the above-named Malayan countries, are yet totally absent from Java; and they include such important forms as the elephant, the tapir, and the Malay bear. It cannot be said that this difference depends on imperfect knowledge, for Java is one of the oldest European settlements in the East, and has been explored by a long succession of Dutch and English naturalists. Every part of it is thoroughly well known, and it would be almost as difficult to find a new mammal of any size in Europe as in Java. Of birds there are twenty-five genera, all typically Malayan, and occurring at least in two, and for the most part in all three, of the Malay countries, which are yet absent from Java. Most of these are large and conspicuous forms, such as jays, gapers, bee-eaters, woodpeckers, hornbills, cuckoos, parrots, pheasants, and partridges, as impossible to have remained undiscovered in Java as the large mammalia above referred to.

Besides these absent genera, there are some curious illustrations of Javan isolation in the species; there being several cases in which the same species occurs in all three of the typical Malay countries, while in Java it is represented by an allied species. Such appear to be the Malayan monkey, *Semnopithecus cristatus*, replaced in Java by *S. maurus;* and the large Malay deer, *Rusa equinus*, represented in Java by *R. hippelaphus*. Among birds there are more numerous examples, no less than seven species which are common to the three great Malay countries being represented in Java by distinct but closely allied species.

From these facts it is impossible to doubt that Java has had a history of its own, quite distinct from that of the other portions of the Malayan area.

Special Relations of the Javan Fauna to that of the Asiatic Continent.—These relations are indicated by comparatively few examples, but they are very clear and of great importance. Among mammalia, the genus Helictis is found in Java, but in no other Malay country, though it inhabits also North India; while two species, *Rhinoceros Javanicus* and *Lepus Kurgosa*, are natives of Indo-Chinese countries and Java, but not of typical Malaya. In birds, there are three genera—Zoothera, Notodela, and Crypsirhina—which inhabit Java and Indo-China; while four others —Brachypteryx, Allotrius, Cochoa, and Psaltria—inhabit Java and the Himalayas, but no intervening country. There are also two species of birds—a trogon (*Harpactes Oreskios*) and the Javanese peacock (*Pavo muticus*)—which inhabit only Java and the Indo-Chinese Peninsula.

Here, then, we find a series of remarkable similarities between Java and the Asiatic continent, quite independent of the typical Malay countries, Borneo, Sumatra, and the Malay Peninsula, which latter have evidently formed one connected land, and thus appear to preclude any independent union of Java and Siam.

The great difficulty in explaining these facts is that all the required changes of sea and land must have occurred within the period of existing species of mammalia. Sumatra, Borneo, and Malacca are, as we have seen, almost precisely alike as regards their species of mammals and birds; while Java, though it differs from them in so curious a manner, has no greater degree of speciality, since its species, when not Malayan, are almost all Indian or Siamese.

There is, however, one consideration which may help us over this difficulty. It seems highly probable that in the equatorial regions species have changed less rapidly than in the north temperate zone, on account of the equality and stability of the equatorial climate. We have seen, in Chapter X., how important an agent in producing extinction and modification of species must have been the repeated changes from cold to warm and from warm to cold conditions, with the inevitable migrations and crowding-together that must have been their necessary consequence. But in the lowlands near the equator these changes would be very little felt, and thus one great cause of specific modification would be wanting. Let us now see whether we can sketch out a series of not improbable changes which may have brought about the existing relations of Java and Borneo to the continent.

Past Geographical Changes of Java and Borneo.—Although Java and Sumatra are mainly volcanic, they are by no means wholly so. Sumatra possesses in its great mountain masses ancient crystalline rocks with much granite, while there are extensive Tertiary deposits of Eocene age, overlying which are numerous beds of coal now raised up many thousand feet above the sea.[1] The volcanoes appear to have burst through these older mountains, and to have partly covered them, as well as great areas of the lowlands, with the product of their eruptions. In Java either the fundamental strata were less extensive and less raised above the sea, or the period of volcanic action has been of longer duration; for here no crystalline rocks have been found except a few boulders of granite in the western part of the island, perhaps a relic of a formation destroyed by denudation, or covered up by volcanic deposits. In the southern part of Java, however, there is an extensive range of low mountains, about 3000 feet high, consisting of basalt with limestone apparently of Miocene age.

During this last-named period, then, Java would have been at least 3000 feet lower than it is now, and such a depression would probably extend to considerable parts of Sumatra and Borneo, so as to reduce them all to a few small islands. At some later period a gradual elevation occurred which ultimately united the whole of the islands with the continent. This may have continued till the glacial period of the Northern Hemisphere, during the severest part of which a few Himalayan species of birds and mammals may have been driven southward, and ranged over suitable portions of the whole area. Java was then separated by subsidence, and these species became imprisoned there; while those in the remaining part of the Malayan area again migrated northward when the cold had passed away from their former home, the equatorial forests of Borneo, Sumatra, and the Malay Peninsula being more especially adapted to the typical Malayan fauna, which is there developed in rich profusion. A little later the subsidence may have extended farther north, isolating Borneo and Sumatra, but probably leav-

---

[1] "On the Geology of Sumatra," by M. R. D. M. Verbeck, *Geological Mag.*, 1877.

ing the Malay Peninsula as a ridge between them as far as the islands of Banca and Biliton. Other slight changes of climate followed, when a further subsidence separated these last-named islands from the Malay Peninsula, and left them with two or three species which have since become slightly modified. We may thus explain how it is that a species is sometimes common to Sumatra and Borneo, while the intervening island (Banca) possesses a distinct form.[1]

In my "Geographical Distribution of Animals," Vol. I., p. 357, I have given a somewhat different hypothetical explanation of the relations of Java and Borneo to the continent, in which I took account of changes of land and sea only; but a fuller consideration of the influence of changes of climate on the migration of animals has led me to the much simpler, and, I think, more probable, explanation above given. The amount of the relationship between Java and Siam, as well as of that between Java and the Himalayas, is too small to be well accounted for by an independent geographical connection in which Borneo and Sumatra did not take part. It is, at the same time, too distinct and indisputable to be ignored; and a change of climate which should drive a portion of the Himalayan fauna southward, leaving a few species in Java (from which they could not return, owing to its subsequent isolation by subsidence), seems to be a cause exactly adapted to produce the kind and amount of affinity between these distant countries that actually exist.

The Philippine Islands.—A sufficiently detailed account of the fauna of these islands, and their relation to the countries which form the subject of this chapter, has been given in my "Geographical Distribution of Animals," Vol. I., pp. 345–349; but since that time considerable additions have been made to their fauna, and these have had the effect of diminishing their isolation from the other islands. Six genera have been added to the terrestrial mammalia—Crocidura, Felis, Tragulus, Hystrix, Pteromys, and Mus, as well as two additional squirrels; while the black ape (*Cynopithecus niger*) has been struck out as not

---

[1] *Pitta megarhynchus* (Banca), allied to *P. brachyurus* (Borneo, Sumatra, Malacca), and *Pitta Bangkanus* (Banca), allied to *P. sordidus* (Borneo and Sumatra).

inhabiting the Philippines. This brings the known mammalia to twenty-one species, and no doubt several others remain to be discovered. The birds have been increased from 219 to 288 species, and the additions include many Malayan genera which were thought to be absent. Such are Phyllornis (green bulbul); Eurylæmus (gaper), Malacopteron, one of the babblers; and Criniger, one of the fruit-thrushes; as well as Batrachostomus, the frog-mouthed goatsucker. There still remain, however, a large number of Malayan genera absent from the Philippines, while there are a few Australian and Indian or Chinese genera which are not Malayan. We must also note that about nine tenths of the mammalia and two thirds of the land birds are peculiar species, a very much larger proportion than is found on any other Malay island.

The origin of these peculiarities is not difficult to trace. The Philippines are almost surrounded by deep sea, but are connected with Borneo by means of two narrow submarine banks, on the northern of which is situated Palawan, and on the southern the Sooloo islands. Two small groups of islands, the Bashees and Babuyanes, have also afforded a partial connection with the continent by way of Formosa. It is evident that the Philippines once formed part of the great Malayan extension of Asia, but that they were separated considerably earlier than Java; and having been since greatly isolated and much broken up by volcanic disturbances, their species have, for the most part, become modified into distinct local species. They have also received a few Chinese types by the route already indicated, and a few Australian forms, owing to their proximity to the Moluccas. The reason of their comparative poverty in genera and species of the higher animals is that they have been subjected to a great amount of submersion in recent times, greatly reducing their area, and causing, no doubt, the extinction of a considerable portion of their fauna. This is not a mere hypothesis, but is supported by direct evidence; for I am informed by Mr. Everett, who has made extensive explorations in the islands, that almost everywhere are found large tracts of elevated coral reefs containing shells similar to those living in the adjacent seas—an indisputable proof of recent elevation.

Concluding Remarks on the Malay Islands.—This completes our sketch of the great Malay islands, the seat of the typical Malayan fauna. It has been shown that the peculiarities presented by the individual islands may be all sufficiently well explained by a very simple and comparatively unimportant series of geographical changes, combined with a limited amount of change of climate towards the northern tropic. Beginning in late Miocene times, when the deposits on the south coast of Java were upraised, we suppose a general elevation of the whole of the extremely shallow seas uniting what are now Sumatra, Java, Borneo, and the Philippines with the Asiatic continent, and forming that extended equatorial area in which the typical Malayan fauna was developed. After a long period of stability, giving ample time for the specialization of so many peculiar types, the Philippines were first separated; then at a considerably later period Java; a little later Sumatra and Borneo; and finally the islands south of Singapore to Banca and Biliton. This one simple series of elevations and subsidences, combined with the changes of climate already referred to, and such local elevations and depressions as must undoubtedly have occurred, appears sufficient to have brought about the curious, and at first sight puzzling, relations of the faunas of Java and the Philippines as compared with those of the larger islands.

We will now pass on to the consideration of two other groups which offer features of special interest, and which will complete our illustrative survey of recent continental islands.

CHAPTER XVIII.

JAPAN AND FORMOSA.

Japan : its Position and Physical Features.—Zoological Features of Japan.—Mammalia.—Birds.—Birds Common to Great Britain and Japan.—Birds Peculiar to Japan.—Japan Birds Recurring in Distant Areas.—Formosa.—Physical Features of Formosa.—Animal Life of Formosa.—Mammalia.—Land Birds Peculiar to Formosa. —Formosan Birds Recurring in India or Malaya. —Comparison of Faunas of Hainan, Formosa, and Japan.—General Remarks on Recent Continental Islands.

JAPAN.

THE Japanese Islands occupy a very similar position on the eastern shore of the great Euro-Asiatic continent to that of the British Islands on the western, except that they are about sixteen degrees farther south, and, having a greater extension in latitude, enjoy a more varied as well as a more temperate climate. Their outline is also much more irregular and their mountains loftier, the volcanic peak of Fusiyama being 14,177 feet high ; while their geological structure is very complex, their soil extremely fertile, and their vegetation in the highest degree varied and beautiful. Like our own islands, too, they are connected with the continent by a marine bank less than a hundred fathoms below the surface—at all events, towards the north and south ; but in the intervening space the Sea of Japan opens out to a width of 600 miles, and in its central portion is very deep ; and this may be an indication that the connection between the islands and the continent is of rather ancient date. At the Strait of Corea the distance from the mainland is about 120 miles, while at the northern extremity of Yesso it is about 200. The island of Saghalien, however, separated from Yesso by a strait only twenty-five miles wide, forms a connection with Amoorland in about 52° N. lat. A southern warm current flowing a little to the eastward of the islands ameliorates their climate much in the same way as the Gulf Stream does ours, and,

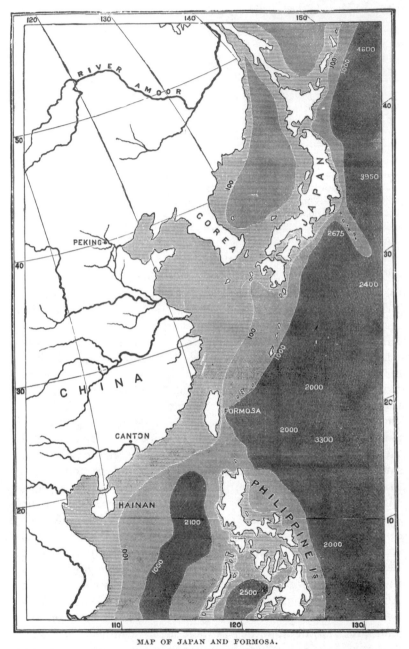

MAP OF JAPAN AND FORMOSA.

Light tint, sea under 100 fathoms; medium tint, under 1000 fathoms; dark tint, over 1000 fathoms. The figures show the depth in fathoms.

added to their insular position, enables them to support a more tropical vegetation and more varied forms of life than are found at corresponding latitudes in China.

Zoological Features of Japan.—As we might expect from the conditions here sketched out, Japan exhibits in all its forms of animal life a close general resemblance to the adjacent continent, but with a considerable element of specific individuality; while it also possesses some remarkable isolated groups. It also exhibits indications of there having been two or more lines of migration at different epochs. The majority of its animals are related to those of the temperate or cold regions of the continent, either as identical or allied species; but a smaller number have a tropical character, and these have in several instances no allies in China, but occur again only in Northern India or the Malay Archipelago. There is also a slight American element in the fauna of Japan, a relic probably of the period when a land communication existed between the two continents over what are now the shallow seas of Japan, Ochotsk, and Kamtschatka. We will now proceed to examine the peculiarities and relations of the fauna.

Mammalia.—The mammalia of Japan at present known are forty in number; not very many when compared with the rich fauna of China and Manchuria, but containing monkeys, bears, deer, wild goats and wild boars, as well as foxes, badgers, moles, squirrels, and hares, so that there can be no doubt whatever that they imply a land connection with the continent. No complete account of Japan mammals has been given by any competent zoologist since the publication of Von Siebold's "Fauna Japonica" in 1844; but by collecting together most of the scattered observations since that period the following list has been drawn up, and will, it is hoped, be of use to naturalists. The species believed to be peculiar to Japan are printed in italics. These are very numerous, but it must be remembered that Corea and Manchuria (the portions of the continent opposite Japan) are comparatively little known, while in very few cases have the species of Japan and of the continent been critically compared. Where this has been done, however, the peculiar species established by the older naturalists have been in many cases found to be correct.

LIST OF THE MAMMALIA OF THE JAPANESE ISLANDS.

1. *Macacus speciosus.* A monkey with rudimentary tail and red face, allied to the Barbary ape. It inhabits the island of Niphon up to 41° N. lat., and has thus the most northern range of any living monkey.

2. *Pteropus dasymallus.* A peculiar fruit-bat, found in Kiusiu Island only (lat. 33° N.), and thus ranging farther north of the equator than any other species of the genus.

3. Rhinolophus ferrum-equinum. The great horseshoe-bat, ranges from Britain across Europe and temperate Asia to Japan. It is the *R. nippon* of the "Fauna Japonica," according to Mr. Dobson's "Monograph of Asiatic Bats."

4. R. minor. Found also in Burmah, Yunnan, Java, Borneo, etc.

5. Vesperugo pipistrellus. From Britain across Europe and Asia.

6. " abramus. Also in India and China.

7. " noctula. From Britain across Europe and Asia.

8. " molossus. Also in China.

9. Vespertilio capaccinii. Philippine Islands and Italy. This is *V. macrodactylus* of the "Fauna Japonica," according to Mr. Dobson.

10. Miniopterus Schreibersii. Philippines, Burmah, Malay Islands. This is *Vespertilio blepotis* of the "Fauna Japonica."

11. *Talpa wogura.* Closely resembles the common mole of Europe, but has six incisors instead of eight in the lower jaw.

12. *Urotrichus talpoides.* A peculiar genus of moles confined to Japan and the northwest coast of North America. The American species has been named *Urotrichus Gibsii ;* but Mr. Lord, after comparing the two, says that he " can find no difference whatever, either generic or specific. In shape, size, and color they are exactly alike.

13. Sorex myosurus. A shrew, found also in India and Malaya.

14. *Sorex dzi-nezumi.*

15. " *umbrinus.*

16. " *platycephalus.*

17. Ursus arctos, var. A peculiar variety of the European brown bear, which inhabits also Amoorland and Kamtschatka. It is the *Ursus ferox* of the "Fauna Japonica."

18. *Ursus Japonicus.* A peculiar species allied to the Himalayan and Formosan species. Named *U. Tibetanus* in the "Fauna Japonica."

19. *Meles anakuma.* Differs from the European and Siberian badgers in the form of the skull.

20. *Mustela brachyura.* A peculiar marten found also in the Kurile Islands.

21. " *melanopus.* The Japanese sable.

22. " *Japonica.* A peculiar marten (see *Proc. Zool. Soc.,* 1865, p. 104).

23. " Sibericus. Also Siberia and China. This is the *M. italsi* of the "Fauna Japonica," according to Dr. Gray.

24. *Lutronectes Whiteleyi.* A new genus and species of otter (*Proc. Zool. Soc.,* 1867, p. 180). In the "Fauna Japonica" named *Lutra vulgaris.*

25. Enhydris marina. The sea-otter of California and Kamtschatka.

26. *Canis hodophylax.* According to Dr. Gray, allied to *Cuon Sumatranus* of the

Malay Islands, and *C. alpinus* of Siberia, if not identical with one of them (*Proc. Zool. Soc.*, 1868, p. 500).

27. *Vulpes Japonica.* A peculiar fox. *Canis vulpes* of "Fauna Japonica."

28. Nyctereutes procyonoides. The raccoon-dog of Northern China and Amoorland.

29. *Lepus brachyurus.* A peculiar hare.

30. *Sciurus lis.* A peculiar squirrel.

31. *Pteromys leucogenys.* The white-cheeked flying squirrel.

32. " *momoga.* Perhaps identical with a Cambojan species (*Proc. Zool. Soc.*, 1861, p. 137).

33. *Myoxus Japonicus.* A peculiar dormouse. *M. elegans* of the "Fauna Japonica;" *M. Javanicus*, Schinz ("Synopsis Mammalium," ii., p. 530).

34. Mus argenteus. China.

35. " *molossinus.*

36. " *nezumi.*

37. " *speciosus.*

38. *Cervus sika.* A peculiar deer allied to *C. pseudaxis* of Formosa and *C. Manchuricus* of Northern China.

39. *Nemorhedus crispa.* A goat-like antelope allied to *N. Sumatranus* of Sumatra, and *N. Swinhoei* of Formosa.

40. *Sus leucomystax.* A wild-boar allied to *S. Taivanus* of Formosa.

We thus find that no less than twenty-six out of the forty Japanese mammals are peculiar; and if we omit the aerial bats (nine in number) as well as the marine sea-otter, we shall have remaining only thirty strictly land mammalia, of which twenty-five are peculiar, or five sixths of the whole. Nor does this represent all their speciality; for we have a mole differing in its dentition from the European mole; another closely allied to an American species; a peculiar genus of otters; and an antelope whose nearest allies are in Formosa and Sumatra. The importance of these facts will be best understood when we shall have examined the corresponding affinities of the birds of Japan.

Birds.—Owing to the recent researches of some English residents, we have probably a fuller knowledge of the birds than of the mammalia; yet the number of true land birds ascertained to inhabit the islands either as residents or migrants is only one hundred and sixty-five, which is less than might be expected considering the highly favorable conditions and the extreme riches of the adjacent continent—Mr. Swinhoe's list of the birds of China containing more than four hundred land species, after deducting all that are peculiar to the adjacent islands. Only sixteen species, or about one tenth of the whole, are now considered to be peculiar to Japan ; but even of these five are classed by Mr. Seebohm as sub-species or slightly modified forms of continental birds, so that eleven only are well-marked species, undoubtedly distinct from those of any other country.

The great majority of the birds are decidedly temperate forms identical with those of Northern Asia and Europe ; while no less than forty of the species are also found in Britain, or are such slight modifications of British species that the difference is only perceptible to a trained ornithologist. The following list of the birds common to Britain and Japan is very interesting, when we consider that these countries are separated by the whole extent of the European and Asiatic continents, or by almost exactly one fourth of the circumference of the globe :

BIRDS COMMON TO GREAT BRITAIN AND JAPAN.[1]

1. Common Creeper (*Certhia familiaris*).
2. Nuthatch (*Sitta Europœa*).
3. Coal Tit (*Parus ater*).
4. Marsh Tit, sub-sp. (*P. Japonicus*).
5. Long-tailed Tit (*Acredula caudata*).
6. Great Gray Shrike (*Lanius excubitor*).
7. Nutcracker (*Nucifraga caryocatactes*).
8. Carrion-crow (*Corvus corone*).
9. Raven (*Corvus corax*).
10. Waxwing (*Ampelis garrulus*).
11. Swallow, sub-sp. (*Hirundo gutturalis*).
12. Sand-martin (*Cotyle riparia*).
13. Brambling (*Fringilla montifringilla*).
14. Siskin (*Chrysomitris spinus*).
15. Lesser Redpole (*Ægiothus linaria*).
16. Tree-sparrow (*Passer montanus*).
17. Pine-grosbeak (*Pyrrhula enucleator*).
18. Reed - bunting, sub - sp. (*Emberiza pyrrhulina*).
19. Snow-bunting (*Plectrophanes nivalis*).
20. Gray Wagtail, sub-sp. (*Motacilla melanope*).
21. Great Spotted Woodpecker (*Picus major*).
22. Great Black Woodpecker (*Dryocopus martius*).
23. Cuckoo (*Cuculus canorus*).
24. Hoopoe (*Upupa epops*).
25. Rock-dove (*Columba livia*).
26. Hen-harrier (*Circus cyaneus*).
27. Goshawk (*Astur palumbarius*).
28. Sparrow-hawk (*Accipiter nisus*).
29. Rough-legged Buzzard (*Buteo lagopus*).
30. Golden Eagle (*Aquila chrysaetos*).
31. White-tailed Eagle (*Haliœtus albicilla*).
32. Kestrel (*Falco tinnunculus*).
33. Hobby (*F. subbuteo*).
34. Merlin (*F. œsalon*).
35. Peregrine Falcon (*F. peregrinus*).
36. Greenland Falcon (*F. candicans*).
37. Osprey (*Pandion haliœtus*).
38. Eagle-owl (*Bubo maximus*).
39. Short - eared Owl (*Asio accipitrinus*).
40. Long-eared Owl (*A. otus*).

[1] Extracted from Messrs. Blakiston and Pryer's "Catalogue of Birds of Japan" (*Ibis*, 1878, p. 209), with Mr. Seebohm's additions and corrections (*Ibis*, 1879, p. 18).

But these forty species by no means fairly represent the amount of *resemblance* between Britain and Japan as regards birds; for there are also wrens, hedge-sparrows, gold-crests, sedge-warblers, pipits, larks, rock-thrushes, jays, and many others, which, though distinct species from our own, have the same general appearance, and give a familiar aspect to the ornithology. There remain, however, a considerable body of Chinese and Siberian species, which link the islands to the neighboring parts of the continent; and there are also a few which are Malayan or Himalayan rather than Chinese, and thus afford us an interesting problem in distribution.

The sixteen species and sub-species which are altogether peculiar to Japan are, for the most part, allied to birds of North China and Siberia, but three are decidedly tropical, and one of them—a fruit-pigeon (*Treron Sieboldi*)—has no close ally nearer than Java and the Himalayas. In the following list the affinities of the species are indicated wherever they have been ascertained :

List of the Species of Land Birds Peculiar to Japan.

1. *Parus Japonicus.* A sub-species of *P. palustris*, very like Siberian varieties.
2. " *varius.* Very distinct. Its nearest ally is in Formosa.
3. *Hypsipetes amaurosis.* A tropical genus. Allied to species of South China and India.
4. *Garrulus Japonicus.* Allied to our European jay. In Niphon only.
5. " *Lidthi.* A very distinct and handsome species. (See *Ibis*, 1873, p. 478.)
6. *Zosterops Japonica.* Allied to a migratory Chinese species.
7. *Chelidon Blakistoni.* Allied to *C. Whiteleyi* of North China.
8. *Chlorospiza Kawarahiba.* Allied to *C. Sinica* of China and Japan.
9. *Emberiza ciopsis.* A sub-species of the *E. cioides* of North China.
10. " *Yessoensis.* Allied to the Siberian *E. passerina.*
11. *Euspiza variabilis.* A very distinct species.
12. *Picus Kisuki.* Allied to *P. pygmœus* of Central Asia.
13. *Gecinus awokera.* Allied to *G. canus* (North China) and *G. viridis* (Europe).
14. *Mulleripicus Richardsi.* Allied to *M. Crawfurdi* of Pegu. In Tzus Sima Island (*Proc. Zool. Soc.*, 1879, p. 386).
15. *Treron Sieboldi.* Allied to *T. sphenura* (Himalayas) and *T. Korthalsi* (Java).
16. *Accipiter gularis.* A sub-species of the Malayan *A. virgatus* (also in Formosa).
17. *Buteo Hemilasius.* A distinct species.
18. *Syrnium rufescens.* A sub-species of *S. Uralense* of East Europe and Siberia.

Japan Birds Recurring in Distant Areas.—The most interest-ing feature in the ornithology of Japan is undoubtedly the pres-ence of several species which indicate an alliance with such re-mote districts as the Himalayas, the Malay Islands, and Europe. Among the peculiar species, the most remarkable of this class are the fruit-pigeon of the genus Treron, entirely unknown in China, but reappearing in Formosa and Japan ; the Hypsipetes, whose nearest ally is in South China at a distance of nearly five hundred miles ; and the jay (*Garrulus Japonicus*), whose close ally (*G. glandarius*) inhabits Europe only, at a distance of 3700 miles. But even more extraordinary are the following non-peculiar species :—*Spizætus orientalis,* a crested eagle, inhabit-ing the Himalayas, Formosa, and Japan, but unknown in Chi-na ; *Ceryle guttata,* a spotted kingfisher, entirely confined to the Himalayas and Japan ; and *Halcyon Coromanda,* a brilliant red kingfisher inhabiting Northern India, the Malay Islands to Cel-ebes, Formosa, and Japan. We have here an excellent illustra-tion of the favorable conditions which islands afford both for species which elsewhere live farther south (*Halcyon Coromanda*), and for the preservation in isolated colonies of species which are verging towards extinction ; for such we must consider the above-named eagle and kingfisher, both confined to a very limit-ed area on the continent, but surviving in remote islands. The spotted kingfisher, indeed, affords us one of the best examples of that rare phenomenon—a species with a discontinuous range ; for although an island is considered, for purposes of distribution, to form part of one continuous area with the adjacent continent (as when a species is found in France and Britain, or in Siam and Borneo, we do not say that the area of distribution is dis-continuous), yet in this case we have to pass over three thousand miles of land after quitting the island, before we come to the continental portion of the area occupied by the species. Re-ferring to our account of the birth, growth, and death of a spe-cies (in Chapter IV.), it can hardly be doubted that the *Ceryle guttata* formerly ranged from the Himalayas to Japan, and has now died out in the intervening area, owing to geographical and physical changes—a subject which will be better discussed when we have examined the interesting fauna of the island of Formosa.

The other orders of animals are not yet sufficiently known to enable us to found any accurate conclusions upon them. The main facts of their distribution have already been given in my "Geographical Distribution of Animals" (Vol. I., pp. 227–231), and they sufficiently agree with the birds and mammalia in showing a mixture of temperate and tropical forms, with a considerable proportion of peculiar species. Owing to the comparatively easy passage from the northern extremity of Japan through the island of Saghalien to the mainland of Asia, a large number of temperate forms of insects and birds are still able to enter the country, and thus diminish the proportionate number of peculiar species. In the case of mammals this is more difficult; and the large proportion of specific difference in their case is a good indication of the comparatively remote epoch at which Japan was finally separated from the continent. How long ago this separation took place we cannot, of course, tell, but we may be sure it was much longer than in the case of our own islands, and therefore probably in the earlier portion of the Pliocene period.

Formosa.

Among recent continental islands, there is probably none that surpasses in interest and instructiveness the Chinese island named by the Portuguese Formosa, or the "Beautiful." Till quite recently it was a *terra incognita* to naturalists, and we owe all our present knowledge of it to a single man, the late Mr. Robert Swinhoe, who, in his official capacity as one of our consuls in China, visited it several times between 1856 and 1866, besides residing on it for more than a year. During this period he devoted all his spare time and energy to the study of natural history, more especially of the two important groups, birds and mammals; and by employing a large staff of native collectors and hunters he obtained a very complete knowledge of its fauna. In this case, too, we have the great advantage of a very thorough knowledge of the adjacent parts of the continent, in great part due to Mr. Swinhoe's own exertions during the twenty years of his service in that country. We possess, too, the further advantage of having the whole of the available materials in these two classes collected together by Mr. Swinhoe himself after full examination and comparison of specimens; so that there is probably no part of the world (if we except Europe, North America, and British India) of whose warm-blooded vertebrates we possess fuller or more accurate knowledge than we do of those of the coast districts of China and its islands.[1]

Physical Features of Formosa.—The island of Formosa is nearly half the size of Ireland, being 220 miles long, and from twenty to eighty miles wide. It is traversed down its centre by a fine mountain-range, which reaches an altitude of about 8000 feet in the south and 12,000 feet in the northern half of the island, and whose higher slopes and valleys are everywhere clothed with magnificent forests. It is crossed by the line of the Tropic of Cancer a little south of its centre; and this position, combined with its lofty mountains, gives it an unusual variety of tropical and temperate climates. These circumstances are all highly favorable to the preservation and development of animal life; and, from what we already know of its productions, it seems probable that few, if any, islands of approximately the same size and equally removed from a continent will be found to equal it in the number and variety of their higher animals. The outline map on page 358 shows that Formosa is connected with the mainland by a submerged bank, the 100-fathom line including it along with Hainan to the southwest and Japan on the northeast; while the line of 200 fathoms includes also the Madjicosima and Loo-choo Islands, and may perhaps mark out proximately the last great extension of the Asiatic continent, the submergence of which isolated these islands from the mainland.

Animal Life of Formosa.—We are at present acquainted with 35 species of mammalia and 128 species of land birds from Formosa, 14 of the former and 43 of the latter being peculiar, while the remainder inhabit also some part of the continent or adjacent islands. This proportion of peculiar species is per-

---

[1] Mr. Swinhoe died in October, 1877, at the early age of forty-two. His writings on natural history are chiefly scattered through the volumes of the *Proceedings of the Zoological Society* and *The Ibis;* the whole being summarized in his "Catalogue of the Mammals of South China and Formosa" (*Proc. Zool. Soc.*, 1870, p. 615), and his "Catalogue of the Birds of China and its Islands" (*Proc. Zool. Soc.*, 1871, p. 337).

haps (as regards the birds) the highest to be met with in any island which can be classed as both continental and recent, and this, in all probability, implies that the epoch of separation is somewhat remote. It was not, however, remote enough to reach back to a time when the continental fauna was very different from what it is now, for we find all the chief types of living Asiatic mammalia represented in this small island. Thus we have monkeys; insectivora; numerous carnivora; pigs, deer, antelopes, and cattle among ungulata; numerous rodents, and the edentate Manis—a very fair representation of Asiatic mammals, all being of known genera, and of species either absolutely identical with some still living elsewhere, or very closely allied to them. The birds exhibit analogous phenomena, with the exception that we have here two peculiar and very interesting genera.

But besides the amount of specific and generic modification that has occurred, we have another indication of the lapse of time in the peculiar relations of a large proportion of the Formosan animals, which show that a great change in the distribution of Asiatic species must have taken place since the separation of the island from the continent. Before pointing these out, it will be advantageous to give lists of the mammalia and peculiar birds of the island, as we shall have frequent occasion to refer to them.

LIST OF THE MAMMALIA OF FORMOSA.

(*The peculiar species are printed in italics.*)

1. *Macacus cyclopis.* A rock-monkey more allied to *M. rhesus* of India than to *M. Sancti-Johannis* of South China.
2. *Pteropus formosus.* A fruit-bat closely allied to the Japanese species. None of the genus are found in China.
3. Vesperugo abramus. China.
4. Vespertilio formosus. Black and orange bat. China.
5. Nyctinomus cestonii. Large-eared bat. China, South of Europe.
6. *Talpo insularis.* A blind mole of a peculiar species.
7. Sorex murinus. Muskrat. China.
8. "   sp. A shrew, undescribed.
9. Erinaceus sp. A hedgehog, undescribed.
10. Ursus Tibetanus. The Thibetan bear. Himalayas and North China.
11. *Helictis subaurantiaca.* The orange-tinted tree-civet. Allied to *H. Nipalensis* of the Himalayas more than to *H. moschata* of China.
12. Martes flavigula, var. The yellow-necked marten. India, China.

13. Felis macroscelis.   The clouded tiger of Siam and Malaya.
14. " viverrina.   The Asiatic wild-cat.   Himalayas and Malacca.
15. " Chinensis.   The Chinese tiger-cat.   China.
16. Viverricula Malaccensis.   Spotted civet.   China, India.
17. Paguma larvata.   Gem-faced civet.   China.
18. Sus Taivanus.   Allied to the wild-pig of Japan.
19. Cervulus Reevesii.   Reeve's muntjac.   China.
20. Cervus pseudaxis.   Formosan spotted deer.   Allied to C. Sika of Japan.
21. " Swinhoei.   Swinhoe's Rusa deer.   Allied to Indian and Malayan species.
22. Nemorhœdus Swinhoei.   Swinhoe's goat-antelope.   Allied to the species of Su-
     matra and Japan.
23. Bos Chinensis.   South China wild-cow.
24. Mus bandicota.   The bandicoot rat.   Perhaps introduced from India.
25. " Indicus.   Indian rat.
26. Mus coxinga.   Spinous country-rat.
27. " canna.   Silken country-rat.
28. " losea.   Brown country-rat.
29. Sciurus castaneoventris.   Chestnut-bellied squirrel.   China and Hainan.
30. " MacClellandi.   MacClelland's squirrel.   Himalayas, China.
31. Sciuropterus Kaleensis.   Small Formosan flying-squirrel.   Allied to S. alboniger
     of Nepal.
32. Pteromys grandis.   Large red flying-squirrel.   Allied to Himalayan and Bor-
     nean species.   From North Formosa.
33. Pteromys pectoralis.   White-breasted flying-squirrel.   From South Formosa.
34. Lepus Sinensis.   Chinese hare.   Inhabits South China.
35. Manis Dalmanni.   Scaly ant-eater.   China and the Himalayas.

The most interesting and suggestive feature connected with
these Formosan mammals is the identity or affinity of several
of them with Indian or Malayan rather than with Chinese species.   We have the rock-monkey of Formosa allied to the rhesus
monkeys of India and Burmah, not to those of South China and
Hainan.   The tree-civet (*Helictis subaurantiaca*) and the small
flying-squirrel (*Sciuropterus Kaleensis*) are both allied to Himalayan species.   Swinhoe's deer and goat-antelope are nearest to
Malayan species, as are the red and white-breasted flying-squir-
rels; while the fruit-bat, the wild-pig, and the spotted deer are
all allied to peculiar Japanese species.   The clouded tiger is a
Malay species unknown in China, while the Asiatic wild-cat is
a native of the Himalayas and Malacca.   It is clear, therefore,
that before Formosa was separated from the mainland the above-
named animals or their ancestral types must have ranged over
the intervening country as far as the Himalayas on the west,

Japan on the north, and Borneo or the Philippines on the south; and that, after that event occurred, the conditions were so materially changed as to lead to the extinction of these species in what are now the coast provinces of China, while they or their modified descendants continued to exist in the dense forests of the Himalayas and the Malay Islands, and in such detached islands as Formosa and Japan. We will now see what additional light is thrown upon this subject by an examination of the birds.

LIST OF THE LAND BIRDS PECULIAR TO FORMOSA.

TURDIDÆ (Thrushes).

1. *Turdus albiceps.* Allied to Chinese species.

SYLVIIDÆ (Warblers).

2. *Cisticola volitans.* Allied to *C. schœnicola* of India and China.
3. *Herbivox cantans.* Sub-species of *H. cantillans* of North China and Japan.
4. *Notodela montium.* Allied to *N. leucura* of the Himalayas; no ally in China.

TIMALIIDÆ (Babblers).

5. *Pomatorhinus musicus.* Allies in South China and the Himalayas.
6.   " " *erythrocnemis.* "   "   "
7. *Garrulax ruficeps.* Allied to *G. albogularis* of North India and East Thibet, not to the species of South China (*G. sannio*).
8. *Janthocincla pœcilorhyncha.* Allied to *J. cœrulata* of the Himalayas. None of the genus in China.
9. *Trochalopteron Taivanus.* Allied to a Chinese species.
10. *Alcippe Morrisoniana.*} Near the Himalayan *A. Nipalensis.* None of the genus
11.   "   *Brunnea.*   }   in China.
12. *Sibia auricularis.* Allied to the Himalayan *S. capistrata.* The genus not known in China.

PANURIDÆ (Bearded Tits, etc.).

13. *Suthora bulomachus.* Allied to the Chinese *S. suffusa.*

CINCLIDÆ (Dippers and Whistling-thrushes).

14. *Myiophoneus insularis.* Allied to *M. Horsfieldi* of South India.

PARIDÆ (Tits).

15. *Parus insperatus.* Sub-species of *P. monticola* of the Himalayas and East Thibet.
16. *P. castaneoventris.* Allied to *P. varius* of Japan.

LIOTRICHIDÆ (Hill Tits).

17. *Liocichla Steerii.* A peculiar genus of a specially Himalayan family, quite unknown in China.

PYCNONOTIDÆ (Bulbuls).

18. *Pycnonotus (Spizixos) cinereicapillus.* Very near *P. cemitorques* of China.
19. *Hypsipetes nigerrimus.* Allied to *H. concolor* of Assam, not to *H. MacClellandi* of China.

ORIOLIDÆ (Orioles).

20. *Analcipus ardens.* Allied to *A. Traillii* of the Himalayas and Tenasserim.

CAMPEPHAGIDÆ (Caterpillar Shrikes).

21. *Graucalus rex-pineti.* Closely allied to the Indian *G. Macei.* No ally in China.

DICRURIDÆ (King Crows).

22. *Chaptia Brauniana.* Closely allied to *C. ænea* of Assam. No ally in China.

MUSCICAPIDÆ (Flycatchers).

23. *Cyornis vivida.* Allied to *C. rubeculoides* of India.

CORVIDÆ (Jays and Crows).

24. *Garrulus Taivanus.* Allied to *G. Sinensis* of South China.
25. *Urocissa cærulea.* A very distinct species from its Indian and Chinese allies.
26. *Dendrocitta Formosæ.* A sub-species of the Chinese *D. Sinensis.*

PLOCEIDÆ (Weaver Finches).

27. *Munia Formosana.* Allied to *M. rubronigra* of India and Burmah.

ALAUDIDÆ (Larks).

28. *Alauda Sala.* ⎫
29. " *Wattersi.* ⎬ Allies in South China.

PITTIDÆ (Pittas).

30. *Pitta oreas.* Allied to *P. cyanoptera* of Malaya and South China.

PICIDÆ (Woodpeckers).

31. *Picus insularis.* Allied to *P. leuconotus* of Japan and Siberia.

MEGALÆMIDÆ.

32. *Megalæma Nuchalis.* Allied to *M. Oortii* of Sumatra and *M. faber* of Hainan. No allies in China.

CAPRIMULGIDÆ (Goatsuckers).

33. *Caprimulgus stictomus.* A sub-species of *C. monticolus* of India and China.

COLUMBIDÆ (Pigeons).

34. *Treron Formosæ.* Allied to Malayan species.
35. *Sphenocercus sororius.* Allied to Malayan species and to *S. Sieboldi* of Japan. No allies of these two birds inhabit China.
36. *Chalcophaps Formosana.* Allied to the Indian species which extends to Tenasserim and Hainan.

TETRAONIDÆ (Grouse and Partridges).

37. *Oreoperdix crudigularis.*   A peculiar genus of partridges.
38. *Bambusicola sonorivox.*   Allied to the Chinese *B. thoracica.*
39. *Areoturnix rostrata.*   Allied to the Chinese *A. Blakistonii.*

PHASIANIDÆ (Pheasants).

40. *Phasianus Formosanus.*   Allied to *P. torquatus* of China.
41. *Euplocamus Swinhoei.*   A very peculiar and beautiful species allied to the tropical fire-backed pheasants, and to the silver pheasant of North China.

STRIGIDÆ (Owls).

42. *Athene pardalota.*   Closely allied to a Chinese species.
43. *Lempigius Hambroekii.*   Allied to a Chinese species.

This list exhibits to us the marvellous fact that more than half the peculiar species of Formosan birds have their nearest allies in such remote regions as the Himalayas, South India, the Malay Islands, or Japan, rather than in the adjacent parts of the Asiatic continent.   Fourteen species have Himalayan allies, and six of these belong to genera which are unknown in China. One has its nearest ally in the Nilgherries, and five in the Malay Islands; and of these six, four belong to genera which are not Chinese.   Two have their only near allies in Japan.   Perhaps more curious still are those cases in which, though the genus is Chinese, the nearest allied species is to be sought for in some remote region.   Thus, we have the Formosan babbler (*Garrulax ruficeps*) not allied to the species found in South China, but to one inhabiting North India and East Thibet; while the black bulbul (*Hypsipetes nigerrimus*) is not allied to the Chinese species, but to an Assamese form.

In the same category as the above we must place eight species not peculiar to Formosa, but which are Indian or Malayan instead of Chinese, so that they offer examples of discontinuous distribution somewhat analogous to what we found to occur in Japan.   These are enumerated in the following list:

SPECIES OF BIRDS COMMON TO FORMOSA AND INDIA OR MALAYA, BUT NOT FOUND IN CHINA.

1. *Siphia superciliaris.*   The rufous-breasted flycatcher of the Southeast Himalayas.
2. *Halcyon Coromanda.*   The great red kingfisher of India, Malaya, and Japan.
3. *Palumbus pulchricollis.*   The Darjeeling wood-pigeon of the Southeast Himalayas.

4. *Turnix Dussumieri.* The larger button-quail of India.
5. *Spizœtus Nipalensis.* The spotted hawk-eagle of Nepal and Assam.
6. *Lophospiza trivirgata.* The crested goshawk of the Malay Islands.
7. *Bulaca Newarensis.* The brown wood-owl of the Himalayas.
8. *Strix candida.* The grass-owl of India and Malaya.

The most interesting of the above are the pigeon and the fly-catcher, both of which are, so far as yet known, strictly confined to the Himalayan mountains and Formosa. They thus afford examples of discontinuous specific distribution exactly parallel to that of the great spotted kingfisher, already referred to as found only in the Himalayas and Japan.

Comparison of the Faunas of Hainan, Formosa, and Japan.
—The island of Hainan, on the extreme south of China, and only separated from the mainland by a strait fifteen miles wide, appears to have considerable similarity to Formosa, inasmuch as it possesses seventeen peculiar land birds (out of 130 obtained by Mr. Swinhoe), two of which are close allies of Formosan species, while two others are identical. We also find four species whose nearest allies are in the Himalayas. Our knowledge of this island and of the adjacent coast of China is not yet sufficient to enable us to form an accurate judgment of its relations, but it seems probable that it was separated from the continent at, approximately, the same epoch as Formosa and Japan, and that the special features of each of these islands are mainly due to their geographical position. Formosa, being more completely isolated than either of the others, possesses a larger proportion of peculiar species of birds ; while its tropical situation and lofty mountain-ranges have enabled it to preserve an unusual number of Himalayan and Malayan forms. Japan, almost equally isolated towards the south, and having a much greater variety of climate as well as a much larger area, possesses about an equal number of mammalia with Formosa, and an even larger proportion of peculiar species. Its birds, however, though more numerous, are less peculiar; and this is probably due to the large number of species which migrate northward in summer, and find it easy to enter Japan through the Kurile Isles or Saghalien. Japan, too, is largely peopled by those northern types which have an unusually wide range, and which, being almost

all migratory, are accustomed to cross over seas of moderate extent. The regular or occasional influx of these species prevents the formation of special insular races, such as are almost always produced when a portion of the population of a species remains for a considerable time completely isolated. We thus have explained the curious fact that while the mammalia of the two islands are almost equally peculiar (those of Japan being most so in the present state of our knowledge), the birds of Formosa show a far greater number of peculiar species than those of Japan.

General Remarks on Recent Continental Islands.—We have now briefly sketched the zoological peculiarities of an illustrative series of recent continental islands, commencing with one of the most recent—Great Britain—in which the process of formation of peculiar species has only just commenced, and terminating with Formosa, probably one of the most ancient of the series, and which accordingly presents us with a very large proportion of peculiar species, not only in its mammalia, which have no means of crossing the wide strait which separates it from the mainland, but also in its birds, many of which are quite able to cross over.

Here, too, we obtain a glimpse of the way in which species die out and are replaced by others, which quite agrees with what the theory of evolution assures us must have occurred. On a continent, the process of extinction will generally take effect on the circumference of the area of distribution, because it is there that the species comes into contact with such adverse conditions or competing forms as prevent it from advancing farther. A very slight change will evidently turn the scale and cause the species to contract its range, and this usually goes on till it is reduced to a very restricted area, and finally becomes extinct. It may conceivably happen (and almost certainly has sometimes happened) that the process of restriction of range by adverse conditions may act in one direction only, and over a limited district, so as ultimately to divide the specific area into two separated parts, in each of which a portion of the species will continue to maintain itself. We have seen that there is reason to believe that this has occurred in a very few cases both in North

America and in Northern Asia (see pp. 62, 64). But the same thing has certainly occurred in a considerable number of cases, only it has resulted in the divided areas being occupied by *representative forms* instead of by the very same species. The cause of this is very easy to understand. We have already shown that there is a large amount of local variation in a considerable number of species, and we may be sure that were it not for the constant intermingling and intercrossing of the individuals inhabiting adjacent localities, this tendency to local variation would soon form distinct races. But as soon as the area is divided into two portions, the intercrossing is stopped, and the usual result is that two closely allied races, classed as representative species, become formed. Such pairs of allied species on the two sides of a continent, or in two detached areas, are very numerous; and their existence is only explicable on the supposition that they are descendants of a parent form which once occupied an area comprising that of both of them; that this area then became discontinuous; and, lastly, that, as a consequence of the discontinuity, the two sections of the parent species became segregated into distinct races or new species.

Now, when the division of the area leaves one portion of the species in an island, a similar modification of the species, either in the island or in the continent, occurs, resulting in closely allied but distinct forms; and such forms are, as we have seen, highly characteristic of island faunas. But islands also favor the occasional preservation of the unchanged species—a phenomenon which very rarely occurs in continents. This is probably due to the absence of competition in islands, so that the parent species there maintains itself unchanged, while the continental portion, by the force of that competition, is driven back to some remote mountain area, where it too obtains a comparative freedom from competition. Thus may be explained the curious fact that the species common to Formosa and India are generally confined to limited areas in the Himalayas, or in other cases are found only in remote islands, as Japan or Hainan.

The distribution and affinities of the animals of continental islands thus throw much light on that obscure subject—the decay and extinction of species; while the numerous and delicate gradations in the modification of the continental species—from perfect identity, through slight varieties, local forms, and insular races, to well-defined species and even distinct genera—afford an overwhelming mass of evidence in favor of the theory of "descent with modification."

We shall now pass on to another class of islands, which, though originally forming parts of continents, were separated from them at very remote epochs. This antiquity is clearly manifested in their existing faunas, which present many peculiarities, and offer some most curious problems to the student of distribution.

CHAPTER XIX.

ANCIENT CONTINENTAL ISLANDS: THE MADAGASCAR GROUP.

Remarks on Ancient Continental Islands.—Physical Features of Madagascar.—Biological Features of Madagascar.—Mammalia.—Reptiles.—Relation of Madagascar to Africa.—Early History of Africa and Madagascar.—Anomalies of Distribution, and how to Explain them.—The Birds of Madagascar as Indicating a Supposed Lemurian Continent.—Submerged Islands between Madagascar and India.—Concluding Remarks on "Lemuria."—The Mascarene Islands.—The Comoro Islands. —The Seychelles Archipelago.—Birds of the Seychelles.—Reptiles and Amphibia. —Fresh-water Fishes.—Land Shells.—Mauritius, Bourbon, and Rodriguez.— Birds.—Extinct Birds and their Probable Origin.—Reptiles.—Flora of Madagascar and the Mascarene Islands.—Curious Relations of Mascarene Plants.—Endemic Genera of Mauritius and Seychelles.—Fragmentary Character of the Mascarene Flora.—Flora of Madagascar Allied to that of South Africa.—Preponderance of Ferns in the Mascarene Flora.—Concluding Remarks on the Madagascar Group.

WE have now to consider the phenomena presented by a very distinct class of islands—those which, although once forming part of a continent, have been separated from it at a remote epoch when its animal forms were very unlike what they are now. Such islands preserve to us the record of a bygone world —of a period when many of the higher types had not yet come into existence, and when the distribution of others was very different from what prevails at the present day. The problem presented by these ancient islands is often complicated by the changes they themselves have undergone since the period of their separation. A partial subsidence will have led to the extinction of some of the types that were originally preserved, and may leave the ancient fauna in a very fragmentary state; while subsequent elevations may have brought it so near to the continent that some immigration even of mammalia may have taken place. If these elevations and subsidences occurred several times over, though never to such an extent as again to unite the island with the continent, it is evident that a very complex result might be produced; for, besides the relics of the ancient fauna, we might have successive immigrations from surrounding lands reaching down to the era of existing species. Bearing in mind these possible changes, we shall generally be able to arrive at a fair conjectural solution of the phenomena of distribution presented by these ancient islands.

Undoubtedly the most interesting of such islands, and that which exhibits their chief peculiarities in the greatest perfection, is Madagascar, and we shall therefore enter somewhat fully into its biological and physical history.

Physical Features of Madagascar.—This great island is situated about 250 miles from the east coast of Africa, and extends from 12° to 25½° S. lat. It is almost exactly 1000 miles long, with an extreme width of 360 and an average width of more than 260 miles. A lofty granitic plateau, from 80 to 160 miles wide, and from 3000 to 5000 feet high, occupies its central portion, on which rise peaks and domes of basalt and granite to a height of nearly 9000 feet; and there are also numerous extinct volcanic cones and craters. All round the island, but especially developed on the south and west, are plains of a few hundred feet elevation, formed of rocks which are shown by their fossils to be of Jurassic age, or, at all events, to belong to somewhere near the middle portion of the Secondary period. The higher granitic plateau consists of bare undulating moors, while the lower Secondary plains are more or less wooded; and there is here also a continuous belt of dense forest, varying from six or eight to fifty miles wide, encircling the whole island, usually at about thirty miles' distance from the coast, but in the northeast coming down to the sea-shore.

The sea around Madagascar, when the shallow bank on which it stands is passed, is generally deep. This 100-fathom bank is only from one to three miles wide on the east side, but on the west it is much broader, and stretches out opposite Mozambique to a distance of about eighty miles. The Mozambique Channel varies from less than 500 to more than 1500 fathoms deep, the shallowest part being where the Comoro Islands and adjacent shoals seem to form stepping-stones to the continent of Africa. The 500-fathom line includes Aldabra and the

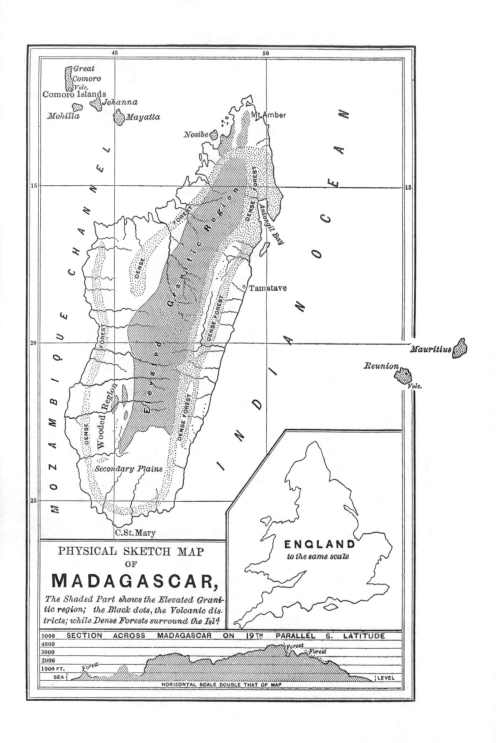

PHYSICAL SKETCH MAP
OF
MADAGASCAR,

The Shaded Part shows the Elevated Granitic region; the Black dots, the Volcanic districts; while Dense Forests surround the Isl^d.

ENGLAND
to the same scale

SECTION ACROSS MADAGASCAR ON 19TH PARALLEL S. LATITUDE

HORIZONTAL SCALE DOUBLE THAT OF MAP

small Farquhar Islands to the north of Madagascar, while to the east the sea deepens rapidly to the 1000-fathom line, and then more slowly, a profound channel of 2400 fathoms separating Madagascar from Bourbon and Mauritius. To the northeast of Mauritius are a series of extensive shoals, forming four large banks less than 100 fathoms below the surface, while the 1000-fathom line includes them all, with an area about half that of Madagascar itself. A little farther north is the Seychelles group, also standing on an extensive 1000-fathom bank, while all around the sea is more than 2000 fathoms deep.

It seems probable, then, that to the northeast of Madagascar there was once a series of very large islands, separated from it by not very wide straits; while eastward across the Indian Ocean we find the Chagos and Maldive coral atolls, marking the position of other large islands, which together would form a line of communication by comparatively easy stages of 400 or 500 miles each between Madagascar and India. These submerged islands, as shown in our map on page 389, are of great importance in explaining some anomalous features in the zoology of this great island.

If the rocks of Secondary age which form a belt around the island are held to indicate that Madagascar was once of less extent than it is now (though this by no means necessarily follows), we have also evidence that it has recently been considerably larger; for along the east coast there is an extensive barrier coral reef about 350 miles in length, and varying in distance from the land from a quarter of a mile to three or four miles. This is good proof of recent subsidence; while we have no record of raised coral rocks inland which would certainly mark any recent elevation, because fringing coral reefs surround a considerable portion of the northern, eastern, and southwestern coasts. We may therefore conclude that during Tertiary times the island was usually as large as, and often probably much larger than, it is now.

Biological Features of Madagascar.—Madagascar possesses an exceedingly rich and beautiful fauna and flora, rivalling in some groups most tropical countries of equal extent, and, even when poor in species, of surpassing interest from the singularity,

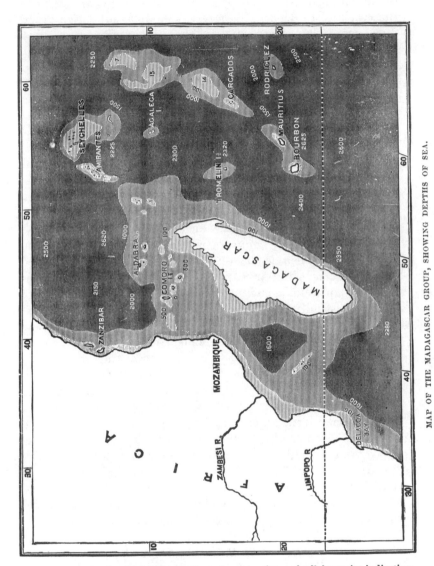

MAP OF THE MADAGASCAR GROUP, SHOWING DEPTHS OF SEA.

In this map the depth of the sea is shown by three tints; the lightest tint indicating
from 0 to 100 fathoms, the medium tint from 100 to 1000 fathoms, the dark tint
more than 1000 fathoms.

the isolation, or the beauty of its forms of life. In order to exhibit the full peculiarity of its natural history and the nature of the problems it offers to the biological student, we must give an outline of its more important animal forms in systematic order.

Mammalia.—Madagascar possesses no less than sixty-six species of mammals—a certain proof in itself that the island has once formed part of a continent; but the character of these animals is very extraordinary and very different from the assemblage now found in Africa or in any other existing continent. Africa is now most prominently characterized by its monkeys, apes, and baboons; by its lions, leopards, and hyenas; by its zebras, rhinoceroses, elephants, buffaloes, giraffes, and numerous species of antelopes. But no one of these animals, nor anything like them, is found in Madagascar, and thus our first impression would be that it could never have been united with the African continent. But as the tigers, the bears, the tapirs, the deer, and the numerous squirrels of Asia are equally absent, there seems no probability of its having been united with that continent. Let us, then, see to what groups the mammalia of Madagascar belong, and where we must look for their probable allies.

First and most important are the lemurs, consisting of six genera and thirty-three species, thus comprising just half the entire mammalian population of the island. This group of lowly organized and very ancient creatures still exists scattered over a wide area; but they are nowhere so abundant as in the island of Madagascar. They are found from West Africa to India, Ceylon, and the Malay Archipelago, consisting of a number of isolated genera and species, which appear to maintain their existence by their nocturnal and arboreal habits, and by haunting dense forests. It can hardly be said that the African forms of lemurs are more nearly allied to those of Madagascar than are the Asiatic, the whole series appearing to be the disconnected fragments of a once more compact and extensive group of animals.

Next, we have about a dozen species of Insectivora, consisting of one shrew, a group distributed over all the great continents; and five genera of a peculiar family, Centetidæ, which family exists nowhere else on the globe except in the two largest West Indian islands, Cuba and Hayti, thus adding still further to our embarrassment in seeking for the original home of the Madagascar fauna.

We then come to the Carnivora, which are represented by a peculiar cat-like animal, Cryptoprocta, forming a distinct family, and having no allies in any part of the globe; and eight civets belonging to four peculiar genera. Here we first meet with some decided indications of an African origin; for the civet family is more abundant in this continent than in Asia, and some of the Madagascar genera seem to be decidedly allied to African groups—as, for example, Eupleres to Suricata and Crossarchus.[1]

The Rodents consist only of four rats and mice of peculiar genera, one of which is said to be allied to an American genus; and, lastly, we have a river-hog of the African genus Potamochærus, and a small sub-fossil hippopotamus, both of which, being semi-aquatic animals, might easily have reached the island from Africa, by way of the Comoros, without any actual land-connection.

Reptiles of Madagascar.—Passing over the birds for the present, as not so clearly demonstrating land-connection, let us see what indications are afforded by the reptiles. The large and universally distributed family of Colubrine snakes is represented in Madagascar not by African or Asiatic genera, but by two American genera—Philodryas and Heterodon, and by Herpetodryas, a genus found in America and China. The other genera are all peculiar, and belong mostly to wide-spread tropical families; but two families—Lycodontidæ and Viperidæ, both abundant in Africa and the Eastern tropics—are absent. Lizards are mostly represented by peculiar genera of African or tropical families, but several African genera are represented by peculiar species, and there are also some species belonging to two American genera of the Iguanidæ, a family which is exclusively American; while a genus of geckoes, inhabiting America and Australia, also occurs in Madagascar.

---

[1] See Dr. J. E. Gray's " Revision of the Viverridæ," in *Proceedings of the Zoological Society*, 1864, p. 507.

Relation of Madagascar to Africa.—These facts, taken all together, are certainly very extraordinary, since they show in a considerable number of cases as much affinity with America as with Africa ; while the most striking and characteristic groups of animals now inhabiting Africa are entirely wanting in Madagascar. Let us first deal with this fact, of the absence of so many of the most dominant African groups. The explanation of this deficiency is by no means difficult, for the rich deposits of fossil mammals of Miocene age in France, Germany, Greece, and Northwest India have demonstrated the fact that all the great African mammals then inhabited Europe and temperate Asia. We also know that a little earlier (in Eocene times) tropical Africa was cut off from Europe and Asia by a sea stretching from the Atlantic to the Bay of Bengal, at which time Africa must have formed a detached island-continent such as Australia is now, and probably, like it, very poor in the higher forms of life. Coupling these two facts, the inference seems clear that all the higher types of mammalia were developed in the great Euro-Asiatic continent (which then included Northern Africa), and that they only migrated into tropical Africa when the two continents became united by the upheaval of the sea-bottom, probably in the latter portion of the Miocene or early in the Pliocene period.[1]

---

[1] This view was, I believe, first advanced by Professor Huxley in his "Anniversary Address to the Geological Society" in 1870. He says, "In fact, the Miocene mammalian fauna of Europe and the Himalayan regions contain, associated together, the types which are at present separately located in the South African and Indian provinces of Arctogæa. Now there is every reason to believe, on other grounds, that both Hindostan south of the Ganges, and Africa south of the Sahara, were separated by a wide sea from Europe and North Asia during the Middle and Upper Eocene epochs. Hence it becomes highly probable that the well-known similarities, and no less remarkable differences, between the present faunæ of India and South Africa have arisen in some such fashion as the following. Some time during the Miocene epoch, the bottom of the nummulitic sea was upheaved and converted into dry land in the direction of a line extending from Abyssinia to the mouth of the Ganges. By this means the Dekkan, on the one hand, and South Africa, on the other, became connected with the Miocene dry land and with one another. The Miocene mammals spread gradually over this intermediate dry land ; and if the condition of its eastern and western ends offered as wide contrasts as the valleys of the Ganges and Arabia do now, many forms which made their way into Africa must have been different

It is clear, therefore, that if Madagascar had once formed part of Africa, but had been separated from it before Africa was united to Europe and Asia, it would not contain any of those kinds of animals which then first entered the country. But, besides the African mammals, we know that some birds now confined to Africa then inhabited Europe, and we may therefore fairly assume that all the more important groups of birds, reptiles, and insects, now abundant in Africa, but absent from Madagascar, formed no part of the original African fauna, but entered the country only after it was joined to Europe and Asia.

Early History of Africa and Madagascar.—We have seen that Madagascar contains an abundance of mammals, and that most of them are of types either peculiar to, or existing also in, Africa; it follows that that continent must have had an earlier union with Europe, Asia, or America, or it could never have obtained any mammals at all. Now these ancient African mammals are Lemurs, Insectivora, and small Carnivora, chiefly Viverridæ; and all these groups are known to have inhabited Europe in Eocene and Miocene times; and that the union was with Europe rather than with America is clearly proved by the fact that even the Insectivorous Centetidæ, now confined to Madagascar and the West Indies, inhabited France in the Lower Miocene period; while the Viverridæ, or civets, which form so important a part of the fauna of Madagascar as well as of Africa, were abundant in Europe throughout the whole Tertiary period, but are not known to have ever lived in any part of the American continent. We here see the application of the principle which we have already fully proved and illustrated (Chapter IV., p. 61), that all extensive groups have a wide range at the period of their maximum development; but as they decay

---

from those which reached the Dekkan, while others might pass into both these subprovinces."

This question is fully discussed in my "Geographical Distribution of Animals" (Vol. I., p. 285), where I expressed views somewhat different from those of Professor Huxley, and made some slight errors which are corrected in the present work. As I did not then refer to Professor Huxley's prior statement of the theory of Miocene immigration into Africa (which I had read, but the reference to which I could not recall), I am happy to give his views here.

their area of distribution diminishes or breaks up into detached fragments, which one after another disappear till the group becomes extinct. Those animal forms which we now find isolated in Madagascar and other remote portions of the globe all belong to ancient groups which are in a decaying or nearly extinct condition, while those which are absent from it belong to more recent and more highly developed types, which range over extensive and continuous areas, but have had no opportunity of reaching the more ancient continental islands.

Anomalies of Distribution, and how to Explain them.—If these considerations have any weight, it follows that there is no reason whatever for supposing any former direct connection between Madagascar and the Greater Antilles merely because the Insectivorous Centetidæ now exist only in these two groups of islands; for we know that the ancestors of this family must once have had a much wider range, which almost certainly extended over the great northern continents. We might as reasonably suppose a land-connection across the Pacific to account for the camels of Asia having their nearest existing allies in the llamas and alpacas of the Peruvian Andes, and another between Sumatra and Brazil, in order that the ancestral tapir of one country might have passed over to the other. In both these cases we have ample proof of the former wide extension of the group. Extinct camels of numerous species abounded in North America in Miocene, Pliocene, and even Post-pliocene times, and one has also been found in Northwestern India, but none whatever among all the rich deposits of mammalia in Europe. We are thus told, as clearly as possible, that from the North American continent as a centre the camel tribe spread westward, over now-submerged land at the shallow Behring Strait and Kamtschatka Sea into Asia, and southward along the Andes into South America. Tapirs are even more interesting and instructive. Their remotest known ancestors appear in Western Europe in the early portion of the Eocene period; in the later Eocene and the Miocene other forms occur both in Europe and North America. These seem to have become extinct in North America, while in Europe they developed largely into many forms of true tapirs, which at a much later period found their

way again to North and thence to South America, where their
remains are found in caves and gravel-deposits. It is an instructive fact that in the Eastern continent, where they were
once so abundant, they have dwindled down to a single species,
existing in small numbers in the Malay Peninsula, Sumatra, and
Borneo only; while in the Western continent, where they are
comparatively recent immigrants, they occupy a much larger
area, and are represented by three or four distinct species. Who
could possibly have imagined such migrations, and extinctions,
and changes of distribution as are demonstrated in the case of
the tapirs if we had only the distribution of the existing species
to found an opinion upon? Such cases as these—and there are
many others equally striking—show us with the greatest distinctness how nature has worked in bringing about the examples of anomalous distribution that everywhere meet us; and
we must, on every ground of philosophy and common-sense,
apply the same method of interpretation to the more numerous
instances of anomalous distribution we discover among such
groups as reptiles, birds, and insects, where we rarely have any
direct evidence of their past migrations through the discovery
of fossil remains. Whenever we can trace the past history of
any group of terrestrial animals, we invariably find that its
actual distribution can be explained by migrations effected by
means of comparatively slight modifications of our existing continents. In no single case have we any direct evidence that the
distribution of land and sea has been radically changed during
the whole lapse of the Tertiary and Secondary periods, while,
as we have already shown in our fifth chapter, the testimony
of geology itself, if fairly interpreted, upholds the same theory
of the stability of our continents and the permanence of our
oceans. Yet so easy and pleasant is it to speculate on former
changes of land and sea with which to cut the Gordian knot
offered by anomalies of distribution that we still continually
meet with suggestions of former continents stretching in every
direction across the deepest oceans, in order to explain the presence in remote parts of the globe of the same genera even of
plants or of insects—organisms which possess such exceptional
facilities both for terrestrial, aerial, and oceanic transport, and of

whose distribution in past ages we generally know absolutely nothing.

The Birds of Madagascar as Indicating a Supposed Lemurian Continent. — Having thus shown how the distribution of the land mammalia and reptiles of Madagascar may be well explained by the supposition of a union with Africa before the greater part of its existing fauna had reached it, we have now to consider whether, as some ornithologists think, the distribution and affinities of the birds present an insuperable objection to this view, and require the adoption of a hypothetical continent—Lemuria—extending from Madagascar to Ceylon and the Malay Islands.

There are about one hundred land birds known from the island of Madagascar, all but four or five being peculiar ; and about half of these peculiar species belong to peculiar genera, many of which are extremely isolated, so that it is often difficult to class them in any of the recognized families, or to determine their affinities to any living birds. Among the other moiety, belonging to known genera, we find fifteen which have undoubted African affinities, while five or six are as decidedly Oriental, the genera or nearest allied species being found in India or the Malay Islands. It is on the presence of these peculiar Indian types that Dr. Hartlaub, in his recent work on the "Birds of Madagascar and the Adjacent Islands," lays great stress, as proving the former existence of "Lemuria;" while he considers the absence of such peculiar African families as the plantain-eaters, glossy-starlings, ox-peckers, barbets, honeyguides, hornbills, and bustards, besides a host of peculiar African genera, as sufficiently disproving the statement in my "Geographical Distribution of Animals" that Madagascar is "more nearly related to the Ethiopian than to any other region," and that its fauna was evidently "mainly derived from Africa."

But the absence of the numerous peculiar groups of African birds is so exactly parallel to the same phenomenon among mammals that we are justified in imputing it to the same cause, the more especially as some of the very groups that are wanting—the plantain-eaters and the trogons, for example—are actually known to have inhabited Europe along with the large mammalia which subsequently migrated to Africa. As to the peculiarly Eastern genera—such as Copsychus and Hypsipetes, with a Dicrurus, Ploceus, a Cisticola, and a Scops, all closely allied to Indian or Malayan species—although very striking to the ornithologist, they certainly do not outweigh the fourteen African genera found in Madagascar. Their presence may, moreover, be accounted for more satisfactorily than by means of an ancient Lemurian continent, which, even if granted, would not explain the very facts adduced in its support.

Let us first prove this latter statement.

The supposed "Lemuria" must have existed, if at all, at so remote a period that the higher animals did not then inhabit either Africa or Southern Asia, and it must have become partially or wholly submerged before they reached those countries; otherwise we should find in Madagascar many other animals besides Lemurs, Insectivora, and Viverridæ, especially such active arboreal creatures as monkeys and squirrels, such hardy grazers as deer or antelopes, or such wide-ranging carnivores as foxes or bears. This obliges us to date the disappearance of the hypothetical continent about the earlier part of the Miocene epoch at latest, for during the latter part of that period we. know that such animals existed in abundance in every part of the great northern continents wherever we have found organic remains. But the Oriental birds in Madagascar, by whose presence Dr. Hartlaub upholds the theory of a Lemuria, are slightly modified forms of *existing Indian genera*, or sometimes, as Dr. Hartlaub himself points out, *species hardly distinguishable from those of India*. Now all the evidence at our command leads us to conclude that, even if these genera and species were in existence in the early Miocene period, they must have had a widely different distribution from what they have now. Along with so many African and Indian genera of mammals, they then probably inhabited Europe, which at that epoch enjoyed a subtropical climate; and this is rendered almost certain by the discovery in the Miocene of France of fossil remains of trogons and jungle-fowl. If, then, these Indian birds date back to the very period during which alone Lemuria could have existed, that continent was quite unnecessary for their introduction into

Madagascar, as they could have followed the same track as the mammalia of Miocene Europe and Asia; while if, as I maintain, they are of more recent date, then Lemuria had ceased to exist, and could not have been the means of their introduction.

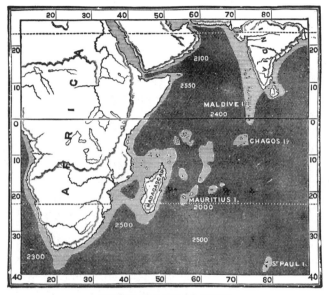

MAP OF THE INDIAN OCEAN.

Showing the position of banks less than 1000 fathoms deep between Africa and the Indian Peninsula.

Submerged Islands between Madagascar and India.—Looking at the accompanying map of the Indian Ocean, we see that between Madagascar and India there are now extensive shoals and coral reefs, such as are always held to indicate subsidence; and we may therefore fairly postulate the former existence here of several large islands, some of them not much inferior to Madagascar itself. These reefs are all separated from each other by very deep sea—much deeper than that which divides Madagascar from Africa, and we have therefore no reason to imagine their former union. But they would, nevertheless, greatly facilitate the introduction of Indian birds into the Mascarene Islands and Madagascar; and these facilities existing, such an immigration would be sure to take place, just as surely as American birds have entered the Galapagos and Juan Fernandez, as Euro pean birds now reach the Azores, and as Australian birds reach such a distant island as New Zealand. This would take place the more certainly because the Indian Ocean is a region of violent periodical storms at the changes of the monsoons, and we have seen in the case of the Azores and Bermuda how important a factor this is in determining the transport of birds across the ocean.

Mr. Darwin's theory of the formation of atolls is now almost universally accepted as the true one, and this theory implies that the areas in question are still, or have very recently been, subsiding. The final disappearance of these now sunken islands does not, therefore, in all probability, date back to a very remote epoch; and this exactly accords with the fact that some of the birds, as well as the fruit-bats of the genus Pteropus, are very closely allied to Indian species, if not actually identical, others being distinct species of the same genera. The fact that not one closely allied species or even genus of Indian or Malayan mammals is found in Madagascar sufficiently proves that it is no land-connection that has brought about this small infusion of Indian birds and bats; while we have sufficiently shown that, when we go back to remote geological times, no land-connection in this direction was necessary to explain the phenomena of the distribution of the Lemurs and Insectivora. A land-connection with *some* continent was undoubtedly necessary, or there would have been no mammalia at all in Madagascar; and the nature of its fauna, on the whole, no less than the moderate depth of the intervening strait and the comparative approximation of the opposite shores, clearly indicate that the connection was with Africa.

Concluding Remarks on " Lemuria."—I have gone into this question in some detail, because Dr. Hartlaub's criticism on my views has been reproduced in a scientific periodical,[1] and the supposed Lemurian continent is constantly referred to by quasi-scientific writers, as well as by naturalists and geologists, as if its

---

[1] *The Ibis*, 1877, p. 334.

existence had been demonstrated by facts, or as if it were absolutely necessary to postulate such a land in order to account for
the entire series of phenomena connected with the Madagascar
fauna, and especially with the distribution of the Lemuridæ.[1]
I think I have now shown, on the other hand, that it was essentially a provisional hypothesis, very useful in calling attention
to a remarkable series of problems in geographical distribution,
but not affording the true solution of those problems, any more
than the hypothesis of an Atlantis solved the problems presented by the Atlantic Islands and the relations of the European
and North American flora and fauna. The Atlantis is now rarely introduced seriously except by the absolutely unscientific, having received its death-blow by the chapter on Oceanic Islands in
the "Origin of Species," and the researches of Professor Asa
Gray on the affinities of the North American and Asiatic floras.
But "Lemuria" still keeps its place—a good example of the survival of a provisional hypothesis which offers what seems an
easy solution of a difficult problem, and has received an appropriate and easily remembered name, long after it has been proved
to be untenable.

It is now more than four years since I first showed, by a careful examination of all the facts to be accounted for, that the

---

[1] In a paper read before the Geological Society in 1874, Mr. H. F. Blanford,
from the similarity of the fossil plants and reptiles, supposed that India and South
Africa had been connected by a continent, "and remained so connected with some
short intervals from the Permian up to the end of the Miocene period," and Mr.
Woodward expressed his satisfaction with "this further evidence derived from the
fossil flora of the Mesozoic series of India in corroboration of the former existence
of an old submerged continent—Lemuria."

Those who have read the preceding chapters of the present work will not need to
have pointed out to them how utterly inconclusive is the fragmentary evidence derived from such remote periods (even if there were no evidence on the other side) as
indicating geographical changes. The notion that a similarity in the productions of
widely separated continents at any past epoch is only to be explained by the existence of a *direct* land-connection, is entirely opposed to all that we know of the wide
and varying distribution of *all* types at different periods, as well as to the great powers of dispersal over moderate widths of ocean possessed by all animals except mammalia. It is no less opposed to what is now known of the general permanency of the
great continental and oceanic areas; while in this particular case it is totally inconsistent (as has been shown above) with the actual facts of the distribution of animals.

hypothesis of a Lemurian continent was alike unnecessary to explain one portion of the facts, and inadequate to explain the remaining portion.[1] Since that time I have seen no attempt even to discuss the question on general grounds in opposition to my views, nor, on the other hand, have those who have hitherto supported the hypothesis taken any opportunity of acknowledging its weakness and inutility. I have therefore here explained my reasons for rejecting it somewhat more fully and in a more popular form, in the hope that a check may thus be placed on the continued restatement of this unsound theory as if it were one of the accepted conclusions of modern science.

The Mascarene Islands.[2]—In the " Geographical Distribution of Animals," a summary is given of all that was known of the zoology of the various islands near Madagascar, which to some extent partake of its peculiarities, and with it form the Malagasy sub-region of the Ethiopian region. As no great additions have since been made to our knowledge of the fauna of these islands, and my object in this volume being more especially to illustrate the mode of solving distributional problems by means of the most suitable examples, I shall now confine myself to pointing out how far the facts presented by these outlying islands support the views already enunciated with regard to the origin of the Madagascar fauna.

The Comoro Islands.—This group of islands is situated nearly midway between the northern extremity of Madagascar and the coast of Africa. The four chief islands vary between sixteen and forty miles in length, the largest being 180 miles from the coast of Africa, while one or two smaller islets are less than 100 miles from Madagascar. All are volcanic, Great Comoro being an active volcano 8500 feet high; and, as already stated, they are situated on a submarine bank with less than 500 fathoms soundings, connecting Madagascar with Africa. There is reason to believe, however, that these islands are of comparatively recent origin, and that the bank has been formed by mat-

---

[1] "Geographical Distribution of Animals," Vol. I., p. 272-292.

[2] The term "Mascarene" is used here in an extended sense to include all the islands near Madagascar which resemble it in their animal and vegetable productions.

ter ejected by the volcanoes or by upheaval. Anyhow, there is no indication whatever of there having been here a land-connection between Madagascar and Africa, while the islands themselves have been mainly colonized from Madagascar, to the 100-fathom bank surrounding which some of them make a near approach.

The Comoros contain two land mammals, a lemur and a civet, both of Madagascar genera and the latter an identical species, and there is also a peculiar species of fruit-bat (*Pteropus Comorensis*), a group which ranges from Australia to Asia and Madagascar, but is unknown in Africa. Of land birds forty-one species are known, of which sixteen are peculiar to the islands, twenty-one are found also in Madagascar, and three found in Africa and not in Madagascar; while of the peculiar species six belong to Madagascar or Mascarene genera.

These facts point to the conclusion that the Comoro Islands have been formerly more nearly connected with Madagascar than they are now, probably by means of intervening islets and the former extension of the latter island to the westward, as indicated by the extensive shallow bank at its northern extremity, so as to allow of the easy passage of birds, and the occasional transmission of small mammalia by means of floating trees.[1]

The Seychelles Archipelago.—This interesting group consists of about thirty small islands situated 700 miles N.N.E. of Madagascar, or almost exactly in the line formed by continuing the central ridge of that great island. The Seychelles stand upon a rather extensive shallow bank, the 100-fathom line around them enclosing an area nearly 200 miles long by 100 miles wide, while the 500-fathom line shows an extension of nearly 100 miles in a southern direction. All the larger islands are of granite, with mountains rising to 3000 feet in Mahé, and to from 1000 to 2000 feet in several of the other islands. We can therefore hardly doubt that they form a portion of the great line of upheaval which produced the central granitic mass of Madagascar, intervening points being indicated by the Amirantes, the Provi-

---

[1] For the birds of the Comoro Islands, see *Proceedings of the Zoological Society*, 1877, p. 295, and 1879, p. 673.

dence, and the Farquhar Islands, which, though all coralline, probably rest on a granitic basis. Deep channels of more than 1000 fathoms now separate these islands from each other, and if they were ever sufficiently elevated to be united, it was probably at a very remote epoch.

The Seychelles may thus have had ample facilities for receiving from Madagascar such immigrants as can pass over narrow seas; and, on the other hand, they were equally favorably situated as regards the extensive Saya de Malha and Cargados banks, which were probably once large islands, and may have supported a rich insular flora and fauna of mixed Mascarene and Indian type. The existing fauna and flora of the Seychelles must therefore be looked upon as the remnants which have survived the partial submergence of a very extensive island; and the entire absence of mammalia may be due either to this island having never been actually united to Madagascar, or to its having since undergone so much submergence as to have led to the extinction of such mammals as may once have inhabited it. The birds and reptiles, however, though few in number, are very interesting, and throw some further light on the past history of the Seychelles.

Birds of the Seychelles.—Fifteen indigenous land birds are known to inhabit the group, thirteen of which are peculiar species,[1] belonging to genera which occur also in Madagascar or Africa. The genera which are more peculiarly Indian are, Copsychus and Hypsipetes, also found in Madagascar; and Palæornis, which has species in Mauritius and Rodriguez, as well as one on the continent of Africa. A black parrot (Coracopsis), congeneric with two species that inhabit Madagascar and with

---

[1] The following is a list of these peculiar birds (see the *Ibis* for 1867, p. 359; and 1879, p. 97):

PASSERES.

Ellisia Seychellensis.
Copsychus Seychellarum.
Hypsipetes crassirostris.
Tchitrea corvina.
Nectarinia Dussumieri.
Zosterops modesta.
"   semiflava.
Foudia Seychellarum.

PSITTACI.

Coracopsis Barklyi.
Palæornis Wardi.

COLUMBÆ.

Alectorænas pulcherrimus.
Turtur rostratus.

ACCIPITRES.

Tinnunculus gracilis.

one that is peculiar to the Comoros; and a beautiful red-headed blue pigeon (*Alectorœnas pulcherrimus*) allied to those of Madagascar and Mauritius, but very distinct, are the most remarkable species characteristic of this group of islands.

Reptiles and Amphibia of the Seychelles.—The reptiles and amphibia are rather numerous and very interesting, indicating clearly that the islands can hardly be classed as oceanic. There are five species of lizards, three being peculiar to the islands, while the two others have a rather wide range. The first is a chameleon—a defenceless slow-moving lizard, especially abundant in Madagascar, from which no less than twenty-one species are now known, about the same number as on the continent of Africa. The Seychelles species (*Chameleo tigris*) is peculiar to the islands. The next is one of the skinks (*Euprepes cyanogaster*), small ground-lizards with a very wide distribution in the Eastern Hemisphere. This species is, however, peculiar to the islands. The other peculiar species is one of the geckoes (*Phelsuma Seychellensis*). An East African species (*P. cepedianus*) is also found in the Seychelles, as well as in the Comoro Islands, Bourbon, Mauritius, Madagascar, and Rodriguez; and there is also a third gecko of another genus (*Peropus mutilatus*) which is found also in Mauritius, Bourbon, Rodriguez, and Ceylon, and even in Penang and the Philippine Islands. These lizards, clinging as they do to trees and timber, are exceedingly liable to be carried in ships from one country to another, and I am told by Dr. Günther that some are found almost every year in the London Docks. It is therefore probable that when species of this family have a very wide range they have been assisted in their migrations by man, though their habit of clinging to trees also renders them likely to be floated with large pieces of timber to considerable distances. Dr. Percival Wright, to whom I am indebted for much information on the productions of the Seychelle Archipelago, informs me that the last-named species varies greatly in color in the different islands, so that he could always tell from which particular island a specimen had been brought. This is analogous to the curious fact of certain lizards on the small islands in the Mediterranean being always very different in color from those of the mainland, usually becoming rich blue or black (see *Nature*, Vol. XIX., p. 97); and we thus learn how readily in some cases differences of color are brought about by local conditions.

Snakes, as is usually the case in small or remote islands, are far less numerous than lizards, only two species being known. One, *Dromicus Seychellensis*, is a peculiar species of the family Colubridæ, the rest of the genus being found in Madagascar and South America. The other, *Boodon geometricus*, one of the Lycodontidæ, or fanged ground-snakes, inhabits also South and West Africa. So far, then, as the reptiles are concerned, there is nothing but what is easily explicable by what we know of the general means of distribution of these animals.

We now come to the amphibia, which are represented in the Seychelles by two tailless and two serpent - like forms. The frogs are, *Rana Mascariensis*, found also in Mauritius, Bourbon, Angola, and Abyssinia, and probably all over tropical Africa; and *Megalixalus infrarufus*, a tree-frog altogether peculiar to the islands, and forming a peculiar genus of the wide-spread tropical family Polypedatidæ. It is found, Dr. Wright informs me, on the Pandani, or screw-pines; and as these form a very characteristic portion of the vegetation of the Mascarene Islands, all the species being peculiar and confined each to a single island or small group, we may perhaps consider it as a relic of the indigenous fauna of that more extensive land of which the present islands are the remains.

The serpentine amphibia are represented by two species of Cæcilia. These creatures externally resemble large worms, except that they have a true head with jaws and rudimentary eyes, while internally they have, of course, a true vertebrate skeleton. They live underground, burrowing by means of the ring-like folds of the skin, which simulate the jointed segments of a worm's body; and when caught they exude a viscid slime. The young have external gills which are afterwards replaced by true lungs, and this peculiar metamorphosis shows that they belong to the amphibia rather than to the reptiles. The Cæcilias are widely but very sparingly distributed through all the tropical regions—a fact which may, as we have seen, be taken as an indication of the great antiquity of the group, and that it is now verging towards extinction. In the Seychelle Islands two species have been found, named respectively *Cæcilia oxyura* and *C. rostrata.* The former also inhabits the Malabar coast of India, while the latter has been found in West Africa and also South America.[1] This is certainly one of the most remarkable cases of the wide and discontinuous distribution of a species known ; and when we consider the habits of life of these animals, and the extreme slowness with which it is likely they can migrate into new areas, we can hardly arrive at any other conclusion than that this species once had an almost world-wide range, and that in the process of dying out it has been left stranded, as it were, in these three remote portions of the globe. The extreme stability and long persistence of specific form which this implies is extraordinary, but not unprecedented, among the lower vertebrates. The crocodiles of the Eocene period differ but slightly from those of the present day, while a small fresh-water turtle from the Miocene deposits of the Siwalik Hills is absolutely identical with a still living Indian species, *Emys tectus.* The mud-fish of Australia, *Ceratodus Forsteri*, is a very ancient type, and may well have remained specifically unchanged since early Tertiary times. It is not, therefore, incredible that the Seychelles Cæcilia may be the oldest land vertebrate now living on the globe—dating back to the early part of the Tertiary period, when the warm climate of the Northern Hemisphere in high latitudes, and the union of the Asiatic and American continents, allowed of the migration of such types over the whole Northern Hemisphere, from which they subsequently passed into the Southern Hemisphere, maintaining themselves only in certain limited areas where the physical conditions were especially favorable, or where they were saved from the attacks of enemies or the competition of higher forms.

Fresh-water Fishes.—The only other vertebrates in the Sey-

---

[1] Specimens are recorded from West Africa in the *Proceedings of the Academy of Natural Science,* Philadelphia, 1857, p. 72, while specimens in the Paris Museum were brought by D'Orbigny from South America. Dr. Wright's specimens from the Seychelles have, as he informs me, been determined to be the same species by Dr. Peters, of Berlin.

chelles are two fresh-water fishes abounding in the streams and rivulets. One, *Haplochilus Playfairii*, is peculiar to the islands, but there are allied species in Madagascar. It is a pretty little fish about four inches long, of an olive color, with rows of red spots, and is very abundant in some of the mountain streams. The fishes of this genus, as I am informed by Dr. Günther, often inhabit both sea and fresh water, so that their migration from Madagascar to the Seychelles and subsequent modification offer no difficulty. The other species is *Fundulus orthonotus*, found also on the east coast of Africa; and as both beiong to the same family—Cyprinodontidæ—this may possibly have migrated in a similar manner.

Land Shells.—The only other group of animals inhabiting the Seychelles which we know with any approach to completeness are the iand and fresh-water mollusca, but they do not furnish any facts of special interest. About forty species are known ; and Mr. Geoffrey Nevill, who has studied them, thinks their meagre number is chiefly owing to the destruction of so much of the forests which once covered the islands. Seven of the species— and among them one of the most conspicuous, *Achatina fulica* —have almost certainly been introduced ; and the remainder show a mixture of Madagascar and Indian forms, with a preponderance of the latter. Five genera—Streptaxis, Cyathoponea, Onchidium, Helicina, and Paludomus—are mentioned as being especially Indian, while only two—Tropidophora and Gibbus— are found in Madagascar, but not in India.[1] About two thirds of the species appear to be peculiar to the islands.

Mauritius, Bourbon, and Rodriguez.—These three islands are somewhat out of place in this chapter because they really belong to the oceanic group, being of volcanic formation, surrounded by deep sea, and possessing no indigenous mammals or amphibia. Yet their productions are so closely related to those of Madagascar, to which they may be considered as attendant satellites, that it is absolutely necessary to associate them together if we wish to comprehend and explain their many interesting features.

---

[1] "Additional Notes on the Land Shells of the Seychelle Islands," by Geoffrey Nevill, C.M.Z.S., in *Proceedings of the Zoological Society*, 1869, p. 61.

Mauritius and Bourbon are lofty volcanic islands, evidently of great antiquity. They are about 100 miles apart, and the sea between them is less than 1000 fathoms deep, while on each side it sinks rapidly to depths of 2400 and 2600 fathoms. We have therefore no reason to believe that they have ever been connected with Madagascar, and this view is strongly supported by the character of their indigenous fauna. Of this, however, we have not a very complete or accurate knowledge, for though both islands have long been occupied by Europeans, the study of their natural products was for a long time greatly neglected, and, owing to the rapid spread of sugar cultivation, the virgin forests, and with them, no doubt, many native animals, have been almost wholly destroyed. There is, however, no good evidence of there ever having been any indigenous mammals or amphibia, though both are now found and are often recorded among the native animals.[1]

The smaller and more remote Rodriguez is also volcanic; but it has, besides, a good deal of coralline rock—an indication of partial submergence, and helping to account for the poverty of its fauna and flora. It stands on a 100-fathom bank of considerable extent, but beyond this the sea rapidly deepens to more than

---

[1] In Maillard's "Notes sur l'Île de Reunion," a considerable number of mammalia are given as "wild," such as *Lemur mongoz* and *Centetes setosus*, both Madagascar species, with such undoubtedly introduced animals as a wild-cat, a hare, and several rats and mice. He also gives two species of frogs, seven lizards, and two snakes. The latter are both Indian species and certainly imported, as are most probably the frogs. Legouat, who resided some years in the island nearly two centuries ago, and who was a close observer of nature, mentions numerous birds, large bats, land tortoises, and lizards, but no other reptiles or venomous animals except scorpions. We may be pretty sure, therefore, that the land mammalia, snakes, and frogs now found wild have all been introduced. Of lizards, on the other hand, there are several species, some peculiar to the island, others common to Africa and the other Mascarene Islands. The following list by Professor Duméril is given in Maillard's work:

Platydactylus cepedianus.	Hemidactylus frenatus.
" ocellatus.	Gongylus Bojerii.
Hemidactylus Peronii.	Ablepharus Peronii.
" mutilatus.	

Four species of chameleon are now recorded from Bourbon and one from Mauritius (J. Reay Greene, M.D., in *Popular Science Review*, April, 1880); but as they are not mentioned by the old writers, it is pretty certain that these creatures are recent introductions, and this is the more probable as they are favorite domestic pets.

2000 fathoms; so that it is truly oceanic, like its larger sister-isles.

Birds.—The living birds of these islands are few in number, and consist mainly of peculiar species of Mascarene types, together with two peculiar genera—Oxynotus belonging to the Campephagidæ or caterpillar-catchers, a family abundant in the Old World tropics; and a dove, Trocazza, forming a peculiar sub-genus. The origin of these birds offers no difficulty, looking at the position of the islands and of the surrounding shoals and islets.

Extinct Birds.—These three islands are, however, pre-eminently remarkable as being the home of a group of large ground-birds, quite incapable of flight, and altogether unlike anything found elsewhere on the globe; and which, though once very abundant, have become totally extinct within the last two hundred years. The best-known of these birds is the dodo, which inhabited Mauritius; while allied species certainly lived in Bourbon and Rodriguez, abundant remains of the species of the latter island—the "solitaire"—having been discovered, corresponding with the figure and description given of it by Leguat, who resided in Rodriguez in 1692. These birds constitute a distinct family, Dididæ, allied to the pigeons, but very isolated. They were quite helpless, and were rapidly exterminated when man introduced dogs, pigs, and cats into the islands, and himself sought them for food. The fact that such perfectly defenceless creatures survived in great abundance to a quite recent period in these three islands only, while there is no evidence of their ever having inhabited any other countries whatever, is itself almost demonstrative that Mauritius, Bourbon, and Rodriguez are very ancient but truly oceanic islands. From what we know of the general similarity of Miocene birds to living genera and families, it seems clear that the origin of so remarkable a type as the dodos must date back to early Tertiary times. If we suppose some ancestral ground-feeding pigeon of large size to have reached the group by means of intervening islands afterwards submerged, and to have thenceforth remained to increase and multiply, unchecked by the attacks of any more powerful animals, we can well understand that the wings, being useless, would in time become almost aborted.[1] It is also not improbable that this process would be aided by natural selection, because the use of wings might be absolutely prejudicial to the birds in their new home. Those that flew up into trees to roost, or tried to cross over the mouths of rivers, might be blown out to sea and destroyed, especially during the hurricanes which have probably always more or less devastated the islands; while, on the other hand, the more bulky and short-winged individuals, who took to sleeping on the ground in the forest, would be preserved from such dangers, and perhaps also from the attacks of birds of prey which may always have visited the islands. But whether or not this was the mode by which these singular birds acquired their actual form and structure, it is perfectly certain that their existence and development depended on complete isolation and on freedom from the attacks of enemies. We have no single example of such defenceless birds having ever existed on a continent at any geological period, whereas analogous though totally distinct forms do exist in New Zealand, where enemies are equally wanting. On the other hand, every continent has always produced abundance of carnivora adapted to prey upon the herbivorous animals inhabiting it at the same period; and we may therefore be sure that these islands have never formed part of a continent during any portion of the time when the dodos inhabited them.

---

[1] That the dodo is really an abortion from a more perfect type, and not a direct development from some lower form of wingless bird, is shown by its possessing a keeled sternum, though the keel is exceedingly reduced, being only three quarters of an inch deep in a length of seven inches. The most terrestrial pigeon—the Didunculus of the Samoan Islands—has a far deeper and better-developed keel, showing that in the case of the dodo the degradation has been extreme. We have also analogous examples in other extinct birds of the same group of islands, such as the flightless rails, Aphanapteryx of Mauritius and Erythromachus of Rodriguez, as well as the large parrot, Lophopsittacus of Mauritius, and the night heron, *Nycticorax megacephala* of Rodriguez, the last two birds probably having been able to fly a little. The commencement of the same process is to be seen in the peculiar dove of the Seychelles, *Turtur rostratus*, which, as Mr. Edward Newton has shown, has much shorter wings than its close ally, *T. picturatus*, of Madagascar. For a full and interesting account of these and other extinct birds, see Professor Newton's article on " Fossil Birds," in the *Encyclopædia Britannica*, 9th ed., Vol. III., p. 732, and that on " The Extinct Birds of Rodriguez," by Dr. A. Günther and Mr. E. Newton, in the Royal Society's volume on the " Transit of Venus Expedition."

It is a remarkable thing that an ornithologist of Dr. Hart-laub's reputation, looking at the subject from a purely ornithological point of view, should yet entirely ignore the evidence of these wonderful and unique birds against his own theory, when he so confidently characterizes Lemuria as "that sunken land which, containing parts of Africa, must have extended far east-ward over Southern India and Ceylon, and the highest points of which we recognize in the volcanic peaks of Bourbon and Mau-ritius, and in the central range of Madagascar itself—the last re-sorts of the mostly extinct Lemurine race which formerly peo-pled it." [1] It is here implied that lemurs formerly inhabited Bourbon and Mauritius, but of this there is not a particle of ev-idence; and we feel pretty sure that had they done so, the dodos would never have been developed there. In Madagascar there are no traces of dodos, while there are remains of extinct gigan-tic struthious birds of the genus Æpyornis, which were, no doubt, as well able to protect themselves against the smaller car-nivora as are the ostriches, emus, and cassowaries in their re-spective countries at the present day.

The whole of the evidence at our command, therefore, tends to establish in a very complete manner the "oceanic" character of the three islands—Mauritius, Bourbon, and Rodriguez, and that they have never formed part of "Lemuria," or of any continent.

Reptiles.—Mauritius, like Bourbon, has lizards, some of which are peculiar species; but no snakes, and no frogs or toads but such as have been introduced. [2] Strange to say, however, a small islet called Round Island, only about a mile across, and situated about fourteen miles northeast of Mauritius, possesses a snake which is not only unknown in Mauritius, but also in any other part of the world, being altogether confined to this minute islet! It belongs to the Python family, and forms a peculiar and very distinct genus, Casarea, whose nearest allies seem to be the Un-galia of Cuba and Bolgeria of Australia. It is hardly possible to believe that this serpent has very long maintained itself on so

---

[1] See *Ibis*, 1877, p. 334.

[2] A common Indian and Malayan toad (*Bufo melanostictus*) has been introduced into Mauritius, and also some European toads, as I am informed by Dr. Günther.

small an island ; and though we have no record of its existence
on Mauritius, it may very well have inhabited the lowland forests without being met with by the early settlers; and the introduction of swine, which soon ran wild and effected the final destruction of the dodo, may also have been fatal to this snake. It
is, however, now almost certainly confined to the one small islet,
and is probably the land vertebrate of most restricted distribution on the globe.

On the same island there is a small lizard, *Thyrus Boyeri*, also
a peculiar species and genus, but this is recorded from Mauritius
and Bourbon as well, though it appears to be rare in both islands.
As Round Island is connected with Mauritius by a bank under
a hundred fathoms below the surface, it has probably been once
joined to it, and when first separated would have been both
much larger and much nearer the main island—circumstances
which would greatly facilitate the transmission of these reptiles
to their present dwelling-place.

Flora of Madagascar and the Mascarene Islands.—The botany of the great island of Madagascar has been perhaps more
thoroughly explored than that of the opposite coasts of Africa,
so that its peculiarities may not be really so great as they now
appear to be. Yet there can be no doubt of its extreme richness
and grandeur, its remarkable speciality, and its anomalous external relations. It is characterized by a great abundance of forest
trees and shrubs of peculiar genera or species, and often adorned
with magnificent flowers. Some of these are allied to African
forms, others to those of Asia, and it is said that of the two affinities the latter preponderates. But there are also, as in the animal world, some decided South American relations, while others
point to Australia, or are altogether isolated.

Among the most prominent characteristics of the Madagascar
flora is the possession of a peculiar and isolated family, Chlæ-
naceæ, allied somewhat to the balsams, but presenting very anomalous characters. It consists of four genera and a number of species all entirely confined to the island. They are handsome trees
or shrubs, mostly with showy red flowers. One of them, *Rho-
dolæna altivola*, is a semi-scandent shrub with magnificent cam-
panulate flowers the size of a camellia, and of a brilliant purple

color.   The genus Chrysopia consists of large forest trees with spreading crowns adorned with umbels or corymbs of large purple flowers.   It belongs to the Clusiaceæ, and is most nearly allied to the South American genus Moronobea.   The Colvillea, a peculiar genus of Leguminosæ, is a tree with splendid scarlet flowers; and there are a large number of other peculiar genera more or less remarkable.   Combretaceæ with splendid flowers abound in Madagascar itself, though they are rare in the Mascarene Islands; while the Ravenala, or " traveller's tree ;" the extraordinary lattice-leaved Ouvirandra; the *Poinciana regia*, one of the most gorgeous of flowering trees ; and the long-spurred *Angræcum sesquipedale*, one of the most elegant and remarkable of orchids, are among its vegetable wonders.[1]

Of the flora of the smaller Madagascarian islands we possess a much fuller account, owing to the recent publication of Mr. Baker's " Flora of the Mauritius and the Seychelles," including also Rodriguez.   The total number of species in this flora is 1058, more than half of which (536) are exclusively Mascarene—that is, found only in some of the islands of the Madagascar group, while nearly a third (304) are endemic or confined to single islands.   Of the wide-spread plants, 66 are found in Africa but not in Asia, and 86 in Asia but not in Africa, showing similar Asiatic preponderance to what is said to occur in Madagascar. With the genera, however, the proportions are different, for I find by going through the whole of the generic distributions as given by Mr. Baker, that out of the 440 genera of wild plants 50 are endemic, 22 are Asiatic but not African, while 28 are African but not Asiatic.   This implies that the more ancient connection has been on the side of Africa, while a more recent immigration, shown by identity of species, has come from the side of Asia ; and it is probable that when the flora of Madagascar is more thoroughly worked out, the same or a still greater African preponderance will be found in that island.

A few Mascarene genera are found elsewhere only in South America, Australia, or Polynesia; and there are also a consid-

---

[1] This sketch of the flora of Madagascar is taken chiefly from a series of articles by M. Émile Blanchard, in the *Revue des Deux Mondes*, Vol. CI. (1872).

erable number of genera whose metropolis is South America, but which are represented by one or more species in Madagascar, and by a single often widely distributed species in Africa. This fact throws light upon the problem offered by those mammals, reptiles, and insects of Madagascar which now have their only allies in South America, since the two cases would be exactly parallel were the African plants to become extinct. Plants, however, are undoubtedly more long-lived specifically than animals, especially the more highly organized groups, and are less liable to complete extinction through the attacks of enemies or through changes of climate or of physical geography ; hence we find comparatively few cases in which groups of Madagascar plants have their *only* allies in such distant regions as America and Australia, while such cases are numerous among animals, owing to the extinction of the allied forms in intervening areas, for which extinction, as we have already shown, ample cause can be assigned.

Curious Relations of Mascarene Plants.—Among the curious affinities of Mascarene plants we have culled the following from Mr. Baker's volume. Trochetia, a genus of Sterculiaceæ, has four species in Mauritius, one in Madagascar, and one in the remote island of St. Helena. Mathurina, a genus of Turneraceæ, consisting of a single species peculiar to Rodriguez, has its nearest ally in another monotypic genus, Erblichia, confined to Central America. Siegesbeckia, one of the Compositæ, consists of two species, one inhabiting the Mascarene Islands, the other Peru. Labourdonasia, a genus of Sapotaceæ, has two species in Mauritius, one in Natal, and one in Cuba. Nesogenes, belonging to the verbena family, has one species in Rodriguez and one in Polynesia. Mespilodaphne, an extensive genus of Lauraceæ, has six species in the Mascarene Islands, and all the rest (about fifty species) in South America. Nepenthes, the well-known pitcher-plants, are found chiefly in the Malay Islands, South China, and Ceylon, with species in the Seychelle Islands and in Madagascar. Milla, a large genus of Liliaceæ, is exclusively American, except one species found in Mauritius and Bourbon. Agauria, a genus of Ericaceæ, is confined to the Mascarene Islands and the Camaroon Mountains in West Africa. An acacia found in Mauritius and Bourbon (*A. heterophylla*) can hardly be separated specifically from *Acacia koa* of the Sandwich Islands. The genus Pandanus, or screw-pine, has sixteen species in the three islands —Mauritius, Rodriguez, and the Seychelles—all being peculiar, and none ranging beyond a single island. Of palms there are fifteen species belonging to ten genera, and all these genera are peculiar to the islands. We have here ample evidence that plants exhibit the same anomalies of distribution in these islands as do the animals, though in a smaller proportion ; while they also exhibit some of the transitional stages by which these anomalies have, in all probability, been brought about, rendering quite unnecessary any other changes in the distribution of sea and land than physical and geological evidence warrants.[1]

---

[1] It may be interesting to botanists and to students of geographical distribution to give here an enumeration of the endemic genera of the " Flora of the Mauritius and the Seychelles," as they are nowhere separately tabulated in that work.

Aphloia (Bixaceæ)............ 1 sp., a shrub, Mauritius, Rodriguez, Seychelles, also Madagascar.
Medusagyne (Ternströmiaceæ). 1 sp., a shrub, Seychelles.
Astiria (Sterculiaceæ).......... 1 sp., a shrub, Mauritius.
Quivisia (Meliaceæ)............ 3 sp., shrubs, Mauritius (2 sp.), Rodriguez (1 sp.), also Bourbon.
Cossignya (Sapindaceæ)....... 1 sp., a shrub, Mauritius, also Bourbon.
Hornea   "   .... 1 sp., a shrub, Mauritius.
Stadtmannia   "   .... 1 sp., a shrub, Mauritius.
Doratoxylon   "   .... 1 sp., a shrub, Mauritius and Bourbon.
Gagnebina (Leguminosæ)...... 1 sp., a shrub, Mauritius, also Madagascar.
Roussea (Saxifragaceæ)........ 1 sp., a climbing shrub, Mauritius and Bourbon.
Tetrataxis (Lythraceæ)......... 1 sp., a shrub, Mauritius.
Psiloxylon   "   .... 1 sp., a shrub, Mauritius and Bourbon.
Mathurina (Turneraceæ)....... 1 sp., a shrub, Rodriguez.
Fœtidia (Myrtaceæ)............ 1 sp., a tree, Mauritius.
Danais   (Rubiaceæ)....... 4 sp., climbing shrubs, Mauritius (1 sp.), Rodriguez (1 sp.), also Bourbon and Madagascar.
Fernelia   "   .... 1 sp., a shrub, Mauritius and Rodriguez.
Pyrostria   "   .... 6 sp., shrubs, Mauritius (3 sp.), also Bourbon and Madagascar.
Scyphochlamys "   .... 1 sp., a shrub, Rodriguez.
Myonima   "   .... 3 sp., shrubs, Mauritius, also Bourbon.
Cylindrocline (Compositæ)..... 1 sp., a shrub, Mauritius.
Monarrhenus   "   ..... 2 sp., shrubs, Mauritius, also Bourbon and Madagascar.
Faujasia   "   ..... 3 sp., shrubs, Mauritius, also Bourbon and Madagascar.
Heterochænia (Campanulaceæ). 1 sp., a shrub, Mauritius, also Bourbon.
Tanulepis (Asclepiadaceæ)..... 1 sp., a·climber, Rodriguez.
Decanema   "   ..... 1 sp., a climber, Mauritius, also Madagascar.

Fragmentary Character of the Mascarene Flora.—Although the peculiar character and affinities of the vegetation of these islands are sufficiently apparent, there can be little doubt that we only possess a fragment of the rich flora which once adorned them. The cultivation of sugar and other tropical products has led to the clearing-away of the virgin forests from all the lowlands, plateaus, and accessible slopes of the mountains, so that remains of the aboriginal woodlands only linger in the recesses of the hills, and numbers of forest-haunting plants must inevitably have been exterminated. The result is that nearly three hundred species of foreign plants have run wild in Mauritius, and have in their turn helped to extinguish the native species.

Nicodemia (Loganiaceæ)....... 2 sp., shrubs, Mauritius (1 sp.), also Comoro Islands and Madagascar.
Bryodes (Serophulariaceæ)..... 1 sp., herb, Mauritius.
Radamæa   "   ..... 2 sp., herb, Seychelles (1 sp.) and Madagascar.
Colea (Rignoniaceæ)...........10 sp., Mauritius (1 sp.), Seychelles (1 sp.), also Bourbon and Madagascar. (Shrubs, trees, or climbers.)
Obetia (Urticaceæ)............ 2 sp., shrubs, Mauritius, Seychelles, and Madagascar.
Bosquiea (Moreæ)............. 3 sp., trees, Seychelles (1 sp.), also Madagascar.
Monimia (Monimiaceæ)....... 3 sp., trees, Mauritius (2 sp.), also Bourbon.
Cynorchis (Orchideæ)........ 3 sp., herb, ter., Mauritius.
Amphorchis   "   .... 1 sp., herb, ter., Mauritius, also Bourbon.
Arnottia   "   .... 2 sp., herb, ter., Mauritius, also Bourbon.
Aplostellis   "   .... 1 sp., herb, ter., Mauritius.
Cryptopus   "   .... 1 sp., herb, Epiphyte, Mauritius, also Bourbon and Madagascar.
Lomatophyllum (Liliaceæ)..... 3 sp., shrubs (succulent), Mauritius, also Bourbon.
Lodoicea   (Palmæ)......... 1 sp., tree, Seychelles.
Latania   "   .... 3 sp., trees, Mauritius (2 sp.), Rodriguez, also Bourbon.
Hyophorbe   "   .... 3 sp., trees, Mauritius (2 sp.), Rodriguez, also Bourbon.
Dictyosperma   "   .... 1 sp., tree, Mauritius, Rodriguez, also Bourbon.
Acanthophœnix   "   .... 2 sp., trees, Mauritius, also Bourbon.
Deckenia   "   .. .. 1 sp., tree, Seychelles.
Nephrosperma   "   .... 1 sp., tree, Seychelles.
Roscheria   "   .... 1 sp., tree, Seychelles.
Verschaffeltia "   .... 1 sp., tree, Seychelles.
Stevensonia   "   .... 1 sp., tree, Seychelles.
Ochropteris (Filices).......... 1 sp., herb, Mauritius, also Bourbon and Madagascar.

Among the curious features in this list are the great number of endemic shrubs in Mauritius, and the remarkable assemblage of five endemic genera of palms in the Seychelle Islands. We may also notice that one palm (*Latania Loddigesii*) is confined to Round Island and two other adjacent islets, offering a singular analogy to the peculiar snake also found there.

In the Seychelles, too, the indigenous flora has been almost entirely destroyed in most of the islands, although the peculiar palms, from their longevity and comparative hardiness, have survived. Mr. Geoffrey Nevill tells us that at Mahé and most of the other islands visited by him, it was only in a few spots near the summits of the hills that he could perceive any remains of the ancient flora. Pineapples, cinnamon, bamboos, and other plants have obtained a firm footing, covering large tracts of country, and killing the more delicate native flowers and ferns. The pineapple, especially, grows almost to the tops of the mountains. Where the timber and shrubs have been destroyed, the water falling on the surface immediately cuts channels, runs off rapidly, and causes the land to become dry and arid; and the same effect is largely seen both in Mauritius and Bourbon, where, originally, dense forest covered the entire surface, and perennial moisture, with its ever-accompanying luxuriance of vegetation, prevailed.

Flora of Madagascar Allied to that of South Africa.—In my "Geographical Distribution of Animals," I have remarked on the relation between the insects of Madagascar and those of south temperate Africa, and have speculated on a great *southern* extension of the continent at the time when Madagascar was united with it. As supporting this view I now quote Mr. Bentham's remarks on the Compositæ. He says, "The connections of the Mascarene endemic Compositæ, especially those of Madagascar itself, are eminently with the southern and subtropical African races; the more tropical races, Plucheineæ, etc., may be rather more of an Asiatic type." He further says that the Composite flora is almost as strictly endemic as that of the Sandwich Islands, and that it is much diversified, with evidences of great antiquity, while it shows insular characteristics in the tendency to tall shrubby or arborescent forms in several of the endemic or prevailing genera.

Preponderance of Ferns in the Mascarene Flora.—A striking character of the flora of these smaller Mascarene islands is the great preponderance of ferns, and next to them of Orchideæ. The following figures are taken from Mr. Baker's "Flora" for Mauritius and the Seychelles, and from an estimate by M. Frap-

pier of the flora of Bourbon given in Maillard's volume already
quoted :

Mauritius, etc.		Bourbon.	
Ferns	168	Ferns	240
Orchideæ	79	Orchideæ	120
Gramineæ	69	Gramineæ	60
Cyperaceæ	62	Compositæ	60
Rubiaceæ	57	Leguminosæ	36
Euphorbiaceæ	45	Rubiaceæ	24
Compositæ	43	Cyperaceæ	24
Leguminosæ	41	Euphorbiaceæ	18

The cause of the great preponderance of ferns in oceanic islands has already been discussed in my book on "Tropical Nature;" and we have seen that Mauritius, Bourbon, and Rodriguez must be classed as such, though from their proximity to
Madagascar they have to be considered as satellites to that great
island. The abundance of orchids may be in part due to analogous causes. Their usually minute and abundant seeds would
be as easily carried by the wind as the spores of ferns, and their
frequent epiphytic habit affords them an endless variety of stations on which to vegetate, and at the same time removes them,
in a great measure, from the competition of other plants. When,
therefore, the climate is sufficiently moist and equable, and there
is a luxuriant forest vegetation, we may expect to find orchids
abundant on such tropical islands as are not too far removed
from other lands or continents from which their seeds might be
conveyed.

*Concluding Remarks on Madagascar and the Mascarene Isl-
ands.*—There is probably no portion of the globe that contains
within itself so many and such varied features of interest connected with geographical distribution, or which so well illustrates the mode of solving the problems it presents, as the comparatively small insular region which comprises the great island
of Madagascar and the smaller islands and island-groups which
immediately surround it. In Madagascar we have a continental
island of the first rank, and undoubtedly of immense antiquity;
we have detached fragments of this island in the Comoros and
Aldabra; in the Seychelles we have the fragments of another
very ancient island, which may perhaps never have been conti-

nental ; in Mauritius, Bourbon, and Rodriguez we have three undoubtedly oceanic islands ; while in the extensive banks and coral reefs of Cargados, Saya de Malha, the Chagos, and the Maldive Isles we have indications of the submergence of many large islands which may have aided in the transmission of organisms from the Indian Peninsula. But between and around all these islands we have depths of 2500 fathoms and upwards, which renders it very improbable that there has ever been here a continuous land surface, at all events during the Tertiary or Secondary period of geology.

It is most interesting and satisfactory to find that this conclusion, arrived at solely by a study of the form of the sea-bottom and the general principle of oceanic permanence, is fully supported by the evidence of the organic productions of the several islands; because it gives us confidence in those principles, and helps to supply us with a practical demonstration of them. We find that the entire group contains just that amount of Indian forms which could well have passed from island to island ; that many of these forms are slightly modified species, indicating that the migration occurred during late Tertiary times; while others are distinct genera, indicating a more ancient connection. But in no one case do we find animals which necessitate an actual land-connection ; while the numerous Indian types of mammalia, reptiles, birds, and insects, which must certainly have passed over had there been such an actual land-connection, are totally wanting. The one fact which has been supposed to require such a connection—the distribution of the lemurs—can be far more naturally explained by a general dispersion of the group from Europe, where we know it existed in Eocene times; and such an explanation applies equally to the affinity of the Insectivora of Madagascar and Cuba, the snakes (Herpetodryas, etc.) of Madagascar and America, and the lizards (Cryptoblepharus) of Mauritius and Australia. To suppose, in all these cases, and in many others, a direct land-connection is really absurd, because we have the evidence afforded by geology of wide differences of distribution directly we pass beyond the most recent deposits; and when we go back to Mesozoic, and still more to Palæozoic, times, the majority of the groups of animals and plants appear to have had a world-wide range. A large number of our European Miocene genera of vertebrates were also Indian or African, or even American ; the South American Tertiary fauna contained many European types ; while many Mesozoic reptiles and mollusca ranged from Europe and North America to Australia and New Zealand.

By direct proof (the occurrence of wide areas of marine deposits of Eocene age), geologists have established the fact that Africa was cut off from Europe and Asia by an arm of the sea in early Tertiary times, forming a large island-continent. By the evidence of abundant organic remains we know that all the types of large mammalia now found in Africa (but which are absent from Madagascar) inhabited Europe and Asia, and many of them also North America, in the Miocene period. At a still earlier epoch Africa may have received its lower types of mammals — lemurs, insectivora, and small carnivora, together with its ancestral struthious birds, and its reptiles and insects of American or Australian affinity ; and at this period it was joined to Madagascar. Before the later continental period of Africa, Madagascar had become an island ; and thus, when the large mammalia from the northern continent overran Africa, they were prevented from reaching Madagascar, which thenceforth was enabled to develop its singular forms of low-type mammalia, its gigantic ostrich-like Æpyornis, its isolated birds, its remarkable insects, and its rich and peculiar flora. From it the adjacent islands received such organisms as could cross the sea ; while they transmitted to Madagascar some of the Indian birds and insects which had reached them.

The method we have followed in these investigations is to accept the results of geological and palæontological science, and the ascertained facts as to the powers of dispersal of the various animal groups ; to take full account of the laws of evolution as affecting distribution, and of the various ocean depths as implying recent or remote union of islands with their adjacent continents ; and the result is that wherever we possess a sufficient knowledge of these various classes of evidence, we find it possible to give a connected and intelligible explanation of all the most striking peculiarities of the organic world. In Madagascar we have undoubtedly one of the most difficult of these problems; but we have, I think, fairly met and conquered most of its difficulties. The complexity of the organic relations of this island is due partly to its having derived its animal forms from two distinct sources—from one continent through a direct landconnection, and from another by means of intervening islands now submerged; but mainly to the fact of its having been separated from a continent which is now, zoologically, in a very different condition from what it was at the time of the separation; and to its having been thus able to preserve a number of types which may date back to the Eocene, or even to the Cretaceous, period. Some of these types have become altogether extinct elsewhere; others have spread far and wide over the globe, and have survived only in a few remote countries, and especially in those which have been more or less secured by their isolated position from the incursions of the more highly developed forms of later times. This explains why it is that the nearest allies of the Madagascar fauna and flora are now so often to be found in South America or Australia—countries in which low forms of mammalia and birds still largely prevail—it being on account of the long-continued isolation of all these countries that similar forms (descendants of ancient types) are preserved in them. Had the numerous suggested continental extensions connecting these remote continents at various geological periods been realities, the result would have been that all these interesting archaic forms, all these helpless insular types, would long ago have been exterminated, and one comparatively monotonous fauna have reigned over the whole earth. So far from explaining the anomalous facts, the alleged continental extensions, had they existed, would have left no such facts to be explained.

CHAPTER XX.

ANOMALOUS ISLANDS: CELEBES.

Anomalous Relations of Celebes. — Physical Features of the Island. — Zoological Character of the Islands around Celebes. —The Malayan and Australian Banks.— Zoology of Celebes: Mammalia.—Probable Derivation of the Mammals of Cele-bes.—Birds of Celebes.—Bird-types Peculiar to Celebes.—Celebes not strictly a Continental Island.—Peculiarities of the Insects of Celebes.—Himalayan Types of Birds and Butterflies in Celebes.—Peculiarities of Shape and Color of Celebe-sian Butterflies.—Concluding Remarks.—Appendix on the Birds of Celebes.

THE only other islands of the globe which can be classed as "ancient continental" are the larger Antilles (Cuba, Hayti, Ja-maica, and Porto Rico), Iceland, and perhaps Celebes. The An-tilles have been so fully discussed and illustrated in my former work, and there is so little fresh information about them, that I do not propose to treat of them here, especially as they fall short of Madagascar in all points of biological interest, and offer no problems of a different character from such as have already been sufficiently explained.

Iceland, also, must apparently be classed as belonging to the "Ancient Continental Islands," for though usually described as wholly volcanic, it is more probably an island of varied geolog-ical structure buried under the lavas of its numerous volcanoes. But of late years extensive Tertiary deposits of Miocene age have been discovered, showing that it is not a mere congeries of volcanoes; it is connected with the British Islands and with Greenland by seas less than 500 fathoms deep; and it possesses a few mammalia, one of which is peculiar, and at least three pe-culiar species of birds. It was therefore almost certainly united with Greenland, and probably with Europe by way of Britain, in the early part of the Tertiary period, and thus afforded one of the routes by which that intermigration of American and European animals and plants was effected which we know occurred during some portion of the Eocene and Miocene periods, and probably also in the Pliocene. The fauna and flora of this island are, however, so poor, and offer so few peculiarities, that it is unnecessary to devote more time to their consideration here.

There remains the great Malay island Celebes, which, owing to its possession of several large and very peculiar mammalia, must be classed, zoologically, as "ancient continental," but whose central position and relations both to Asia and to Australia render it very difficult to decide in which of the primary zoological regions it ought to be placed, or whether it has ever been united with either of the great continents. Although I have pretty fully discussed its zoological peculiarities and past history in my "Geographical Distribution of Animals," it seems advisable to review the facts on the present occasion, more especially as the systematic investigation of the characteristics of continental islands we have now made will place us in a better position for determining its true zoo-geographical relations.

Physical Features of Celebes.—This large and still comparatively unexplored island is interesting to the geographer on account of its remarkable form, but much more so to the zoologist for its curious assemblage of animal forms. The geological structure of Celebes is almost unknown. The extremity of the northern peninsula is volcanic; while in the southern peninsula there are extensive deposits of a crystalline limestone, in some places overlying basalt. Gold is found in the northern peninsula and in the central mass, as well as iron, tin, and copper in small quantities; so that there can be little doubt that the mountain-ranges of the interior consist of ancient stratified rocks.

It is not yet known whether Celebes is completely separated from the surrounding islands by a deep sea, but the facts at our command render it probable that it is so. The northern and eastern portions of the Celebes Sea have been ascertained to be from 2000 to 2600 fathoms deep, and such depths may extend over a considerable portion of it, or even be much exceeded in the centre. In the Molucca passage a single sounding on the Gilolo side gave 1200 fathoms, and a large part of the Molucca and Banda Seas probably exceeds 2000 fathoms. The southern portion of the Strait of Macassar is full of coral reefs, and a

shallow sea of less than 100 fathoms extends from Borneo to
within about forty miles of the western promontory of Celebes;
but farther north there is deep water close to the shore, and it

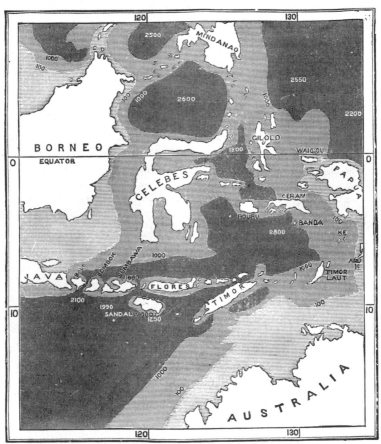

MAP OF CELEBES AND THE SURROUNDING ISLANDS.

The depth of sea is shown by three tints; the lightest indicating less than 100 fathoms, the medium tint less than 1000 fathoms, and the dark tint more than 1000
fathoms.   The figures show depths in fathoms.

seems probable that a deep channel extends quite through the
strait, which has, no doubt, been much shallowed by the deposits
from the great Bornean rivers as well as by those of Celebes
itself.   Southward again, the chain of volcanic islands from Bali

to Timor appear to rise out of a deep ocean, the few soundings we possess showing depths of from 670 to 1300 fathoms almost close to their northern shores. We seem justified, therefore, in concluding that Celebes is entirely surrounded by a deep sea, which has, however, become partially filled up by river deposits, by volcanic upheaval, or by coral reefs. Such shallows, where they exist, may therefore be due to antiquity and isolation, instead of being indications of a former union with any of the surrounding islands.

Zoological Character of the Islands around Celebes.—In order to have a clear conception of the peculiar character of the Celebesian fauna, we must take into account that of the surrounding countries from which we may suppose it to have received immigrants. These we may divide broadly into two groups, those on the west belonging to the Oriental region of our zoological geography, and those on the east belonging to the Australian region. Of the first group Borneo is a typical representative; and from its proximity and the extent of its opposing coasts it is the island which we should expect to show most resemblance to Celebes. We have already seen that the fauna of Borneo is essentially the same as that of Southern Asia, and that it is excessively rich in all the Malayan types of mammalia and birds. Java and Bali closely resemble Borneo in general character, though somewhat less rich and with several peculiar forms; while the Philippine Islands, though very much poorer, and with a greater amount of speciality, yet exhibit essentially the same character. These islands, taken as a whole, may be described as having a fauna almost identical with that of Southern Asia; for no family of mammalia is found in the one which is absent from the other, and the same may be said, with very few and unimportant exceptions, of the birds; while hundreds of genera and of species are common to both.

In the islands east and south of Celebes—the Moluccas, New Guinea, and the Timor group from Lombok eastward—we find, on the other hand, the most wonderful contrast in the forms of life. Of twenty-seven families of terrestrial mammals found in the great Malay islands, all have disappeared but four, and of these it is doubtful whether two have not been introduced by man. We also find here four families of Marsupials, all totally unknown in the western islands. Even birds, though usually more widely spread, show a corresponding difference, about eleven Malayan families being quite unknown east of Celebes, where six new families make their appearance which are equally unknown to the westward.[1]

We have here a radical difference between two sets of islands not very far removed from each other, the one set belonging zoologically to Asia, the other to Australia. The Asiatic or Malayan group is found to be bounded strictly by the eastward limits of the great bank (for the most part less than fifty fathoms below the surface) which stretches out from the Siamese and Malayan peninsulas as far as Java, Sumatra, Borneo, and the Philippines. To the east another bank unites New Guinea and the Papuan Islands as far as Aru, Mysol, and Waigiou, with Australia; while the Molucca and Timor groups are surrounded by much deeper water, which forms, in the Banda and Celebes seas, and perhaps in other parts of this area, great basins of enormous depths (2000 to 3000 fathoms, or even more) enclosed by tracts under 1000 fathoms, which separate the basins from each other and from the adjacent Pacific and Indian oceans (see map). This peculiar formation of the sea-bottom probably indicates that this area has been the seat of great local upheavals and subsidences; and it is quite in accordance with this view that we find the Moluccas, while closely agreeing with New Guinea in their forms of life, yet strikingly deficient in

---

[1] Families of Malayan Birds not Found in Islands East of Celebes.

Troglodytidæ.
Sittidæ.
Paridæ.
Liotrichidæ.
Phyllornithidæ.
Eurylæmidæ.
Picidæ.
Indicatoridæ.
Megalæmidæ.
Trogonidæ.
Phasianidæ.

Families of Moluccan Birds not Found in Islands West of Celebes.

Paradiseidæ.
Meliphagidæ.
Cacatuidæ.
Platycercidæ.
Trichoglossidæ.
Nestoridæ.

many important groups, and exhibiting an altogether poverty-stricken appearance as regards the higher animals. It is a suggestive fact that the Philippine Islands bear an exactly parallel relation to Borneo, being equally deficient in many of the higher groups; and here too, in the Sooloo Sea, we find a similar enclosed basin of great depth. Hence we may in both cases connect, on the one hand, the extensive area of land surface and of adjacent shallow sea with a long period of stability and a consequent rich development of the forms of life; and, on the other hand, a highly broken land surface, with the adjacent seas of great but very unequal depths, with a period of disturbance, probably involving extensive submersions of the land, resulting in a scanty and fragmentary vertebrate fauna.

Zoology of Celebes.—The zoology of Celebes differs so remarkably from that of both the great divisions of the archipelago above indicated that it is very difficult to decide in which to place it. It possesses only about sixteen species of terrestrial mammalia, so that it is at once distinguished from Borneo and Java by its extreme poverty in this class. Of this small number four belong to the Moluccan and Australian fauna—there being two marsupials of the genus Cuscus, and two forest rats said to be allied to Australian types.

The remaining twelve species are, generally speaking, of Malayan or Asiatic types, but some of them are so peculiar that they have no near allies in any part of the world; while the rest are of the ordinary Malay type, or even identical with Malayan species, and some of these may be recent introductions through human agency. These twelve species of Asiatic type will be now enumerated. They consist of five peculiar squirrels —a group unknown farther east; a peculiar species of wild-pig; a deer so closely allied to the *Cervus hippelaphus* of Borneo that it may well have been introduced by man both here and in the Moluccas; a civet, *Viverra tangalunga,* common in all the Malay islands, and also perhaps introduced; the curious Malayan tarsier (*Tarsius spectrum*), said to be only found in a small island off the coast; and, besides these, three remarkable animals, all of large size, and all quite unlike anything found in the Malay Islands or even in Asia. These are a black and almost tailless

baboon-like ape (*Cynopithecus nigrescens*); an antelopean buffalo (*Anoa depressicornis*), and the strange babirusa (*Babirusa
alfurus*).

None of these three animals last mentioned have any close
allies elsewhere, and their presence in Celebes may be considered
the crucial fact which must give us the clew to the past history
of the island. Let us, then, see what they teach us. The ape
is apparently somewhat intermediate between the great baboons
of Africa and the short-tailed macaques of Asia, but its cranium
shows a nearer approach to the former group in its flat project-
ing muzzle, large superciliary crests, and maxillary ridges. The
anoa, though anatomically allied to the buffaloes, externally
more resembles the bovine antelopes of Africa; while the
babirusa is altogether unlike any other living member of the
swine family, the canines of the upper jaws growing directly
upward like horns, forming a spiral curve over the eyes, instead
of downward, as in all other mammalia. An approach to this
peculiarity is made by the African wart-hogs, in which the upper
tusk grows out laterally and then curves up; but these animals
are not otherwise closely allied to the babirusa.

Probable Derivation of the Mammals of Celebes.—It is clear
that we have here a group of extremely peculiar, and in all
probability very ancient, forms, which have been preserved to
us by isolation in Celebes, just as the monotremes and marsupials have been preserved in Australia, and so many of the
lemurs and Insectivora in Madagascar. And this compels us
to look upon the existing island as a fragment of some ancient
land, once perhaps forming part of the great northern continent,
but separated from it far earlier than Borneo, Sumatra, and Java.
The exceeding scantiness of the mammalian fauna, however, remains to be accounted for. We have seen that Formosa, a much
smaller island, contains more than twice as many species; and
we may be sure that at the time when such animals as apes
and buffaloes existed, the Asiatic continent swarmed with varied
forms of mammals to quite as great an extent as Borneo does
now. If the portion of separated land had been anything like
as large as Celebes now is, it would certainly have preserved a
far more abundant and varied fauna. To explain the facts we

have the choice of two theories—either that the original island has since its separation been greatly reduced by submersion, so as to lead to the extinction of most of the higher land animals; or that it originally formed part of an independent land stretching eastward, and was only united with the Asiatic continent for a short period, or perhaps even never united at all, but so connected by intervening islands separated by narrow straits that a few mammals might find their way across. The latter supposition appears best to explain the facts. The three animals in question are such as might readily pass over narrow straits from island to island; and we are thus better enabled to understand the complete absence of the arboreal monkeys, of the Insectivora, and of the very numerous and varied Carnivora and Rodents of Borneo, all of which are entirely unrepresented in Celebes by any peculiar and ancient forms except the squirrels.

The question at issue can only be finally determined by geological investigations. If Celebes has once formed part of Asia, and participated in its rich mammalian fauna which has been since destroyed by submergence, then some remains of this fauna must certainly be preserved in caves or late Tertiary deposits, and proofs of the submergence itself will be found when sought for. If, on the other hand, the existing animals fairly represent those which have ever reached the island, then no such remains will be discovered, and there need be no evidence of any great and extensive subsidence in late Tertiary times.

Birds of Celebes.—Having thus clearly placed before us the problem presented by the mammalian fauna of Celebes, we may proceed to see what additional evidence is afforded by the birds, and any other groups of which we have sufficient information. About 164 species of true land birds are now known to inhabit the island of Celebes itself. Considerably more than half of these (94 species) are peculiar to it; 29 are found also in Borneo and the other Malay islands, to which they specially belong; while 16 are common to the Moluccas or other islands of the Australian region; the remainder being species of wide range, and not characteristic of either division of the archipelago. We have here a large preponderance of Western over Eastern species of birds inhabiting Celebes, though not to quite so great an extent as in the mammalia; and the inference to be drawn from this fact is, simply, that more birds have migrated from Borneo than from the Moluccas—which is exactly what we might expect, both from the greater extent of the coast of Borneo opposite that of Celebes, and also from the much greater richness in species of the Bornean than the Moluccan bird fauna.

It is, however, to the relations of the peculiar species of Celebesian birds that we must turn in order to ascertain the origin of the fauna in past times; and we must look to the source of the generic types which they represent to give us this information. The ninety-four peculiar species above noted belong to about sixty-six genera, of which about twenty-three are common to the whole archipelago, and have therefore little significance. Of the remainder, twelve are altogether peculiar to Celebes; twenty-one are Malayan, but not Moluccan or Australian; while ten are Moluccan or Australian, but not Malayan. This proportion does not differ much from that afforded by the non-peculiar species; and it teaches us that, for a considerable period, Celebes has been receiving immigrants from all sides, many of which have had time to become modified into distinct representative species. These evidently belong to the period during which Borneo, on the one side, and the Moluccas, on the other, have occupied very much the same relative position as now. There remain the twelve peculiar Celebesian genera, to which we must look for some further clew as to the origin of the older portion of the fauna; and as these are especially interesting, we must examine them somewhat closely.

Bird-types Peculiar to Celebes.—First we have Artamides, one of the Campephaginæ, or caterpillar-shrikes—a not very well-marked genus, and which may have been derived either from the Malayan or the Moluccan side of the archipelago. Two peculiar genera of kingfishers—Monachalcyon and Cittura—seem allied, the former to the wide-spread Todiramphus and to the Caridonax of Lombok, the latter to the Australian Melidora. Another kingfisher, Ceycopsis, combines the characters of the Malayan Ceyx and the African Ispidina, and thus forms an example of an ancient generalized form analogous to what occurs among the mammalia. Streptocitta is a peculiar form allied to the magpies; while Basilornis (found also in Ceram), Enodes, and Scissirostrum are very peculiar starlings, the latter altogether unlike any other bird, and perhaps forming a distinct sub-family. Meropogon is a peculiar bee-eater, allied to the Malayan Nyctiornis; Rhamphococcyx is a modification of Phæ-nicophaes, a Malayan genus of cuckoos; Prioniturus (found also in the Philippines) is a genus of parrots distinguished by raquet-formed tail-feathers, altogether unique in the order; while Mega-cephalon is a remarkable and very isolated form of the Aus-tralian Megapodiidæ, or mound-builders.

Omitting those whose affinity may be pretty clearly traced to groups still inhabiting the islands of the western or the eastern half of the archipelago, we find four birds which have no near allies at all, but appear to be either ancestral forms, or extreme modifications, of Asiatic or African birds—Basilornis, Enodes, Scissirostrum, Ceycopsis. These may fairly be associated with the baboon-ape, anoa, and babirusa, as indicating extreme antiq-uity and some communication with the Asiatic continent at a period when the forms of life and their geographical distribu-tion differed considerably from what they are at the present time.

But here again we meet with exactly the same difficulty as in the mammalia, in the comparative poverty of the types of birds now inhabiting Celebes. Although the preponderance of affin-ity, especially in the case of its more ancient and peculiar forms, is undoubtedly with Asia rather than with Australia, yet, still more decidedly than in the case of the mammalia, are we forbid-den to suppose that it ever formed a part of the old Asiatic con-tinent, on account of the *total* absence of so many important and extensive groups of Asiatic birds. It is not single species or even genera, but whole families, that are thus absent, and among them families which are pre-eminently characteristic of all trop-ical Asia. Such are the Timaliidæ, or babblers, of which there are twelve genera in Borneo and nearly thirty genera in the Oriental Region, but of which one species only, hardly distin-guishable from a Malayan form, inhabits Celebes; the Phyllor-nithidæ, or green bulbuls, and the Pycnonotidæ, or bulbuls, both absolutely ubiquitous in tropical Asia and Malaya, but un-

known in Celebes; the Eurylæmidæ, or gapers, found everywhere in the great Malay islands; the Megalæmidæ, or barbets;
the Trogonidæ, or trogons; and the Phasianidæ, or pheasants;
all pre-eminently Asiatic and Malayan, but all absent from Celebes, with the exception of the common jungle-fowl, which,
owing to the passion of Malays for cock-fighting, may have been
introduced. To these important *families* may be added Asiatic
and Malayan *genera* by the score; but, confining ourselves to
these seven ubiquitous families, we must ask, Is it possible that,
at the period when the ancestors of the peculiar Celebes mammals entered the island, and when the forms of life, though distinct, could not have been quite unlike those now living, it could
have actually formed a part of the continent without possessing
representatives of the greater part of these extensive and important families of birds? To get rid altogether of such varied and
dominant types of bird-life by any subsequent process of submersion is more difficult than to exterminate mammalia; and we
are therefore again driven to our former conclusion—that the
present land of Celebes has never (in Tertiary times) been united
to the Asiatic continent, but has received its population of Asiatic forms by migration across narrow straits and intervening islands. Taking into consideration the amount of affinity, on the
one hand, and the isolation, on the other, of the Celebesian fauna,
we may probably place the period of this earlier migration in
the early part of the latter half of the Tertiary period; that is,
in middle or late Miocene times.

Celebes not Strictly a Continental Island.—A study of the
mammalian and of the bird fauna of Celebes thus leads us in
both cases to the same conclusion, and forbids us to rank it as a
strictly continental island on the Asiatic side. But facts of a
very similar character are equally opposed to the idea of a former land-connection with Australia or New Guinea, or even
with the Moluccas. The numerous marsupials of those countries are all wanting in Celebes, except the phalangers of the
genus Cuscus, and these arboreal creatures are very liable to be
carried across narrow seas on trees uprooted by earthquakes or
floods. The terrestrial cassowaries are equally absent; and thus
we can account for the presence of all the Moluccan or Austra-

lian types actually found in Celebes without supposing any land-connection on this side during the Tertiary period. The presence of the Celebes ape in the island of Batchian, and of the babirusa in Bouru, can be sufficiently explained by a somewhat closer approximation of the respective lands, or by a few intervening islands which have since disappeared, or it may even be due to human agency.

If the explanation now given of the peculiar features presented by the fauna of Celebes be the correct one, we are fully justified in classing it as an " anomalous island," since it possesses a small but very remarkable mammalian fauna, without ever having been directly united with any continent or extensive land ; and, both by what it has and what it wants, occupies such an exactly intermediate position between the Oriental and Australian regions that it will perhaps ever remain a mere matter of opinion with which it should properly be associated.   Forming, as it does, the western limit of such typical Australian groups as the marsupials among mammalia, and the Trichoglossidæ and Meliphagidæ among birds, and being so strikingly deficient in all the more characteristic Oriental families and genera of both classes, I have always placed it in the Australian Region ; but it may perhaps with equal propriety be left out of both till a further knowledge of its geology enables us to determine its early history with more precision.

Peculiarities of the Insects of Celebes.—The only other class of animals in Celebes of which we have a tolerable knowledge is that of insects, among which we meet with peculiarities of a very remarkable kind, and such as are found in no other island on the globe.   Having already given a full account of some of these peculiarities in a paper read before the Linnæan Society, republished in my " Contributions to the Theory of Natural Selection," while others have been discussed in my " Geographical Distribution of Animals" (Vol. I., p. 434), I will here only briefly refer to them in order to see whether they accord with, or receive any explanation from, the somewhat novel view of the past history of the island here advanced.

The general distribution of the two best-known groups of insects—the butterflies and the beetles—agrees very closely with that of the birds and mammalia, inasmuch as Celebes forms the eastern limit of a number of Asiatic and Malayan genera, and at the same time the western limit of several Moluccan and Australian genera—the former perhaps preponderating, as in the higher animals.

Himalayan Types of Birds and Butterflies in Celebes.—A curious fact of distribution, exhibited both among butterflies and birds, is the occurrence in Celebes of species and genera unknown to the adjacent islands, but only found again when we reach the Himalayan Mountains or the Indian Peninsula. Among birds we have a small yellow flycatcher (*Myialestes helianthea*), a flower-pecker (*Pachyglossa aureolimbata*), a finch (*Munia brunneiceps*), and a roller (*Coracias Temminckii*), all closely allied to Indian (not Malayan) species—all the genera except Munia being, in fact, unknown in any Malay island. Exactly parallel cases are two butterflies of the genera Dichorrhagia and Euripus, which have very close allies in the Himalayas, but nothing like them in any intervening country. These facts call to mind the similar case of Formosa, where some of its birds and mammals occurred again, under identical or closely allied forms, in the Himalayas; and in both instances they can only be explained by going back to a period when the distribution of these forms was very different from what it is now.

Peculiarities of Shape and Color in Celebesian Butterflies.— Even more remarkable are the peculiarities of shape and color in a number of Celebesian butterflies of different genera. These are found to vary all in the same manner, indicating some general cause of variation able to act upon totally distinct groups, and produce upon them all a common result. Nearly thirty species of butterflies, belonging to three different families, have a common modification in the shape of their wings, by which they can be distinguished at a glance from their allies in any other island or country whatever; and all these are larger than the representative forms inhabiting most of the adjacent islands.[1] No such remarkable local modification as this is known to occur

---

[1] For outline figures of the chief types of these butterflies, see my "Malay Archipelago," Vol. I., p. 441, or p. 281 of the second edition.

in any other part of the globe; and whatever may have been its cause, that cause must certainly have been long in action, and have been confined to a limited area. We have here, therefore, another argument in favor of the long-continued isolation of Celebes from all the surrounding islands and continents — an hypothesis which we have seen to afford the best, if not the only, explanation of its peculiar vertebrate fauna.

Concluding Remarks.—If the view here given of the origin of the remarkable Celebesian fauna is correct, we have in this island a fragment of the great eastern continent which has preserved to us, perhaps from Miocene times, some remnants of its ancient animal forms. There is no other example on the globe of an island so closely surrounded by other islands on every side, yet preserving such a marked individuality in its forms of life; while, as regards the special features which characterize its insects, it is, so far as yet known, absolutely unique. Unfortunately, very little is known of the botany of Celebes, but it seems probable that its plants will to some extent partake of the speciality which so markedly distinguishes its animals; and there is here a rich field for any botanist who is able to penetrate to the forest-clad mountains of its interior.

APPENDIX TO CHAPTER XX.

The following list of the land birds of Celebes and the adjacent islands which partake of its zoological peculiarities, in which are incorporated all the species discovered up to the present year, has been drawn up from the following sources:

1. A List of the Birds Known to Inhabit the Island of Celebes. By Arthur, Viscount Walden, F.R.S. (*Trans. Zool. Soc.*, 1872, Vol. VIII., pt. ii.)
2. Intorno al Genere Hermotimia. (Rchb.) Note di Tommaso Salvadori. (*Atti della Reale Academia delle Scienze di Torino*, Vol. X., 1874.)
3. Intorno a Due Collezioni di Ucelli di Celebes. Note di Tommaso Salvadori. (*Annali del Mus. Civ. di St. Nat. di Genova*, Vol. VII., 1875.)
4. Beiträge zur Ornithologie von Celebes und Sangir. Von Dr. Friedrich Brüggemann. Bremen, 1876.
5. Intorno a Due Piccole Collezioni di Ucelli di Isole Sanghir e di Tifore. Note di Tommaso Salvadori. (*Annali del Mus. Civ. di St. Nat. di Genova*, Vol. IX., 1876-77.)
6. Intorno alle Specie di Nettarinie delle Molucche e del Gruppo di Celebes. Note di Tommaso Salvadori. (*Atti della Reale Acad. delle Scienze di Torino*, Vol. XII., 1877.)
7. Descrizione di Tre Nuove Specie di Ucelli, e Note intorno ad altre poco conosciute delle Isole Sanghir. Per Tommaso Salvadori. (*Loc. cit.*, Vol. XIII., 1878.)
8. Field Notes on the Birds of Celebes. By A. B. Meyer, M.D., etc. (*Ibis*, 1879.)
9. On the Collection of Birds made by Dr. Meyer during his Expedition to New Guinea and some Neighboring Islands. By R. Boulder Sharpe. (*Mitth. d. kgl. Zool. Mus. Dresden*, 1878. Heft 3.) New species from the Sula and Sanghir Islands are described.
10. List of Birds from the Sula Islands (East of Celebes), with Descriptions of the New Species. By Alfred Russel Wallace, F.Z.S. (*Proc. Zool. Soc.*, 1862, p. 333.)

LIST OF LAND BIRDS OF CELEBES.

N.B.—The Species marked with an * are not included in Viscount Walden's list. For these only, an authority is usually given.

	Celebes.	Sula Is.	Sanghir Is.	Range and Remarks.
TURDIDÆ.				
1. Geocichla erythronota......	×			
2. Monticola solitaria........	×		×	Phil., China, Japan
SYLVIIDÆ.				
3. Cisticola cursitans........	×			Assam
4. " Grayi...........	×			
5. Acrocephalus orientalis.....	×			China, Japan
*6. " insularis.....	—	—	× (Salv.)	Moluccas
7. Pratincola caprata.........	×			Asia, Java, Timor
*8. Gerygone flaveola (Cab.)...	×(Meyer)			(Near *G. sulphurea*, Timor)
TIMALIIDÆ.				
9. Trichostoma Celebense.....	×			
PYCNONOTIDÆ.				
*10. Criniger longirostris (Wall.)		×		Oriental genus (near Bouru sp.)
*11. " aureus (Wald.)....	×			
ORIOLIDÆ.				
12. Oriolus Celebensis.........	×			(Var. of *O. coronatus*, Java)
*13. " formosus (Cab.)....	—	—	×(Brügg.)	(Var. of Philipp. sp.)
*14. " frontalis (Wall.)....	—	×		
CAMPEPHAGIDÆ.				
15. Graucalus atriceps.........	×			Ceram, Flores
16. " leucopygius......	×			
17. " Temminckii......	×	×		
18. Campephaga morio........	×			
*19. " melanotis.....	—	× (Wall.)		Moluccas
*20. " Salvadorii (Sharpe)	—	—	×	
21. Lalage leucopygialis........	×			
*22. " Dominica..........	×(Meyer)	—	—	Java
23. Artamides bicolor..........	×			
*24. " schistaceus (Sharpe)	—	×		

	Celebes.	Sula Is.	Sanghir Is.	Range and Remarks.
DICRURIDÆ.				
25. Dicrurus leucops...........	×			
*26. " axillaris (Salv.)...	—	—	×	
*27. " pectoralis (Wall.)		× •		
MUSCICAPIDÆ.				
28. Cyornis rufigula............	×			
29. " banyumas.........	×			Java and Borneo
30. Myialestes helianthea......	×			(Indian ally)
31. Hypothymis puella.........	×	×		
32. " Menadensis?..	×			
*33. Monarcha commutata (Brügg.).................	×			
*34. Monarcha cinerascens......	—	× (Wall.)		Moluccas
PACHYCEPHALIDÆ.				
35. Hylocharis sulfuriventra....	×			
*36. Pachycephala lineolata (Wall.).................	—	×	—	Bouru
*37. Pachycephala rufescens (Wall.).	—	×	—	Bouru
*38. Pachycephala Clio (Wall.).	—	×	—	Bouru
LANIIDÆ.				
*39. Lanius magnirostris (Meyer)	×	—	—	Java
CORVIDÆ.				
40. Corvus enca...............	×	× var.		Java
*41. " annectens (Brügg.).	×			
42. " (Gazzola) typica....	×			
43. Streptocitta Caledonica	×			
44. " torquata...... .	×			
*45. (Charitornis)Albertiæ(Schl.)	—	×		
MELIPHAGIDÆ.				
46. Myzomela chloroptera......	×			(Nearest M. sanguinolenta of Aus.)
NECTARINIIDÆ.				
47. Anthreptes Malaccensis.... (Celebensis. Shelley)	×	×	×	Siam, Malaya
48. Chalcostethia porphyolæma	×			
*49. " auriceps............	—	× (Wall.)	—	Ternate
*50. " Sangirensis (Meyer).	—	—	×	
51. Arachnecthra frenata......	×	×	—	Moluccas and N. Guinea.
52. Nectarophila Grayi	×			
53. Æthopyga flavostriata......	×			(An Oriental genus)
*54. " Beccarii (Salv.).....	×			
*55. " Duyvenbodei (Schl.).	—	—	×	
DICÆIDÆ.				
56. Zosterops intermedia.......	×			Lombok
57. " atrifrons.........	×			
58. Dicæum Celebicum........	×	×		

	Celebes.	Sula Is.	Sanghir Is.	Range and Remarks.
*59. Dicæum Sanghirense (Salv.)	—	—	×	
60. Pachyglossa aureolimbata..	×	—	×	
HIRUNDINIDÆ.				
61. Hirundo gutturalis........	×		×	Indian Region
62. " Javanica.........	×	×		Indo-Malaya
PLOCEIDÆ.				
63. Munia oryzivora..........	×			Java
64. " nisoria.............	×			Java
65. " Molucca...........	×			Moluccas
66. " brunneiceps........	×			(Near *M. rubroni-*
				gra, India)
*67. " Jagori.............	× (Meyer)			Philippines
STURNIDÆ.				
68. Basilornis Celebensis.......	×			
69. Acridotheres cinereus......	×			
70. Sturnia pyrrhogenys.......	×			Malaya
71. Calornis neglecta..........	×	×	× var.	
*72. " metallica.........	×(Brügg.)	× (Wall.)	—	Moluccas
73. Enodes erythrophrys.......	×			
74. Scissirostrum Pagei........	×			
ARTAMIDÆ.				
75. Artamus monachus........	×	×		
76. " leucorhynchus....	×			Malay Archipel.
MOTACILLIDÆ.				
77. Corydalla Gustavi.........	×			
78. Budytes viridis............	×			Java, Moluccas
*79. Calobates melanope=(Motac. sulfurea. Brügg.)...	×			China, Philipp.
PITTIDÆ.				
80. Pitta Forsteni.............	×			
*81. " Sanghirana (Schl.)..	—	—	×	
82. " Celebensis..........	×			
*83. " palliceps (Brügg.)....	—	—	×	
*84. " cæruleitorques (Salv.)	—	—	×	
*85. " irena (=crassirostris)	—	× (Wall.)	—	Timor, Ternate?
PICIDÆ.				
86. Mulleripicus fulvus........	×			
87. Yungipicus Temminckii....	×			
CUCULIDÆ.				
88. Rhamphococcyx calorhynchus...................	×			
89. Centropus Celebensis......	×			
90. " affinis..........	×			Java
91. " Javanensis......	×			Java, Borneo
92. Cuculus canorus...........	×			
93. Cacomantes lanceolatus....	×			Java

	Celebes.	Sula Is.	Sanghir Is.	Range and Remarks.
94. Cacomantes sepulchralis...	×			
95. Hierococcyx crassisostris..	×			
96. Eudynamis melanorhyncha	×			
*97. " facialis(Wall.).	—	×		
*98. " orientalis......	—		×(Brügg.)	Moluccas?
99. Scythrops Novæhollandiæ.	×			Moluccas, etc.
CORACIIDÆ.				
100. Coracias Temminckii.....	×			
101. Eurystomus orientalis.....	×	×	×	Asia
MEROPIDÆ.				
102. Meropogon Forsteni......	×			
103. Merops Philippinus.......	×			Oriental Region
104. " ornatus..........	×	×		Java, Australia
ALCEDINIDÆ.				
105. Alcedo Moluccensis.......	×		×	Moluccas
106. " Asiatica..........	×			Indo-Malaya
107. Pelargopsis melanorhyncha	×	×		
*108. Ceyx Wallacei (Sharpe)..	—	×		(Allied to Mol. sp.)
109. Ceycopsis fallax..........	×			
110. Halcyon chloris..........	×	×	×	All Archipel.
111. " sancta..........	×	×		All Archipel.
112. " Forsteni........	×			
113. " rufa............	×	×		
114. Monachalcyon princeps...	×			
*115. " cyanocephala (Brügg.)	×			
116. Cittura cyanotis..........	×			
*117. " Sanghirensis (Schl.)	--	—	×	
BUCEROTIDÆ.				
118. Hydrocissa exarata.......	×			
119. Cranorhinus cassidix......	×			
CAPRIMULGIDÆ.				
120. Caprimulgus affinis.......	×			
121. " sp..........	×			
122. Lyncornis macropterus....	×			
CYPSELIDÆ.				
123. Dendrochelidon Wallacei..	×	×		
124. Collocalia esculenta.......	×			Mol. to Aru Is.
125. " fuciphaga.......	×			India, Java
126. Chætura gigantea.........	×			India, Java
PSITTACI.				
127. Cacatua sulphurea........	×			Lombok, Flores
128. Prioniturus platurus......	×			
129. " flavicans......	×			
*130. Platycercus dorsalis, var...	—	× (Wall.)		N. Guinea?
131. Tanygnathus Mulleri.....	×	×		
*132. " megalorhynchus	×	—	×	Moluccas. An is. n. Menado (Meyer)

	Celebes.	Sula Is.	Sanghir Is.	Range and Remarks.
*133. Tanygnathus Luzoniensis.	—	—	×(Brügg.)	
134. Loriculus stigmatus.......	×			
*135. " quadricolor (Wald.)	×			Togian Is., Gulf of Tomini.
136. " Sclateri...........	?	×		
137. " exilis..............	×			
*138. " catamene (Schl.)...	—	—	×	
139. Trichoglossus ornatus.....	×			
*140. " flavoviridis (Wall.)	—	×		
141. " Meyeri...........	×			
*142. Eos histrio = E. coccinea	—	—	×	
COLUMBÆ.				
143. Treron vernans..........	×			Malac.,Java,Philip.
144. " griseicauda.......	×	×	× var. Sanghirensis.	
145. Ptilopus formosus........	×			
146. " melanocephalus..	×	×	× var. Xanthorrhoa. Salv.	Java, Lombok
147. " gularis...........	×			
*148. " Fischeri (Brügg.).	×			
149. Carpophaga paulina.......	×	×		
*150. " pulchella (Wald.)..	×			Togian Is.(Ann.and Mag. Nat. Hist., 1874.)
151. " concinna..........	—	—	× (Salv.)	Ké Goram
152. " rosacea...........	×			Gilolo, Timor
*153. " pæcilorrhoa (Brügg.)	×			
154. " luctuosa...........	×	×		
*155. " bicolor...........	×(Meyer)	—	×	N. Guinea, Moluc.
156. " radiata.............	×	—	×	
157. " Forsteni..........	×			
158. Macropygia albicapilla....	×	×		
159. " Macassariensis	×			
*160. " Sanghirensis(Salv.)	—	—		
161. Turacœna Menadensis.....	×	×	×	
*162. Reinwardtænas Reinwardti	× Meyer			Moluccas and N.G.
163. Turtur tigrina...........	×			Malaya, Moluccas
164. Chalcophaps Stephani.....	×			New Guinea
165. " Indica.......	×	× var.	×	India and Archipel.
166. Phlogænas tristigmata.....	×			
167. Geopelia striata..........	×			China, Java, Lomb.
168. Calænas Nicobarica.......	×			Malacca and New Guinea
GALLINÆ.				
169. Gallus Bankiva..........	×			Java, Timor
170. Coturnix minima.........	×			(Var.of C. Chinensis)
171. Turnix rufilatus..........	×			
*172. " Beccarii (Salv.)...	×			
173. Megapodius Gilberti......	×			
174. Megacephalon malleo.....	×	—	×	

	Celebes.	Sula Is.	Sanghir Is.	Range and Remarks.
ACCIPITRES.				
175. Circus assimilis...........	×			Australia
176. Astur griseiceps...........	×			
*177. " tenuirostris (Brügg.)	×			
178. " rhodogastra........	×			
179. " trinotata...........	×			
*180. Accipiter Sulaensis (Schl.)	—	×		
181. " Soloensis........	×			Malacca and New Guinea
182. Neopus Malayensis........	×			Nepaul,Sum., Java, Moluccas
183. Spizætus lanceolatus......	×	×		
184. Haliætus leucogaster......	×			Oriental Region
185. Spilornis rufipectus........	×	×		
186. Butastur liventer..........	×			Java, Timor
187. " Indicus..........	×	—	×	India, Java
188. Haliastur leucosternus.....	×			Moluccas,N.Guinea
189. Milvus affinis.............	×			Australia
190. Elanus hypoleucus........	×			? Java, Borneo
191. Pernis ptilorhyncha.......	×			(Var. Java, etc.)
(var. Celebensis)				
192. Baza erythrothorax.......	×	×		
193. Falco severus.............	×			All Archipel.
194. Cerchneis Moluccensis....	×			Java, Moluccas
195. Polioætus humilis........	×			India, Malaya
STRIGIDÆ.				
196. Athene punctulata........	×			
197. " ochracea..	×			
198. Scops magicus............	×			Amboyna, etc. ?
199. " Menadensis........	×			Flores,Madagascar
200. Ninox Japonicus	×			China, Japan
*201. " scutulata..........	—	—	× (Salv.)	Malacca
202. Strix Rosenbergi..........	×			

CHAPTER XXI.

ANOMALOUS ISLANDS: NEW ZEALAND.

Position and Physical Features of New Zealand.—Zoological Character of New Zealand.—Mammalia.—Wingless Birds Living and Extinct.—Recent Existence of the Moa.—Past Changes of New Zealand Deduced from its Wingless Birds.—Birds and Reptiles of New Zealand.—Conclusions from the Peculiarities of the New Zealand Fauna.

THE fauna of New Zealand has been so recently described, and its bearing on the past history of the islands so fully discussed in my large work already referred to, that it would not be necessary to introduce the subject again, were it not that we now approach it from a somewhat different point of view, and with some important fresh material, which will enable us to arrive at more definite conclusions as to the nature and origin of this remarkable fauna and flora. The present work is, besides, addressed to a wider class of readers than my former volumes, and it would be manifestly incomplete if all reference to one of the most remarkable and interesting of insular faunas were omitted.

The two great islands which mainly constitute New Zealand are together about as large as the kingdom of Italy. They stretch over thirteen degrees of latitude in the warmer portion of the south temperate zone, their extreme points corresponding to the latitudes of Vienna and Cyprus. Their climate throughout is mild and equable, their vegetation is luxuriant, and deserts or uninhabitable regions are as completely unknown as in our own islands.

The geological structure of these islands has a decidedly continental character. Ancient sedimentary rocks, granite, and modern volcanic formations abound; gold, silver, copper, tin, iron, and coal are plentiful; and there are also some considerable deposits of early or late Tertiary age. The Secondary rocks alone are very scantily developed, and such fragments as exist are chiefly of Cretaceous age, often not clearly separated from the succeeding Eocene beds.

The position of New Zealand in the great Southern Ocean, about 1200 miles distant from the Australian continent, is very isolated. It is surrounded by a moderately deep ocean; but the form of the sea-bottom is peculiar, and may help us in the solu-

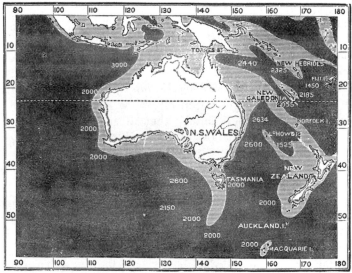

MAP SHOWING DEPTHS OF SEA AROUND AUSTRALIA AND NEW ZEALAND.

The light tint indicates a depth of less than 1000 fathoms.
The dark　"　　　"　　　"　more　　"　　　"

tion of some of the anomalies presented by its living productions. The line of 200 fathoms encloses the two islands and extends their area considerably; but the 1000-fathom line, which indicates the land-area that would be produced if the sea-bottom were elevated 6000 feet, has a very remarkable conformation, extending in a broad mass westward, and then sending out two great arms, one reaching to beyond Lord Howe's Island, while the other stretches over Norfolk Island to the great barrier reef, thus forming a connection with tropical Australia and New

Guinea. Temperate Australia, on the other hand, is divided
from New Zealand by an oceanic gulf about 700 miles wide and
between 2000 and 3000 fathoms deep. The 2000-fathom line
embraces all the islands immediately round New Zealand; and
a submarine plateau at a depth somewhere between one and two
thousand fathoms stretches southward to the antarctic continent. Judging from these indications, we should say that the
most probable ancient connections of New Zealand were with
tropical Australia and New Guinea, and perhaps, at a still more
remote epoch, with the great southern continent by means of
intervening lands and islands ; and we shall find that a land-connection or near approximation in these two directions at remote
periods will serve to explain many of the remarkable anomalies
which these islands present.

Zoological Character of New Zealand.—We see, then, that
both geologically and geographically New Zealand has more of
the character of a "continental" than of an "oceanic" island ;
yet its zoological characteristics are such as almost to bring it
within the latter category, and it is this which gives it its
anomalous character. It is usually considered to possess no indigenous mammalia; it has no snakes, and only one frog; it
possesses (living or quite recently extinct) an extensive group of
birds incapable of flight; and its productions generally are wonderfully isolated, and seem to bear no predominant or close relation to those of Australia or any other continent. These are the
characteristics of an oceanic island ; and thus we find that the
inferences from its physical structure and those from its forms
of life directly contradict each other. Let us see how far a closer
examination of the latter will enable us to account for this apparent contradiction.

Mammalia of New Zealand.—The only undoubtedly indigenous mammalia appear to be two species of bats, one of which
(*Scotophilus tuberculatus*) is, according to Mr. Dobson, identical
with an Australian form, while the other (*Mystacina tubercu-
lata*) forms a very remarkable and isolated genus of Emballonu-
ridæ, a family which extends throughout all the tropical regions
of the globe. The genus Mystacina was formerly considered to
belong to the American Phyllostomidæ, but this has been shown

to be an error.[1] The poverty of New Zealand in bats is very remarkable when compared with our own islands, where there are at least twelve distinct species, though having a far less favorable climate.

Of the existence of truly indigenous land mammals in New Zealand there is at present no positive evidence, but there is some reason to believe that one, if not two, species may be found there. The Maoris say that before Europeans came to their country a forest rat abounded and was largely used for food. They believe that their ancestors brought it with them when they first came to the country; but it has now become almost, if not quite, exterminated by the European brown rat. What this native animal was is still somewhat doubtful. Several specimens have been caught at different times which have been declared by the natives to be the true *Kiore Maori*, as they term it; but these have usually proved, on examination, to be either the European black rat or some of the native Australian rats which now often find their way on board ships. But within the last few years many skulls of a rat have been obtained from the old Maori cooking-places, and from a cave associated with moa bones; and Captain Hutton, who has examined them, states that they belong to a true Mus, but differ from the *Mus rattus*. This animal might have been on the islands when the Maoris first arrived, and in that case would be truly indigenous; while the Maori legend of their "ancestors" bringing the rat from their Polynesian home may be altogether a myth invented to account for its presence in the islands, because the only other land mammal which they knew—the dog—was certainly so brought. The question can only be settled by the discovery of remains of a rat in some deposit of an age decidedly anterior to the first arrival of the Maori race in New Zealand.[2]

Much more interesting is the reported existence in the moun-

---

[1] Dobson, "On the Classification of Chiroptera," *Annals and Magazine of Natural History*, November, 1875.

[2] See Buller, "On the New Zealand Rat," *Transactions of the New Zealand Institute*, 1870, Vol. III., p. 1, and Vol. IX., p. 348; and Hutton, "On the Geographical Relations of the New Zealand Fauna," *Transactions of the New Zealand Institute*, 1872, p. 229.

tains of the South Island of a small otter-like animal. Dr. Haast
has seen its tracks, resembling those of our European otter, at a
height of 3000 feet above the sea in a region never before trod-
den by man; and the animal itself was seen by two gentlemen
near Lake Heron, about seventy miles due west of Christchurch.
It was described as being dark-brown and the size of a large rabbit.
On being struck at with a whip, it uttered a shrill yelping sound
and disappeared in the water.[1] An animal seen so closely as to
be struck at with a whip could hardly have been mistaken for a
dog—the only other animal that it could possibly be supposed
to have been—and a dog would certainly not have "disappeared
in the water." This account, as well as the footsteps, points to
an aquatic animal; and if it now frequents only the high alpine
lakes and streams, this might explain why it has never yet been
captured.   Hochstetter also states that it has a native name—
Waitoteke—a striking evidence of its actual existence; while a
gentleman who lived many years in the district assures me that
it is universally believed in by residents in that part of New
Zealand. The actual capture of this animal, and the determination of its characters and affinities, could not fail to aid us
greatly in our speculations as to the nature and origin of the
New Zealand fauna.[2]

  Wingless Birds, Living and Extinct.—Almost equally valuable

---

[1] Hochstetter's "New Zealand," p. 161, note.

[2] The animal described by Captain Cook as having been seen at Pickersgill Harbor in Dusky Bay (Cook's "Second Voyage," Vol. I., p. 98) may have been the same
creature. He says, " A four-footed animal was seen by three or four of our people ; but
as no two gave the same description of it, I cannot say what kind it is. All, however, agreed that it was about the size of a cat, with short legs, and of a mouse-color.
One of the seamen, and he who had the best view of it, said it had a bushy tail, and
was the most like a jackal of any animal he knew." It is suggestive that so far as
the points on which "all agreed"—the size and the dark color—this description
would answer well to the animal so recently seen, while the "short legs" correspond to the otter-like tracks, and the thick tail of an otter-like animal may well have
appeared "bushy" when the fur was dry. It has been suggested that it was only
one of the native dogs; but as none of those who saw it took it for a dog, and the
points on which they all agreed are not dog-like, we can hardly accept this explanation ; while the actual existence of an unknown animal in New Zealand of corresponding size and color is confirmed by this account of a similar animal having been
seen about a century ago.

with mammalia in affording indications of geographical changes are the wingless birds for which New Zealand is so remarkable. These consist of four species of Apteryx, called by the natives "kiwis"—creatures which hardly look like birds, owing to the apparent absence (externally) of tail or wings and the dense covering of hair-like feathers. They vary in size from that of a small fowl up to that of a turkey, and have a long slightly curved bill, somewhat resembling that of the snipe or ibis. Two species appear to be confined to the South Island and one to the North Island; but all are becoming scarce, and they will, no doubt, gradually become extinct. These birds are generally classed with the Struthiones, or ostrich tribe, but they form a distinct family, and in many respects differ greatly from all other known birds.

But, besides these, a number of other wingless birds, called "moas," inhabited New Zealand during the period of human occupation, and have only recently become extinct. These were much larger birds than the kiwis, and some of them were even larger than the ostrich, a specimen of *Dinornis maximus* mounted in the British Museum in its natural attitude being eleven feet high. They agreed, however, with the living Apteryx in having four toes, and in the character of the pelvis and some other parts of the skeleton; while in their short bill and in some important structural features they resembled the emu of Australia and the cassowaries of New Guinea.[1] No less than eleven distinct species of these birds have now been discovered; and their remains exist in such abundance—in recent fluviatile deposits, in old native cooking-places, and even scattered on the surface of the ground, that complete skeletons of several of them have been put together, illustrating various periods of growth from the chick up to the adult bird. Feathers have also been found attached to portions of the skin, as well as the stones swallowed by the birds to assist digestion, and eggs, some containing portions of the embryo bird; so that everything con-

---

[1] Owen, "On the Genus Dinornis," *Transactions of the Zoological Society*, Vol. X., p. 184; Mivart, "On the Axial Skeleton of the Struthionidæ," *Transactions of the Zoological Society*, Vol. X., p. 51.

firms the statements of the Maoris—that their ancestors found
these birds in abundance on the islands; that they hunted them
for food; and that they finally exterminated them only a short
time before the arrival of Europeans.[1] Bones of Apteryx are
also found fossil, but apparently of the same species as the living birds. How far back in geological time these creatures or
their ancestral types lived in New Zealand we have as yet no
evidence to show. Some specimens have been found under a
considerable depth of fluviatile deposits which may be of Quaternary or even of Pliocene age; but this evidently affords us
no approximation to the time required for the origin and development of such highly peculiar insular forms.

*Past Changes of New Zealand Deduced from its Wingless
Birds.*—It has been well observed by Captain Hutton, in his interesting paper already referred to, that the occurrence of such a
number of species of Struthious birds living together in so small
a country as New Zealand is altogether unparalleled elsewhere
on the globe. This is even more remarkable when we consider that the species are not equally divided between the two
islands, for remains of no less than ten out of the eleven known
species of Dinornis have been found in a single swamp in the
South Island, where also three of the species of Apteryx oc-

---

[1] The recent existence of the moa, and its having been exterminated by the Maoris,
appears to be at length set at rest by the statement of Mr. John White, a gentleman
who has been collecting materials for a history of the natives for thirty-five years, who
has been initiated by their priests into all the mysteries, and is said to "know more
about the history, habits, and customs of the Maoris than they do themselves." His
information on this subject was obtained from old natives long before the controversy
on the subject arose. He says that the histories and songs of the Maoris abound
in allusions to the moa, and that they were able to give full accounts of "its habits,
food, the season of the year it was killed, its appearance, strength, and all the numerous ceremonies which were enacted by the natives before they began the hunt; the
mode of hunting, how cut up, how cooked, and what wood was used in the cooking,
with an account of its nest, and how the nest was made, where it usually lived, etc."
Two pages are occupied by these details, but they are only given from memory, and
Mr. White promises a full account from his MSS. Many of the details given correspond with facts ascertained from the discovery of native cooking-places with moa's
bones; and it seems quite incredible that such an elaborate and detailed account
should be all invention (see *Transactions of the New Zealand Institute*, Vol. VIII.,
p. 79).

cur. The New Zealand Struthiones, in fact, very nearly equal in number those of all the rest of the world, and nowhere else do more than three species occur in any one continent or island, while no more than two ever occur in the same district. Thus there appear to be two closely allied species of ostriches inhabiting Africa and Southwestern Asia respectively. South America has three species of Rhea, each in a separate district. Australia has an eastern and a western variety of emu, and a cassowary in the north; while eight other cassowaries are known from the islands north of Australia—one from Ceram, two from the Aru Islands, one from Jobie, one from New Britain, and three from New Guinea—but of these last, one is confined to the northern and another to the southern part of the island.

This law, of the distribution of allied species in separate areas —which is found to apply more or less accurately to all classes of animals—is so entirely opposed to the crowding together of no less than fifteen species of wingless birds in the small area of New Zealand that the idea is at once suggested of great geographical changes. Captain Hutton points out that if the islands from Ceram to New Britain were to become joined together, we should have a large number of species of cassowary (perhaps several more than are yet discovered) in one land area. If, now, this land were gradually to be submerged, leaving a central elevated region, the different species would become crowded together in this portion, just as the moas and kiwis were in New Zealand. But we also require, at some remote epoch, a more or less complete union of the islands now inhabited by the separate species of cassowaries, in order that the common ancestral form which afterwards became modified into these species could have reached the places where they are now found; and this gives us an idea of the complete series of changes through which New Zealand is believed to have passed in order to bring about its abnormally dense population of wingless birds. First, we must suppose a land-connection with some country inhabited by Struthious birds, from which the ancestral forms might be derived; secondly, a separation into many considerable islands, in which the various distinct species might become differentiated; thirdly, an elevation bringing about the union of these islands to unite the distinct species in one area; and, fourthly, a subsidence of a large part of the area, leaving the present islands with the various species crowded together.

If New Zealand has really gone through such a series of changes as here suggested, some proofs of it might perhaps be obtained in the outlying islands which were once, presumably, joined with it. And this gives great importance to the statement of the aborigines of the Chatham Islands that the Apteryx formerly lived there, but was exterminated about 1835. It is to be hoped that some search will be made here, and also in Norfolk Island, in both of which it is not improbable remains of Apteryx or Dinornis might be discovered.

So far we find nothing to object to in the speculations of Captain Hutton, with which, on the contrary, we almost wholly concur; but we cannot follow him when he goes on to suggest an antarctic continent uniting New Zealand and Australia with South America, and probably also with South Africa, in order to explain the existing distribution of Struthious birds. Our best anatomists, as we have seen, agree that both Dinornis and Apteryx are more nearly allied to the cassowaries and emus than to the ostriches and rheas; and we see that the form of the sea-bottom suggests a former connection with North Australia and New Guinea — the very region where these types most abound, and where in all probability they originated. The suggestion that all the Struthious birds of the world sprang from a common ancestor at no very remote period, and that their existing distribution is due to direct land communication between the countries they *now* inhabit, is one utterly opposed to all sound principles of reasoning in questions of geographical distribution; for it depends upon two assumptions, both of which are at least doubtful, if not certainly false—the first, that their distribution over the globe has never in past ages been very different from what it is now; and the second, that the ancestral forms of these birds never had the power of flight. As to the first assumption, we have found in almost every case that groups now scattered over two or more continents formerly lived in intervening areas of existing land. Thus, the marsupials of South America and Australia are connected by forms which lived in North America and Europe; the camels of Asia and the llamas of the Andes had many extinct common ancestors in North America; the lemurs of Africa and Asia had their ancestors in Europe, as did the trogons of South America, Africa, and tropical Asia. But, besides this general evidence, we have direct proof that the Struthious birds had a wider range in past times than now. Remains of extinct rheas have been found in Central Brazil, and those of ostriches in North India; while remains believed to be of Struthious birds are found in the Eocene deposits of England; and the Cretaceous rocks of North America have yielded the extraordinary toothed bird Hesperornis, which Professor O. Marsh declares to have been "a carnivorous swimming ostrich."

As to the second point, we have the remarkable fact that all known birds of this group have not only the rudiments of wing-bones, but also the rudiments of wings; that is, an external limb bearing rigid quills or largely developed plumes. In the cassowary these wing-feathers are reduced to long spines like porcupine-quills, while even in the Apteryx the minute external wing bears a series of nearly twenty stiff quill-like feathers.[1] These facts render it probable that the Struthious birds do not owe their imperfect wings to a direct evolution from a reptilian type, but to a retrograde development from some low form of winged birds, analogous to that which has produced the dodo and the solitaire from the more highly developed pigeon-type. Professor Marsh has proved that, so far back as the Cretaceous period, the two great forms of birds—those with a keeled sternum and fairly developed wings, and those with a convex keelless sternum and rudimentary wings—already existed side by side; while in the still earlier Archæopteryx of the Jurassic period we have a bird with well-developed wings, and therefore probably with a keeled sternum. We are evidently, therefore, very far from a knowledge of the earlier stages of bird-life, and our acquaintance with the various forms that have existed is scanty in the extreme; but we may be sure that birds acquired wings

---

[1] See figure in *Transactions of the New Zealand Institute*, Vol. III., Plate 12 *b*, Fig. 2.

and feathers, and some power of flight, before they developed a keeled sternum, since we see that bats with no such keel fly very well. Since, therefore, the Struthious birds all have perfect feathers, and all have rudimentary wings which are anatomically those of true birds, not the rudimentary forelegs of reptiles, and since we know that in many higher groups of birds —as the pigeons and the rails—the wings have become more or less aborted, and the keel of the sternum greatly reduced in size by disuse, it seems probable that the very remote ancestors of the rhea, the cassowary, and the Apteryx were true flying birds, although not perhaps provided with a keeled sternum, or possessing very great powers of flight. But, in addition to the possible ancestral power of flight, we have the undoubted fact that the rhea and the emu both swim freely, the former having been seen swimming from island to island off the coast of Patagonia. This, taken in connection with the wonderful aquatic ostrich of the Cretaceous period discovered by Professor Marsh, opens up fresh possibilities of migration ; while the immense antiquity thus given to the group, and their universal distribution in past time, render all suggestions of special modes of communication between the parts of the globe in which their scattered remnants *now* happen to exist altogether superfluous and misleading.

The bearing of this argument on our present subject is that, so far as accounting for the presence of wingless birds in New Zealand is concerned, we have nothing whatever to do with any possible connection by way of a southern continent or antarctic islands with South America and South Africa, because the nearest allies of its moas and kiwis are the cassowaries and emus ; and we have distinct indications of a former land-extension towards North Australia and New Guinea, which is exactly what we require for the original entrance of the Struthious type into the New Zealand area.

Winged Birds and Lower Vertebrates of New Zealand.—Having given a pretty full account of the New Zealand fauna elsewhere,[1] I need only here point out its bearing on the hypothesis now advanced, of the former land-connection having been with

---

[1] " Geographical Distribution of Animals," Vol. I., p. 450.

North Australia, New Guinea, and the Western Pacific islands, rather than with the temperate regions of Australia. Of the Australian genera of birds which are found also in New Zealand, almost every one ranges also into New Guinea or the Pacific islands, while the few that do not extend beyond Australia are found in its northern districts. As regards the peculiar New Zealand genera, all whose affinities can be traced are allied to birds which belong to the tropical parts of the Australian Region; while the starling family, to which four of the most remarkable New Zealand birds belong (the genera Creadion, Heterolocha, and Callæas), is totally wanting in temperate Australia and is comparatively scarce in the entire Australian Region, but is abundant in the Oriental Region, with which New Guinea and the Moluccas are in easy communication. It is certainly a most suggestive fact that there are more than sixty genera of birds peculiar to the Australian continent (with Tasmania), many of them almost or quite confined to its temperate portions, and that no single one of these should be represented in temperate New Zealand.[1] The affinities of the living and more highly organized no less than those of the extinct and wingless birds strikingly accord with the line of communication indicated by the deep submarine bank connecting these temperate islands with the tropical parts of the Australian Region.

The reptiles, so far as they go, are quite in accordance with the birds. The lizards belong to three genera—Hinulia and Mocoa, which have a wide range in the Eastern tropics and the Pacific and Malayan regions, as well as Australia; and Naultinus, a genus peculiar to New Zealand, but belonging to a family (Geckotidæ) spread over the whole of the warmer parts of the world. Australia, on the other hand, has three small but peculiar families, and no less than thirty-six peculiar genera of lizards, many of which are confined to its temperate regions, but no one of them extends to temperate New Zealand. The ex-

---

[1] In my "Geographical Distribution of Animals," Vol. I., p. 541, I have given two peculiar Australian genera (*Orthonyx* and *Tribonyx*) as occurring in New Zealand. But the former has been found in New Guinea, while the New Zealand bird is considered to form a distinct genus, *Clitonyx;* and the latter inhabits Tasmania, and was recorded from New Zealand through an error (see *Ibis*, 1873, p. 427).

traordinary lizard-like *Hatteria punctata* of New Zealand forms
of itself a distinct order of reptiles, in some respects intermediate between lizards and crocodiles, and having therefore no affinity with any living animal.

The only representative of the Amphibia in New Zealand is
a solitary frog of a peculiar genus (*Liopelma Hochstetteri*); but
it has no affinity for any of the Australian frogs, which are numerous, and belong to eleven distinct families; while the Liopelma belongs to a very different family (Bombinatoridæ), confined to Europe and temperate South America.

Of the fresh-water fishes we need only say here that none
belong to peculiar Australian types, but are related to those of
temperate South America or of Asia.

The Invertebrate classes are comparatively little known, and
their modes of dispersal are so varied and exceptional that the
facts presented by their distribution can add little weight to
those already adduced. We will therefore now proceed to the
conclusions which can fairly be drawn from the general facts of
New Zealand natural history already known to us.

Deductions from the Peculiarities of the New Zealand Fauna.
—The total absence (or extreme scarcity) of mammals in New
Zealand obliges us to place its union with North Australia and
New Guinea at a very remote epoch. We must either go back
to a time when Australia itself had not yet received the ancestral forms of its present marsupials and monotremes, or we must
suppose that the portion of Australia with which New Zealand
was connected was then itself isolated from the mainland, and
was thus without a mammalian population. We shall see in
our next chapter that there are certain facts in the distribution
of plants, no less than in the geological structure of the country,
which favor the latter view. But we must on any supposition
place the union very far back, to account for the total want of
identity between the winged birds of New Zealand and those
peculiar to Australia, and a similar want of accordance in the
lizards, the fresh-water fishes, and the more important insect-
groups of the two countries. From what we know of the long
geological duration of the generic types of these groups, we
must certainly go back to the earlier portion of the Tertiary

period at least, in order that there should be such a complete
disseverance as exists between the characteristic animals of the
two countries; and we must further suppose that, since their
separation, there has been no subsequent union or sufficiently
near approach to allow of any important intermigration, even
of winged birds, between them. It seems probable, therefore,
that the Bampton shoal, west of New Caledonia, and Lord Howe's
Island farther south, formed the western limits of that extensive land in which the great wingless birds and other isolated
members of the New Zealand fauna were developed. Whether
this early land extended eastward to the Chatham Islands and
southward to the Macquaries, we have no means of ascertaining;
but as the intervening sea appears to be not more than about
1500 fathoms deep, it is quite possible that such an amount of
subsidence may have occurred. It is possible, too, that there
may have been an extension northward to the Kermadec Islands,
and even farther to the Tonga and Fiji Islands, though this is
hardly probable, or we should find more community between
their productions and those of New Zealand.

A southern extension towards the antarctic continent at a
somewhat later period seems more probable, as affording an easy
passage for the numerous species of South American and antarctic plants, and also for the identical and closely allied fresh-
water fishes of these countries.

The subsequent breaking-up of this extensive land into a
number of separate islands in which the distinct species of moa
and kiwi were developed, their union at a later period, and the
final submergence of all but the existing islands, are pure hypotheses, which seem necessary to explain the occurrence of so
many species of these birds in a small area, but of which we have
no independent proof. There are, however, some other facts
which would be explained by it, as the presence of three peculiar
but allied genera of starlings, the three species of parrots of the
genus Nestor, and the six distinct rails of the genus Ocydromus,
as well as the numerous species in some of the peculiar New
Zealand genera of plants, which seem less likely to have been
developed in a single area than when isolated, and thus preserved from the counteracting influence of intercrossing.

In the present state of our knowledge, these seem all the conclusions we can arrive at from a study of the New Zealand fauna; but as we fortunately possess a very full and accurate knowledge of the flora of New Zealand, as well as of that of Australia and the south temperate lands generally, it will be well to see how far these conclusions are supported by the facts of plant-distribution, and what further indications they afford us of the early history of these most interesting countries. This inquiry is of sufficient importance to occupy a separate chapter.

CHAPTER XXII.

THE FLORA OF NEW ZEALAND: ITS AFFINITIES AND PROBABLE ORIGIN.

Relations of the New Zealand Flora to that of Australia.—General Features of the Australian Flora.—The Floras of Southeastern and Southwestern Australia.—Geological Explanation of the Differences of these two Floras.—The Origin of the Australian Element in the New Zealand Flora.—Tropical Character of the New Zealand Flora Explained.—Species Common to New Zealand and Australia mostly Temperate Forms.—Why Easily Dispersed Plants have often Restricted Ranges.—Summary and Conclusion on the New Zealand Flora.

ALTHOUGH plants have means of dispersal far exceeding those possessed by animals, yet as a matter of fact comparatively few species are carried for very great distances, and the flora of a country taken as a whole usually affords trustworthy indications of its past history. Plants, too, are more numerous in species than the higher animals, and are almost always better known; their affinities have been more systematically studied; and it may be safely affirmed that no explanation of the origin of the fauna of a country can be sound which does not also explain, or at least harmonize with, the distribution and relations of its flora.

The relations of the flora of New Zealand to that of Australia have long formed an insoluble enigma for botanists. Sir Joseph Hooker, in his most instructive and masterly essay on the flora of Australia, says, " Under whatever aspect I regard the flora of Australia and of New Zealand, I find all attempts to theorize on the possible causes of their community of feature frustrated by anomalies in distribution such as I believe no two other similarly situated countries in the globe present. Everywhere else I recognize a parallelism or harmony in the main common features of contiguous floras, which conveys the impression of their generic affinity, at least, being affected by migration from centres of dispersion in one of them, or in some adjacent country. In this case it is widely different. Regarding the question from the Australian point of view, it is impossible, in the present state of science, to reconcile the fact of Acacia, Eucalyptus, Casuarina, Callitris, etc., being absent in New Zealand with any theory of transoceanic migration that may be adopted to explain the presence of other Australian plants in New Zealand; and it is very difficult to conceive of a time or of conditions that could explain these anomalies, except by going back to epochs when the prevalent botanical as well as geographical features of each were widely different from what they are now. On the other hand, if I regard the question from the New Zealand point of view, I find such broad features of resemblance, and so many connecting-links that afford irresistible evidence of a close botanical connection, that I cannot abandon the conviction that these great differences will present the least difficulties to whatever theory may explain the whole case." I will now state, as briefly as possible, what are the facts above referred to as being of so anomalous a character, and there is little difficulty in doing so, as we have them fully set forth, with admirable clearness, in the essay. above alluded to, and in the same writer's "Introduction to the Flora of New Zealand," only requiring some slight modifications, owing to the later discoveries which are given in the "Handbook of the New Zealand Flora."

Confining ourselves always to flowering plants, we find that the flora of New Zealand is a very poor one, considering the extent of surface, and the favorable conditions of soil and climate. It consists of 935 species, our own islands possessing about 1500; but a very large proportion of these are peculiar, there being no less than 677 endemic species and 32 endemic genera.

Out of the 258 species not peculiar to New Zealand, no less than 222 are Australian, but a considerable number of these are also antarctic, South American, or European; so that there are only about 100 species absolutely confined to New Zealand and Australia; and, what is important as indicating a somewhat recent immigration, only six of these belong to genera which are peculiar to the two countries, and hardly any to the larger and more important Australian genera. Many, too, are rare species in both countries, and are often alpines.

Far more important are the relations of the genera and families of the two countries. All the natural orders of New Zealand are found in Australia except three — Coriariæ, a widely scattered group found in South Europe, the Himalayas, and the Andes ; Escallonieæ, a widely distributed group; and Chloranthaceæ, found in tropical Asia, Japan, Polynesia, and South America. Out of a total of 303 New Zealand genera, no less than 251 are Australian, and 60 of these are almost peculiar to the two countries, only 32, however, being absolutely confined to them. In the three large orders Compositæ, Orchideæ, and Gramineæ, the genera are almost identical in the two countries, while the species—in the two former especially—are mostly distinct.

Here, then, we have apparently a wonderful resemblance between the New Zealand flora and that of Australia, indicated by more than two thirds of the non-peculiar species, and more than nine tenths of the non-peculiar genera (255) being Australian. But now let us look at the other side of the question.

There are in Australia seven great genera of plants, each containing more than 100 species, all widely spread over the country, and all highly characteristic Australian forms—Acacia, Eucalyptus, Melaleuca, Leucopogon, Stylidium, Grevillea, and Hakea. These are entirely absent from New Zealand, except one species of Leucopogon, a genus which also has representatives in the Malayan and Pacific islands. Sixteen more Australian genera have over fifty species each, and of these eight are totally absent from New Zealand, five are represented by one or two species, and only two are fairly represented ; but these two— Drosera and Helichrysum—are very wide-spread genera, and might have reached New Zealand from other countries than Australia.

But this by no means exhausts the differences between New Zealand and Australia. No less than seven important Australian natural orders — Dilleniaceæ, Buettneriaceæ, Polygaleæ, Tremandreæ, Casuarineæ, Hæmodoraceæ, and Xyrideæ—are entirely wanting in New Zealand ; and several others which are excessively abundant and highly characteristic of the former country are very poorly represented in the latter. Thus, Leguminosæ are extremely abundant in Australia, where there are over 1000 species belonging to about 100 genera, many of them altogether peculiar to the country; yet in New Zealand this great order is most scantily represented, there being only five genera and thirteen species; and only two of these genera, Swainsonia and Clianthus, are Australian; and as the latter consists of but two species, it may as well have passed from New Zealand to Australia as the other way, or more probably from some third country to them both. Goodeniaceæ, with twenty genera and 230 species Australian, has but two species in New Zealand, and one of these is a salt-marsh plant found also in Tasmania and in Chili; and four other large Australian orders —Rhamneæ, Myoporineæ, Proteaceæ, and Santalaceæ — have very few representatives in New Zealand.

We find, then, that the great fact we have to explain and account for is the undoubted affinity of the New Zealand flora to that of Australia, but an affinity almost exclusively confined to the least predominant and least peculiar portion of that flora, leaving the most predominant, most characteristic, and most widely distributed portion absolutely unrepresented. We must, however, be careful not to exaggerate the amount of affinity with Australia, apparently implied by the fact that nearly six sevenths of the New Zealand genera are also Australian, for, as we have already stated, a very large number of these are European, antarctic, South American, or Polynesian genera, whose presence in the two contiguous areas only indicates a common origin. About one eighth only are absolutely confined to Australia and New Zealand (thirty-two genera), and even of these several are better represented in New Zealand than in Australia, and may therefore have passed from the former to the latter. No less than 174 of the New Zealand genera are temperate South American, many being also antarctic or European; while others, again, are especially tropical or Polynesian; yet undoubtedly a larger proportion of the natural orders and genera are common to Australia than to any other country, so that we may say that the basis of the flora is Australian with a large intermixture of northern and southern temperate forms and others which have remote world-wide affinities.

General Features of the Australian Flora, and its Probable Origin.—Before proceeding to point out how the peculiarities of the New Zealand flora may be best accounted for, it is necessary to consider briefly what are the main peculiarities of Australian vegetation, from which so important a part of that of New Zealand has evidently been derived.

The actual Australian flora consists of two great divisions—a temperate and a tropical, the temperate being again divisible into an eastern and a western portion. Everything that is characteristic of the Australian flora belongs to the temperate division (though these often overspread the whole continent), in which are found almost all the remarkable Australian types of vegetation and the numerous genera peculiar to this part of the world. Contrary to what occurs in most other countries, the tropical is far less rich in species and genera than the temperate region, and, what is still more remarkable, it contains comparatively few peculiar species and very few peculiar genera. Although the area of tropical Australia is about equal to that of the temperate portions, and it has now been pretty well explored botanically, it has less than half as many species.[1] Nearly 500 of its species are identical with Indian or Malayan plants, or are

---

[1] Sir Joseph Hooker informs me that the number of tropical Australian plants discovered within the last twenty years is very great, and that the statement as above made may have to be modified. Looking, however, at the enormous disproportion of the figures given in the "Introductory Essay" in 1859 (2200 tropical to 5800 temperate species), it seems hardly possible that a great difference should not still exist, at all events as regards species. Sir Joseph Hooker also doubts the generally greater richness of tropical over temperate floras which I have taken as almost an axiom. He says, "Taking similar areas to Australia in the Western World—*e.g.* tropical Africa north of 20° as against temperate Africa and Europe up to 47°—I suspect that the latter would present more genera and species than the former." This, however, appears to me to be hardly a case in point, because Europe is a distinct continent from Africa, and has had a very different past history. A closer parallel may perhaps be found in equal areas of Brazil and south temperate America, or of Mexico and the Southern United States, in both of which cases I suppose there can be little doubt that the tropical areas are far the richest. Temperate South Africa is, no doubt, always quoted as richer than an equal area of tropical Africa, or perhaps than any part of the world of equal extent, but this is admitted to be an exceptional case.

very close representatives of them; while there are more than 200 Indian genera confined, for the most part, to the tropical portion of Australia. The remainder of the tropical flora consists of certain species and genera of temperate Australia which range over the whole continent, but these form a very small portion of the peculiarly Australian genera.

These remarkable facts clearly point to one conclusion—that the flora of tropical Australia is, comparatively, recent and derivative. If we imagine the greater part of North Australia to have been submerged beneath the ocean, from which it rose in the middle or latter part of the Tertiary period, offering an extensive area ready to be covered by such suitable forms of vegetation as could first reach it, something like the present condition of things would inevitably arise. From the north widespread Indian and Malay plants would quickly enter; while from the south the most dominant forms of temperate Australia, and such as were best adapted to the tropical climate and arid soil, would intermingle with them. Even if numerous islands had occupied the area of Northern Australia for long periods anterior to the final elevation, very much the same state of things would result.

The existence in North and Northeast Australia of enormous areas covered with Cretaceous and other Secondary deposits, as well as extensive Tertiary formations, lends support to the view that during very long epochs temperate Australia was cut off from all close connection with the tropical and northern lands by a wide extent of sea; and this isolation is exactly what was required in order to bring about the wonderful amount of specialization and the high development manifested by the typical Australian flora. Before proceeding further, however, let us examine this flora itself, so far as regards its component parts and probable past history.

The Floras of Southeastern and Southwestern Australia.— The peculiarities presented by the southeastern and southwestern subdivisions of the flora of temperate Australia are most interesting and suggestive, and are, perhaps, unparalleled in any other part of the world. Southwest Australia is far less extensive than the southeastern division—less varied in soil and climate, with no lofty mountains, and much sandy desert; yet, strange to say, it contains an equally rich flora and a far greater proportion of peculiar species and genera of plants. As Sir Joseph Hooker remarks, "What differences there are in conditions would, judging from analogy with other countries, favor the idea that Southeastern Australia, from its far greater area, many large rivers, extensive tracts of mountainous country and humid forests, would present much the most extensive flora, of which only the dryer types could extend into Southwestern Australia. But such is not the case; for though the far greater area is much the best explored, presents more varied conditions, and is tenanted by a larger number of natural orders and genera, these contain fewer species by several hundreds."[1]

The fewer genera of Southwestern Australia are due almost wholly to the absence of the numerous European, antarctic, and South American types found in the southeastern region; while in purely Australian types it is far the richer, for, while it contains most of those found in the east, it has a large number altogether peculiar to it; and Sir Joseph Hooker states that "there are about 180 genera, out of 600, in Southwestern Australia that are either not found at all in Southeastern, or that are represented there by a very few species only, and these 180 genera include nearly 1100 species."

Geological Explanation of the Differences of these Two Floras. —These facts again clearly point to the conclusion that Southwestern Australia is the remnant of the more extensive and more isolated portion of the continent in which the peculiar Australian flora was principally developed. The existence there of a very large area of granite—800 miles in length by nearly 500 in maximum width—indicates such an extension; for this granitic mass was certainly once buried under piles of stratified

---

[1] Sir Joseph Hooker thinks that later discoveries in the Australian Alps and other parts of East and South Australia may have greatly modified, or perhaps reversed, the above estimate. But even if this should be the case, the small area of Southwest Australia will still be, proportionally, far the richer of the two. It is much to be desired that the enormous mass of facts contained in Mr. Bentham's "Flora Australiensis" should be tabulated and compared by some competent botanist, so as to exhibit the various relations of its wonderful vegetation in the same manner as was done by Sir Joseph Hooker with the materials available twenty-one years ago.

rock, since denuded, and then formed the nucleus of the old Western Australian continent. If we take the 1000-fathom line around the southern part of Australia to represent the probable extension of this old land, we shall see that it would give a wide additional area south of the Great Australian Bight, and form a continent which, even if the greater part of tropical Australia were submerged, would be sufficient for the development of a peculiar and abundant flora. We must also remember that an elevation of 6000 feet, added to the vast amount which has been taken away by denudation, would change the whole country, including what are now the deserts of the interior, into a mountainous and well-watered region.

But, while this rich and peculiar flora was in process of formation, the eastern portion of the continent must either have been widely separated from the western, or had perhaps not yet risen from the ocean. The whole of this part of the country consists of Palæozoic and Secondary formations, with granite and metamorphic rocks—the Secondary deposits being largely developed on both sides of the central range, extending the whole length of the continent from Tasmania to Cape York, and constituting the greater part of the plateau of the Blue Mountains and other lofty ranges. During some portion of the Secondary period, therefore, this side of Australia must have been almost wholly submerged beneath the ocean; and if we suppose that during this time the western part of the continent was at nearly its maximum extent and elevation, we shall have a sufficient explanation of the great difference between the flora of Western and Eastern Australia, since the latter would only have been able to receive immigrants from the former at a later period, and in a more or less fragmentary manner.

If we examine the geological map of Australia (given in Stanford's " Compendium of Geography and Travel," volume " Australasia"), we shall see good reason to conclude that the eastern and the western divisions of the country first existed as separate islands, and only became united at a comparatively recent epoch. This is indicated by an enormous stretch of Cretaceous and Tertiary formations extending from the Gulf of Carpentaria completely across the continent to the mouth of the Murray River.

During the Cretaceous period, therefore, and probably through-out a considerable portion of the Tertiary epoch,[1] there must have been a wide arm of the sea occupying this area, dividing the great mass of land on the west—the true seat and origin of the typical Australian flora—from a long but narrow belt of land on the east, indicated by the continuous mass of Secondary and Palæozoic formations already referred to, which extend uninter-ruptedly from Tasmania to Cape York.   Whether this formed one continuous land, or was broken up into islands, cannot be positively determined ; but the fact that no marine Tertiary beds occur in the whole of this area renders it probable that it was almost, if not quite, continuous, and that it not improbably ex-tended across to what is now New Guinea.   At this epoch, then (as shown in the accompanying map), Australia would consist of a very large and fertile western island, almost or quite extra-tropical, and extending from the Silurian rocks of the Flinders range in South Australia to about 150 miles west of the present west coast, and southward to about 350 miles south of the Great Australian Bight.   To the east of this, at a distance of from 250 to 400 miles, extended in a north and south direction a long but comparatively narrow island, stretching from far south of Tasmania to New Guinea ; while the crystalline and Secondary formations of central North Australia probably in-dicate the existence of one or more large islands in that direc-tion.

The eastern and the western islands—with which we are now chiefly concerned—would then differ considerably in their vege-tation and animal life.   The western and more ancient land al-ready possessed, in its main features, the peculiar Australian flora, and also the ancestral forms of its strange marsupial fauna, both of which it had probably received at some earlier epoch by

---

[1] From an examination of the fossil corals of the Southwest of Victoria, Professor P. M. Duncan concludes " that at the time of the formation of these deposits the central area of Australia was occupied by sea, having open water to the north, with reefs in the neighborhood of Java." The age of these fossils is not known, but as almost all are extinct species, and some are almost identical with European Pliocene and Miocene species, they are supposed to belong to a corresponding period (*Journal of the Geological Society*, 1870).

a temporary union with the Asiatic continent over what is now
the Java Sea.   Eastern Australia, on the other hand, possessed
only the rudiments of its existing mixed flora, derived from
three distinct sources.   Some important fragments of the typical Australian vegetation had reached it across the marine strait,
and had spread widely, owing to the soil, climate, and general

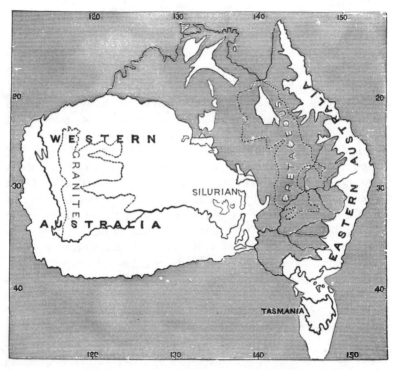

MAP SHOWING THE PROBABLE CONDITION OF AUSTRALIA DURING THE CRETACEOUS
PERIOD.

The white portions represent land ; the shaded parts sea.
The existing land of Australia is shown in outline.

conditions being exactly suited to it ; from the north and northeast a tropical vegetation of Polynesian type had occupied suitable areas in the north ; while the extension southward of the
Tasmanian Peninsula, accompanied, probably, as now, with lofty
mountains, favored the immigration of south temperate forms

from whatever antarctic lands or islands then existed. The marsupial fauna had not yet reached this eastern land, which was, however, occupied in the north by some ancestral Struthious birds, which had entered it by way of New Guinea through some very ancient continental extension, and of which the emu, the cassowaries, the extinct Dromornis of Queensland, and the moas and kiwis of New Zealand are the modified descendants.

The Origin of the Australian Element in the New Zealand Flora.—We have now brought down the history of Australia, as deduced from its geological structure and the strongly marked features of its flora, to the period when New Zealand was first brought into close connection with it by means of a great northwestern extension of that country, which, as already explained in our last chapter, is so clearly indicated by the form of the sea-bottom (see map, p. 435). The condition of New Zealand previous to this event is very obscure. That it had long existed as a more or less extensive land is indicated by its ancient sedimentary rocks; while the very small areas occupied by Jurassic and Cretaceous deposits imply that much of the present land was then also above the sea-level. The country had probably at that time a scanty vegetation of mixed antarctic and Polynesian origin;[1] but now, for the first time, it would be open to the free immigration of such Australian types as were suitable to its climate, and which *had already reached the tropical and subtropical portions of the eastern Australian island.* It is

---

[1] In Dr. Hector's address as President of the Wellington Philosophical Society, in 1872, he refers to the fluviatile deposits of early Tertiary or Cretaceous age as containing valuable deposits of coal, and adds, "In the associated sandstones and shales the flora of the period has been in many cases well preserved, and shows that at a period anterior to the deposit of the marine stratum the New Zealand area was clothed with a mixed vegetation of dicotyledonous leaves and ferns that in general character represent those which now constitute the flora of the country. It would appear from the recent surveys of Dr. Haast that the large saurian reptiles in the Amuri and Waipara beds, the collections of which have been added to largely during the past year by the exertions of Mr. Henry Travers, lived during the formation of these coal-seams, and coeval with them was a species of the kauri-tree, the leaves of which have been found imbedded with the reptilian bones." He goes on to suggest that "even at this remote period, New Zealand formed part of an area that possessed an insular flora, the peculiar characters of which have been preserved to the present time."—*Transactions of the New Zealand Institute,* Vol. V., p. 423.

here that we obtain the clew to those strange anomalies and con-
tradictions presented by the New Zealand flora in its relation to
Australia, which have been so clearly set forth by Sir Joseph
Hooker, and which have so puzzled botanists to account for. But
these apparent anomalies cease to present any difficulty when we
see that the Australian plants in New Zealand were acquired, not
directly, but, as it were, at second-hand, by union with an island
which itself had as yet only received a portion of the flora. And
then further difficulties were placed in the way of New Zealand
receiving such an adequate representation of that portion of the
flora which had reached East Australia as its climate and position entitled it to, by the fact of the union being, not with the
temperate, but with the tropical and subtropical portions of that
island, so that only those groups could be acquired which were
less exclusively temperate and had already established themselves in the warmer portion of their new home.

It is therefore no matter of surprise, but exactly what we
should expect, that the great mass of pre-eminently temperate
Australian genera should be absent from New Zealand, including the whole of such important families as Dilleniaceæ, Tremandreæ, Buettneriaceæ, Polygaleæ, Casuarineæ, and Hæmodoraceæ; while others, such as Rutaceæ, Stackhousieæ, Rhamneæ,
Myrtaceæ, Proteaceæ, and Santalaceæ, are represented by only
a few species. Thus, too, we can explain the absence of *all* the
peculiar Australian Leguminosæ; for these were still mainly
confined to the great western island, along with peculiar Acacias
and Eucalypti, which at a later period spread over the whole
continent. It is equally accordant with the view we are maintaining that among the groups which Sir Joseph Hooker enumerates as "keeping up the features of extratropical Australia in
its tropical quarter," several should have reached New Zealand,
such as Drosera, some Pittosporeæ and Myoporineæ, with a few
Proteaceæ, Loganiaceæ, and Restiaceæ; for most of these are
not only found in tropical Australia, but also in the Malayan and
Pacific islands.

Tropical Character of the New Zealand Flora Explained.—In
this origin of the New Zealand fauna by a northwestern route
from Northeastern Australia, we find also an explanation of the

remarkable number of tropical groups of plants found there; for though, as Sir Joseph Hooker has shown, a moist and uniform climate favors the extension of tropical forms in the temperate zone, yet some means must be afforded them for reaching a temperate island. On carefully going through the "Handbook," and comparing its indications with those of Bentham's "Flora Australiensis," I find that there are in New Zealand thirty-eight thoroughly tropical genera, thirty-three of which are found in Australia—mostly in the tropical portion of it, though a few are temperate, and these may have reached it through New Zealand.[1] To these we must add thirty-two more genera, which, though chiefly developed in temperate Australia, extend into the tropical or subtropical portion of it, and may well have reached New Zealand by the same route.

---

[1] The following are the tropical genera common to New Zealand and Australia :

1. *Melicope.* Queensland, Pacific islands.
2. *Eugenia.* Tropical Australia, Asia, and America.
3. *Passiflora.* Queensland, tropics of Old World, and America.
4. *Myrsine.* Tropical and temperate Australia, tropical and subtropical regions.
5. *Sapota.* Australia, Norfolk Islands, tropics.
6. *Cyathodes.* Australia and Pacific islands.
7. *Parsonsia.* Tropical Australia and Asia.
8. *Geniostoma.* Queensland, Polynesia, Asia.
9. *Mitrasacme.* Tropical and temperate Australia, India.
10. *Ipomœa.* Tropical Australia, tropics.
11. *Mazus.* Temperate Australia, India, China.
12. *Vitex.* Tropical Australia, tropical and subtropical.
13. *Pisonia.* Tropical Australia, tropical and subtropical.
14. *Alternanthera.* Tropical Australia, India, and South America.
15. *Tretranthera.* Tropical Australia, tropics.
16. *Santalum.* Tropical and subtropical Australia, Pacific, Malay Islands.
17. *Carumbium.* Tropical and subtropical Australia, Pacific islands.
18. *Elatostemma.* Subtropical Australia, Asia, Pacific islands.
19. *Peperomia.* Tropical and subtropical Australia, tropics.
20. *Piper.* Tropical and subtropical Australia, tropics.
21. *Dacrydium.* Tasmania, Malay, and Pacific islands.
22. *Dammara.* Tropical Australia, Malay, and Pacific islands.
23. *Dendrobium.* Tropical Australia, Eastern tropics.
24. *Bolbophyllum.* Tropical and subtropical Australia. tropics.
25. *Sarcochilus.* Tropical and subtropical Australia, Fiji, and Malay Islands.
26. *Freycinetia.* Tropical Australia, tropical Asia.
27. *Cordyline.* Tropical Australia, Pacific islands.
28. *Dianella.* Australia, India, Madagascar, Pacific islands.
29. *Cyperus.* Australia, tropical regions mainly.
30. *Fimbristylis.* Tropical Australia, tropical regions.
31. *Paspalum.* Tropical and subtropical grasses.
32. *Isachne.* Tropical and subtropical grasses.
33. *Sporobolus.* Tropical and subtropical grasses.

On the other hand, we find but few New Zealand genera certainly derived from Australia which are especially temperate, and it may be as well to give a list of such as do occur, with a few remarks. They are sixteen in number, as follows:

1. Pennantia (1 sp.). This genus has a species in Norfolk Island, indicating perhaps its former extension to the northwest.
2. Pomaderris (3 sp.). Two species are common to temperate Australia and New Zealand, indicating recent transoceanic migration.
3. Quintinia (2 sp.). This genus has winged seeds, facilitating migration.
4. Olearia (20 sp.). Seeds with pappus.
5. Craspedia (2 sp.). Seeds with pappus. Alpine; identical with Australian species, and therefore of comparatively recent introduction.
6. Celmisia (25 sp.). Seeds with pappus. Only three Australian species, two of which are identical with New Zealand forms; probably, therefore, derived from New Zealand.
7. Ozothamnus (5 sp.). Seeds with pappus.
8. Epacris (4 sp.). Minute seeds. Some species are subtropical, and they are all found in the northern (warmer) island of New Zealand.
9. Archeria (2 sp.). Minute seeds. Tasmania and New Zealand only.
10. Logania (3 sp.). Small seeds. Alpine plants.
11. Hedycarya (1 sp.).
12. Chiloglottis (1 sp.). Minute seeds. In Auckland Islands. Alpine in Australia.
13. Prasophyllum (1 sp.). Minute seeds. Identical with Australian species.
14. Orthoceras (1 sp.). Minute seeds. Close to an Australian species.
15. Alepyrum (1 sp.). Alpine, moss-like. An antarctic type.
16. Dichelachne (3 sp.). Identical with Australian species. An awned grass.

We thus see that there are special features in most of these plants that would facilitate transmission across the sea between temperate Australia and New Zealand, or to both from some antarctic island; and the fact that in several of them the species are absolutely identical shows that such transmission has occurred in geologically recent times.

Species Common to New Zealand and Australia mostly Temperate Forms.—Let us now take the species which are common to New Zealand and Australia, but found nowhere else, and which must therefore have passed from one country to the other at a more recent period than the mass of genera with which we have hitherto been dealing. These are ninety-six in number, and they present a striking contrast to the similarly restricted genera in being wholly temperate in character, the entire list presenting only a single species which is confined to subtropical East Aus-

tralia—a grass (*Apera arundinacea*) only found in a few localities on the New Zealand coast.

Now it is clear that the larger portion, if not the whole, of
these plants must have reached New Zealand from Australia (or
in few cases Australia from New Zealand) by transmission across
the sea, because we know there has been no land-connection during the Tertiary period, as proved by the absence of all the Australian mammalia and almost all the most characteristic Australian birds, insects, and plants. The form of the sea-bed shows
that the distance could not have been less than 600 miles, even
during the greatest extension of Southern New Zealand and
Tasmania; and we have no reason to suppose it to have been
less, because in other cases an equally abundant flora of identical
species has reached islands at a still greater distance—notably
in the case of the Azores and Bermuda. The character of the
plants is also just what we should expect; for about two thirds
of them belong to genera of world-wide range in the temperate
zones, such as Ranunculus, Drosera, Epilobium, Gnaphalium,
Senecio, Convolvulus, Atriplex, Luzula, and many sedges and
grasses whose exceptionally wide distribution shows that they
possess exceptional powers of dispersal and vigor of constitution, enabling them not only to reach distant countries, but also
to establish themselves there. Another set of plants belong to
especially antarctic or south temperate groups, such as Coloban-
thus, Acæna, Gaultheria, Pernettya, and Muhlenbeckia, and
these may in some cases have reached both Australia and New
Zealand from some now submerged antarctic island. Again,
about one fourth of the whole are alpine plants, and these possess two advantages as colonizers. Their lofty stations place
them in the best position to have their seeds carried away by
winds; and they would in this case reach a country which, having derived the earlier portion of its flora from the side of the
tropics, would be likely to have its higher mountains and favorable alpine stations to a great extent unoccupied, or occupied by
plants unable to compete with specially adapted alpine groups.

Fully one third of the exclusively Australo-New Zealand species belong to the two great orders of the sedges and the grasses;
and there can be no doubt that these have great facilities for

dispersion in a variety of ways. Their seeds, often enveloped in chaffy glumes, would be carried long distances by storms of wind, and even if finally dropped into the sea would have so much less distance to reach the land by means of surface currents; and Mr. Darwin's experiments show that even cultivated oats germinated after a hundred days' immersion in sea-water. Others have hispid awns by which they would become attached to the feathers of birds, and there is no doubt this is an effective mode of dispersal. But a still more important point is, probably, that these plants are generally, if not always, wind-fertilized, and are thus independent of any peculiar insects, which might be wanting in the new country.

Why Easily Dispersed Plants have often Restricted Ranges. —This last consideration throws light on a very curious point, which has been noted as a difficulty by Sir Joseph Hooker, that plants which have most clear and decided powers of dispersal by wind or other means have *not* generally the widest specific range; and he instances the small number of Compositæ common to New Zealand and Australia. But in all these cases it will, I think, be found that although the species have not a wide range, the genera often have. In New Zealand, for instance, the Compositæ are very abundant, there being no less than 148 species, almost all belonging to Australian genera; yet only nine species, or less than one sixteenth of the whole, are identical in the two countries. The explanation of this is not difficult. Owing to their great powers of dispersal, the Australian Compositæ reached New Zealand at a very remote epoch, and such as were adapted to the climate and the means of fertilization established themselves; but being highly specialized plants with great flexibility of organization, they soon became modified in accordance with the new conditions, producing many special forms in different localities; and these, spreading widely, soon took possession of all suitable stations. Henceforth immigrants from Australia had to compete with these indigenous and well-established plants, and only in a few cases were able to obtain a footing; whence it arises that we have many Australian types, but few Australian species, in New Zealand, and both phenomena are directly traceable to the combination of great powers of dispersal with a high degree of specialization. Exactly the same thing occurs with the still more highly specialized Orchideæ. These are not proportionally so numerous in New Zealand (thirty - eight species), and this is no doubt due to the fact that so many of them require insect-fertilization, often by a particular family or genus (whereas almost any insect will fertilize Compositæ), and insects of all orders are remarkably scarce in New Zealand. This would at once prevent the establishment of many of the orchids which may have reached the islands, while those which did find suitable fertilizers and other favorable conditions would soon become modified into new species. It is thus quite intelligible why only three species of orchids are identical in Australia and New Zealand, although their minute and abundant seeds must be dispersed by the wind almost as readily as the spores of ferns.

Another specialized group, the Scrophularineæ, abounds in New Zealand, where there are sixty-two species; but though almost all the genera are Australian, only three species are so. Here, too, the seeds are usually very small, and the powers of dispersal great, as shown by several European genera—Veronica, Euphrasia, and Limosella being found in the Southern Hemisphere.

Looking at the whole series of these Australo-New Zealand plants, we find the most highly specialized groups—Compositæ, Scrophularineæ, Orchideæ—with a small proportion of identical species (one thirteenth to one twentieth), the less highly specialized—Ranunculaceæ, Onagrariæ, and Ericeæ—with a higher proportion (one ninth to one sixth), and the least specialized—Junceæ, Cyperaceæ, and Gramineæ—with the high proportion in each case of one fourth. These nine are the most important New Zealand orders which contain species common to that country and Australia and confined to them; and the marked correspondence they show between high specialization and want of specific identity, while the generic identity is in all cases approximately equal, points to the conclusion that the means of diffusion are in almost all plants ample when long periods of time are concerned, and that diversities in this respect are not so important in determining the peculiar character of a derived flora as

adaptability to varied conditions, great powers of multiplica-
tion, and inherent vigor of constitution. This point will have
to be more fully discussed in treating of the origin of the
antarctic and north temperate members of the New Zealand
flora.

Summary and Conclusion on the New Zealand Flora.—Confining ourselves strictly to the direct relations between the plants
of New Zealand and of Australia, as I have done in the preceding discussion, I think I may claim to have shown that the
union between the two countries in the latter part of the Secondary epoch, at a time when Eastern Australia was widely separated from Western Australia (as shown by its geological formation and by the contour of the sea-bottom), does sufficiently
account for all the main features of the New Zealand flora. It
shows why the basis of the flora is fundamentally Australian
both as regards orders and genera, for it was due to a direct
land-connection between the two countries. It shows also why
the great mass of typical Australian forms are unrepresented;
for the Australian flora is typically *western* and *temperate*, and
New Zealand received its immigrants from the *eastern* island,
which had itself received only a fragment of this flora, and from
the *tropical* end of this island, and thus could only receive such
forms as were not exclusively temperate in character. It shows,
further, why New Zealand contains such a very large proportion
of tropical forms, for we see that it derived the main portion of
its flora directly from the tropics. Again, this hypothesis shows
us why, though the specially Australian genera in New Zealand
are largely tropical or subtropical, the specially Australian species are wholly temperate or alpine; for as these are comparatively recent arrivals, they must have migrated across the sea in
the temperate zone, and these temperate and alpine forms are
exactly such as would be best able to establish themselves in a
country already stocked mainly by tropical forms and their modified descendants. This hypothesis further fulfils the conditions
implied in Sir Joseph Hooker's anticipation that "these great
differences [of the floras] will present the least difficulties to
whatever theory may explain the whole case;" for it shows
that these differences are directly due to the history and devel-

opment of the Australian flora itself, while the resemblances depend upon the most certain cause of all such broad resemblances —actual land-connection.

One objection will undoubtedly be made to the above theory —that it does not explain why some species of the prominent Australian genera Acacia, Eucalyptus, Melaleuca, Grevillea, etc., have not reached New Zealand in recent times along with the other temperate forms that have established themselves. But it is doubtful whether any detailed explanation of such a negative fact is possible, while general explanations sufficient to cover it are not wanting. Nothing is more certain than that numerous plants never run wild and establish themselves in countries where they, nevertheless, grow freely if cultivated; and the explanation of this fact given by Mr. Darwin—that they are prevented doing so by the competition of better-adapted forms —is held to be sufficient. In this particular case, however, we have some very remarkable evidence of the fact of their non-adaptation. The intercourse between New Zealand and Europe has been the means of introducing a host of common European plants—more than 150 in number as enumerated at the end of the second volume of the "Handbook;" yet, although the intercourse with Australia has probably been greater, only two or three Australian plants have similarly established themselves. More remarkable still, Sir Joseph Hooker states, "I am informed that the late Mr. Bidwell habitually scattered Australian seeds during his extensive travels in New Zealand." We may be pretty sure that seeds of such excessively common and characteristic groups as Acacia and Eucalyptus would be among those so scattered, yet we have no record of any plants of these or other peculiar Australian genera ever having been found wild, still less of their having spread and taken possession of the soil in the way that many European plants have done. We are, then, entitled to conclude that the plants above referred to have not established themselves in New Zealand (although their seeds may have reached it) because they could not successfully compete with the indigenous flora, which was already well established and better adapted to the conditions of climate and of the organic environment. This explanation is so perfectly in accordance with a large body of well-known facts, including that which is known to every one—how few of our oldest and hardiest garden-plants ever run wild—that the objection above stated will, I feel convinced, have no real weight with any naturalists who have paid attention to this class of questions.

CHAPTER XXIII.

ON THE ARCTIC ELEMENT IN SOUTH TEMPERATE FLORAS.

European Species and Genera of Plants in the Southern Hemisphere.—Aggressive Power of the Scandinavian Flora.—Means by which Plants have Migrated from North to South.—Newly Moved Soil as Affording Temporary Stations to Migrating Plants.—Elevation and Depression of the Snow-line as Aiding the Migration of Plants.—Changes of Climate Favorable to Migration.—The Migration from North to South has been long going on.—Geological Changes as Aiding Migration.—Proofs of Migration by Way of the Andes.—Proofs of Migration by Way of the Himalayas and Southern Asia.—Proofs of Migration by Way of the African Highlands.—Supposed Connection of South Africa and Australia.—The Endemic Genera of Plants in New Zealand.—The Absence of Southern Types from the Northern Hemisphere.—Concluding Remarks on the New Zealand and South Temperate Floras.

WE have now to deal with another portion of the New Zealand flora which presents perhaps equal difficulties—that which appears to have been derived from remote parts of the north and south temperate zones; and this will lead us to inquire into the origin of the northern or arctic element in all the south temperate floras.

More than one third of the entire number of New Zealand genera (115) are found also in Europe, and even 58 species are identical in these remote parts of the world. Temperate South America has 74 genera in common with New Zealand, and there are even 11 species identical in the two countries, as well as 32 which are close allies or representative species. A considerable number of these northern or antarctic plants, and many more which are representative species, are found also in Tasmania and in the mountains of temperate Australia; and Sir Joseph Hooker gives a list of 38 species very characteristic of Europe and Northern Asia, but almost or quite unknown in the warmer regions, which yet reappear in temperate Australia. Other genera seem altogether antarctic—that is, confined to the extreme southern

lands and islands; and these often have representative species
in Southern America, Tasmania, and New Zealand, while others
occur only in one or two of these areas.   Many north temperate
genera also occur in the mountains of South Africa.   On the
other hand, few, if any, of the peculiar Australian or antarctic
types have spread northward, except some of the former which
have reached the mountains of Borneo, and a few of the latter
which spread along the Andes to Mexico.

On these remarkable facts, of which I have given but the
barest outline, Sir Joseph Hooker makes the following suggestive observations:

"When I take a comprehensive view of the vegetation of the
Old World, I am struck with the appearance it presents of there
being a continuous current of vegetation (if I may so fancifully
express myself) from Scandinavia to Tasmania; along, in short,
the whole extent of that arc of the terrestrial sphere which presents the greatest continuity of land.   In the first place, Scandinavian genera, and even species, reappear everywhere from
Lapland and Iceland to the tops of the Tasmanian Alps—in rapidly diminishing numbers, it is true, but in vigorous development throughout.   They abound on the Alps and Pyrenees, pass
on to the Caucasus and Himalaya; thence they extend along the
Khasia Mountains, and those of the peninsulas of India to those
of Ceylon and the Malayan Archipelago (Java and Borneo), and
after a hiatus of 30° they appear on the Alps of New South
Wales, Victoria, and Tasmania, and beyond these again on those
of New Zealand and the antarctic islands, many of the species
remaining unchanged throughout!   It matters not what the
vegetation of the bases and flanks of these mountains may be;
the northern species may be associated with alpine forms of
Germanic, Siberian, Oriental, Chinese, American, Malayan, and
finally Australian and antarctic types; but whereas these are all,
more or less, local assemblages, the Scandinavian asserts his pre-
rogative of ubiquity from Britain to beyond its antipodes."[1]

It is impossible to place the main facts more forcibly before
the reader than in the above striking passage.   It shows clearly

---

[1] Introductory essay "On the Flora of Australia," p. 103.

that this portion of the New Zealand flora is due to wide-spread causes which have acted with even greater effect in other south temperate lands, and that in order to explain its origin we must grapple with the entire problem of the transfer of the north temperate flora to the Southern Hemisphere. Taking, therefore, the facts as given by Sir Joseph Hooker in the works already referred to, I shall discuss the whole question broadly, and shall endeavor to point out the general laws and subordinate causes that, in my opinion, have been at work in bringing about the anomalous phenomena of distribution he has done so much to make known and to elucidate.

Aggressive Power of the Scandinavian Flora.—The first important fact bearing upon this question is the wonderful aggressive and colonizing power of the Scandinavian flora, as shown by the way in which it establishes itself in any temperate country to which it may gain access. About 150 species have thus established themselves in New Zealand, often taking possession of large tracts of country; about the same number are found in Australia, and nearly as many in the Atlantic states of America, where they form the commonest weeds. Whether or not we accept Mr. Darwin's explanation of this power as due to development in the most extensive land area of the globe where competition has been most severe and long-continued, the fact of the existence of this power remains, and we can see how important an agent it must be in the formation of the floras of any lands to which these aggressive plants have been able to gain access.

But not only are these plants pre-eminently capable of holding their own in any temperate country in the world, but they also have exceptional powers of migration and dispersal over seas and oceans. This is especially well shown by the case of the Azores, where no less than 400 out of a total of 478 flowering plants are identical with European species. These islands are more than 800 miles from Europe, and, as we have already seen in Chapter XII., there is no reason for supposing that they have ever been more nearly connected with it than they are now, since an extension of the European coast to the 1000-fathom line would very little reduce the distance. Now it is a

most interesting and suggestive fact that more than half the
European genera which occur in the Australian flora occur also
in the Azores, and in several cases even the species are identical
in both.[1] The importance of such a case as this cannot be exaggerated, because it affords a demonstration of the power of the
very plants in question to pass over wide areas of sea—some, no
doubt, wholly through the air, carried by storms in the same
way as the European birds and insects which annually reach the
Azores; others by floating on the waters, or by a combination of
the two methods; while some may have been carried by aquatic
birds, to whose feathers many seeds have the power of attaching
themselves. We have in such facts as these a complete disproof
of the necessity for those great changes of sea and land which
are continually appealed to by those who think land-connection
the only efficient means of accounting for the migration of animals or plants; but, at the same time, we do not neglect to make
the fullest use of such moderate changes as all the evidence at
our command leads us to believe have actually occurred, and especially of the former existence of intermediate islands, so often
indicated by shoals in the midst of the deepest oceans.

Means by which Plants have Migrated from North to South.
—But if plants can thus pass in considerable numbers and variety
over wide seas and oceans, it must be yet more easy for them to
traverse continuous areas of land, wherever mountain-chains offer suitable stations at moderate intervals on which they might
temporarily establish themselves. The facilities afforded for
the transmission of plants by mountains has hardly received
sufficient attention. The numerous land-slips, the fresh surfaces
of broken rock and precipice, the débris of torrents, and the
moraines deposited by glaciers afford numerous unoccupied stations on which wind-borne seeds have a good chance of germinating. It is a well-known fact that fresh surfaces of soil or
rock, such as are presented by railway cuttings and embank-
ments, often produce plants strange to the locality, which survive for a few years, and then disappear as the normal vegeta-

---

[1] Hooker, "On the Flora of Australia," p. 95; H. C. Watson, in Godman's
"Azores," pp. 278-286.

tion gains strength and permanence.[1]   But such a surface will, in the meantime, have acted as a fresh centre of dispersal; and thus a plant might pass on step by step, by means of stations

---

[1] As this is a point of great interest in its bearing on the dispersal of plants by means of mountain-ranges, I have endeavored to obtain a few illustrative facts :

1. Mr. William Mitten, of Hurst Pierpont, Sussex, informs me that when the London and Brighton Railway was in progress in his neighborhood, *Melilotus vulgaris* made its appearance on the banks, remained for several years, and then altogether disappeared. Another case is that of *Diplotaxis muralis*, which formerly occurred only near the sea-coast of Sussex and at Lewes; but since the railway was made has spread along it, and still maintains itself abundantly on the railway banks, though rarely found anywhere else.

2. A correspondent in Tasmania informs me that whenever the virgin forest is cleared in that island there invariably comes up a thick crop of a plant locally known as fire-weed—a species of Senecio, probably *S. australis*. It never grows except where the fire has gone over the ground, and is unknown except in such places. My correspondent adds : "This autumn I went back about thirty-five miles through a dense forest, along a track marked by some prospectors the year before, and in one spot where they had camped, and the fire had burned the fallen logs, etc., there was a fine crop of 'fire-weed.' All around for many miles was a forest of the largest trees and dense scrub." Here we have a case in which burnt soil and ashes favor the germination of a particular plant, whose seeds are easily carried by the wind, and it is not difficult to see how this peculiarity might favor the dispersal of the species for enormous distances, by enabling it temporarily to grow and produce seeds on burnt spots.

3. In answer to an inquiry on this subject, Mr. H. C. Watson has been kind enough to send me a detailed account of the progress of vegetation on the railway banks and cuttings about Thames Ditton. This account is written from memory; but as Mr. Watson states that he took a great interest in watching the process year by year, there can be no reason to doubt the accuracy of his memory. I give a few extracts which bear especially on the subject we are discussing :

"One rather remarkable biennial plant appeared early (the second year, as I recollect) and renewed itself in either two or three years; namely, *Isatis tinctoria*—a species usually supposed to be one of our introduced, but pretty well naturalized, plants. The nearest stations then or since known to me for this Isatis are on chalk about Guildford, twenty miles distant. There were two or three plants of it at first, never more than half a dozen. Once since I saw a plant of Isatis on the railway bank near Vauxhall.

"Close by Ditton Station three species appeared which may be called interlopers. The biennial *Barbarea precox*, one of these, is the least remarkable, because it might have come as seed in the earth from some garden, or possibly in the Thames gravel (used as ballast). At first it increased to several plants, then became less numerous, and will soon, in all probability, become extinct, crowded out by other plants. The biennial *Petroselinum segetum* was at first one very luxuriant plant on the slope of the embankment. It increased by seed into a dozen or a score, and is now nearly, if not quite, extinct. The third species is *Linaria purpurea*, not strictly

temporarily occupied, till it reached a district where, the general conditions being more favorable, it was able to establish itself as a permanent member of the flora.   Such, generally speaking,

---

a British plant, but one established in some places on old walls.   A single root of it appeared on the chalk facing of the embankment by Ditton Station.   It has remained there several years and grown into a vigorous specimen.   Two or three smaller examples are now seen by it, doubtless sprung from some of the hundreds or thousands of seeds shed by the original one plant.   The species is not included in Salmon and Brewer's "Flora of Surrey."

"The main line of the railway has introduced into Ditton parish the perennial *Arabis hirsuta*, likely to become a permanent inhabitant.   The species is found on the chalk and greensand miles away from Thames Ditton ; but neither in this parish nor in any adjacent parish, so far as known to myself or to the authors of the flora of the county, does it occur.   Some years after the railway was made a single root of this *Arabis* was observed in the brickwork of an arch by which the railway is carried over a public road.   A year or two afterwards there were three or four plants.   In some later year I laid some of the ripened seed-pods between the bricks in places where the mortar had partly crumbled out.   Now there are several scores of specimens in the brickwork of the arch.   It is presumable that the first seed may have been brought from Guildford.   But how could it get on to the perpendicular face of the brickwork ?

"The bee orchis (*Ophrys apifera*), plentiful on some of the chalk lands in Surrey, is not a species of Thames Ditton, or (as I presume) of any adjacent parish.   Thus, I was greatly surprised some years back to see about a hundred examples of it in flower in one clayey field either on the outskirts of Thames Ditton or just within the limits of the adjoining parish of Cobham.   I had crossed this same field in a former year without observing the Ophrys there.   And on finding it in the one field, I closely searched the surrounding fields and copses, without finding it anywhere else.   Gradually the plants became fewer and fewer in that one field, and some six or eight years after its first discovery there the species had quite disappeared again.   I guessed it had been introduced with chalk, but could obtain no evidence to show this."

4. Mr. A. Bennett, of Croydon, has kindly furnished me with some information on the temporary vegetation of the banks and cuttings on the railway from Yarmouth to Caistor, in Norfolk, where it passes over extensive sandy dunes with a sparse vegetation.   The first year after the railway was made, the banks produced abundance of *Œnothera odorata* and *Delphinium Ajacis* (the latter only known thirty miles off in corn-fields in Cambridgeshire), with *Atriplex patula* and *A. deltoidea*.   Gradually the native sand-plants—Carices, grasses, *Galium verum*, etc.—established themselves, and year by year covered more ground, till the new introductions almost completely disappeared.   The same phenomenon was observed in Cambridgeshire between Chesterton and Newmarket, where, the soil being different, *Stellaria media* and other annuals appeared in large patches ; but these soon gave way to a permanent vegetation of grasses, composites, etc., so that in the third year no Stellaria was to be seen.

5. Mr. T. Kirk (writing in 1878) states that "in Auckland, where a dense sward

was probably the process by which the Scandinavian flora has made its way to the Southern Hemisphere; but it could hardly have done so to any important extent without the aid of those powerful causes explained in our eighth chapter—causes which acted as a constantly recurrent motive power to produce that "continuous current of vegetation" from north to south across the whole width of the tropics referred to by Sir Joseph Hooker. Those causes were, the repeated changes of climate which, during all geological time, appear to have occurred in both hemispheres, culminating at rare intervals in glacial epochs, and which have been shown to depend upon changes of eccentricity of the earth's orbit and the occurrence of summer or winter in aphelion, in conjunction with the slower and more irregular changes of geographical conditions; these combined causes acting chiefly through the agency of heat - bearing oceanic currents, and of snow- and ice-collecting highlands. Let us now briefly consider how such changes would act in favoring the dispersal of plants.

Elevation and Depression of the Snow-line as Aiding the

of grass is soon formed, single specimens of the European Milk Thistle (*Carduus marianus*) have been known for the past fifteen years; but although they seeded freely, the seeds had no opportunity of germinating, so that the thistle did not spread. A remarkable exception to this rule occurred during the formation of the Onehunga Railway, where a few seeds fell on disturbed soil, grew up, and flowered. The railway works being suspended, the plant increased rapidly, and spread wherever it could find disturbed soil."

Again, "The fiddle-dock (*Rumex pulcher*) occurs in great abundance on the formation of new streets, etc., but soon becomes comparatively rare. It seems probable that it was one of the earliest plants naturalized here, but that it partially died out, its buried seeds retaining their vitality."

Medicago sativa and *Apium graveolens* are also noted as escapes from cultivation which maintain themselves for a time, but soon die out.*

The preceding examples of the *temporary* establishment of plants on newly exposed soil, often at considerable distances from the localities they usually inhabit, might, no doubt, by further inquiry be greatly multiplied; but, unfortunately, the phenomenon has received little attention, and is not even referred to in the elaborate work of De Candolle ("Géographie Botanique Raisonnée"), in which almost every other aspect of the dispersion and distribution of plants is fully discussed. Enough has been advanced, however, to show that it is of constant occurrence, and from the point of view here advocated it becomes of great importance in explaining the almost world-wide distribution of many common plants of the north temperate zone.

* *Transactions of the New Zealand Institute*, Vol. X., p. 367.

Migration of Plants.—We have endeavored to show (in an earlier portion of this volume) that wherever geographical or physical conditions were such as to produce any considerable amount of perpetual snow, this would be increased whenever a high degree of eccentricity concurred with winter in aphelion, and diminished during the opposite phase. On all mountain-ranges, therefore, which reached above the snow-line there would be a periodical increase and decrease of snow; and when there were extensive areas of plateau at about the same level, the lowering of the snow-line might cause such an increased accumulation of snow as to produce great glaciers and ice-fields, such as we have seen occurred in South Africa during the last period of high eccentricity. But along with such depression of the line of perpetual snow there would be a corresponding depression of the alpine and sub-alpine zones suitable for the growth of an arctic and temperate vegetation, and, what is perhaps more important, the depression would necessarily produce a great *extension* of the area of these zones on all high mountains, thus affording a number of new stations suitable for such temperate plants as might first reach them. But just above and below the snow-line is the area of most powerful disintegration and denudation, from the alternate action of frost and sun, of ice and water; and thus the more extended area would be subject to the constant occurrence of land-slips, berg-falls, and floods, with their accompanying accumulations of débris and of alluvial soil, affording innumerable stations in which solitary wind-borne seeds might germinate and temporarily establish themselves.

This lowering and rising of the snow-line each 10,500 years during periods of high eccentricity would occur in the Northern and Southern hemispheres alternately; and where there were high mountains within the tropics the two would probably overlap each other, so that the northern depression would make itself felt in a slight degree even across the equator some way into the Southern Hemisphere, and *vice versa*; and even if the difference of the height of perpetual snow at the two extremes did not average more than a few hundred feet, this would be amply sufficient to supply the new and unoccupied stations needful to facilitate the migration of plants.

But the differences of temperature in the two hemispheres caused by the sun being in perihelion in the winter of the one while it was in aphelion during the same season in the other would necessarily lead to increased aerial and marine currents, as already explained; and whenever geographical conditions were such as to favor the production of glaciation in any area, these effects would become more powerful, and would further aid in the dispersal of the seeds of plants.

Changes of Climate Favorable to Migration.—It is clear, then, that during periods when no glacial epochs were produced in the Northern Hemisphere, and even when a mild climate extended over the whole polar area, alternate changes of climate favoring the dispersal of plants would occur on all high mountains, and with particular force on such as rise above the snow-line. But, during that long-continued, though comparatively recent, phase of high eccentricity which produced an extensive glaciation in the Northern Hemisphere and local glaciations in the Southern, these risings and lowerings of the snow-line on all mountain-ranges would have been at a maximum, and would have been increased by the depression of the ocean which must have arisen from such a vast bulk of water being locked up in land-ice, and which depression would have produced the same effect as a general elevation of all the continents. At this time, too, aerial currents would have attained their maximum of force in both hemispheres; and this would greatly facilitate the dispersal of all wind-borne seeds as well as of those carried in the plumage or in the stomachs of birds, since we have seen how vastly the migratory powers of birds are increased by a stormy atmosphere.

Migration from North to South has been long going on.—Now, if each phase of colder and warmer mountain-climate, each alternate depression and elevation of the snow-line, only helped on the migration of a few species some stages of the long route from the north to the south temperate regions, yet during the long course of the Tertiary period there might well have arisen that representation of the northern flora in the Southern Hemisphere which is now so conspicuous. For it is very important to remark that it is not the existing flora alone that is represented, such as might have been conveyed during the last glacial epoch only; but we find a whole series of northern types evidently of varying degrees of antiquity, while even some genera character-istic of the Southern Hemisphere appear to have been originally derived from Europe. Thus Eucalyptus and Metrosideros have been determined by Dr. Ettinghausen from their fruits in the Eocene beds of Sheppey; while Pimelea, Leptomeria, and four genera of Proteaceæ have been recognized by Professor Heer in the Miocene of Switzerland; and the former writer has detected fifty-five Australian forms in the Eocene plant-beds of Häring (? Belgium).[1] Then we have such peculiar genera as Pachychla-don and Notothlaspi of New Zealand said to have affinities with arctic plants; while Stilbocarpa, another peculiar New Zealand genus, has its nearest allies in the Himalayan and Chinese Ara-lias. Following these are a whole host of very distinct species of northern genera which may date back to any part of the Terti-ary period, and which occur in every south temperate land. Then we have closely allied representative species of European or arc-tic plants, and, lastly, a number of identical species; and these two classes are probably due entirely to the action of the last great glacial epoch, whose long continuance, and the repeated fluctuations of climate with which it commenced and termi-nated, rendered it an agent of sufficient power to have brought about this result.

---

[1] Sir Joseph Hooker informs me that he considers these identifications worthless, and Mr. Bentham has also written very strongly against the value of similar identifi-cations by Heer and Unger. Giving due weight to the opinions of these eminent botanists, we must admit that Australian genera have not yet been *demonstrated* to have existed in Europe during the Tertiary period; but, on the other hand, the evi-dence that they did so appears to have some weight, on account of the improbability that the numerous resemblances to Australian plants which have been noticed by different observers should *all* be illusory; while the well-established fact of the former wide distribution of many tropical or now restricted types of plants and ani-mals, so frequently illustrated in the present volume, removes the antecedent improb-ability which is supposed to attach to such identifications. I am myself the more inclined to accept them because, according to the views here advocated, such migra-tions must have taken place at remote as well as at recent epochs; and the preserva-tion of some of these types in Australia while they have become extinct in Europe is exactly paralleled by numerous facts in the distribution of animals which have been already referred to in Chapter XIX. and elsewhere in this volume, and also re-peatedly in my larger work.

Here, then, we have that constant or constantly recurrent process of dispersal acting throughout long periods with varying power—that "continuous current of vegetation," as it has been termed—which the facts demand; and the extraordinary phenomenon of the species and genera of European and even of arctic plants being represented abundantly in South Africa, Australia, and New Zealand thus adds another to the long series of phenomena which are rendered intelligible by frequent alternations of warmer and colder climates in either hemisphere, culminating, at long intervals and in favorable situations, in actual glacial epochs.

Geological Changes as Aiding Migration.—It will be well also to notice here that there is another aid to dispersion, dependent upon the changes effected by denudation during the long periods included in the duration of the species and genera of plants. A considerable number of the plants of Europe of the Miocene period were so much like existing species that although they have generally received fresh names they may well have been identical; and a large proportion of the vegetation during the whole Tertiary period consisted of genera which are still living.[1] But from what is now known of the rate of subaerial denudation we are sure that during each division of this period many mountain-chains must have been considerably lowered, while we know that some of the existing ranges have been greatly elevated. Ancient volcanoes, too, have been destroyed by denudation, and new ones have been built up, so that we may be quite sure that ample means for the transmission of temperate plants across the tropics may have existed in countries where they are now no longer to be found. The great mountain masses of Guiana and Brazil, for example, must have been far more lofty before the sedimentary covering was denuded from their granitic bosses and metamorphic peaks, and may have aided the southern

---

[1] Out of forty-two genera from the Eocene of Sheppey, enumerated by Dr. Ettinghausen in the *Geological Magazine* for January, 1880, only two or three appear to be extinct, while there is a most extraordinary intermixture of tropical and temperate forms—Musa, Nipa, and Victoria, with Corylus, Prunus, Acer, etc. The rich Miocene flora of Switzerland, described by Professor Heer, presents a still larger proportion of living genera.

migration of plants before the final elevation of the Andes. And if Africa presents us with an example of a continent of vast antiquity we may be sure that its great central plateaus once bore far loftier mountain-ranges ere they were reduced to their present condition by long ages of denudation.

Proofs of Migration by Way of the Andes.—We are now prepared to apply the principles above laid down to the explanation of the character and affinities of the various portions of the north temperate flora in the Southern Hemisphere, and especially in Australia and New Zealand.

At the present time the only unbroken chain of highlands and mountains connecting the arctic and north temperate with the antarctic lands is to be found in the American continent, the only break of importance being the comparatively low Isthmus of Panama, where there is a distance of about 300 miles occupied by rugged forest-clad hills, between the lofty peaks of Veragua and the northern extremity of the Andes of New Granada. Such distances are, as we have already seen, no barrier to the diffusion of plants; and we should accordingly expect that this great continuous mountain-chain has formed the most effective agent in aiding the southward migration of the arctic and north temperate vegetation. We do find, in fact, not only that a large number of northern genera and many species are scattered all along this line of route, but that at the end of the long journey, in Southern Chili and Fuegia, they have established themselves in such numbers as to form an important part of the flora of those countries. From the lists given in the works already referred to, it appears that there are between sixty and seventy northern genera in Fuegia and Southern Chili, while about forty of the species are absolutely identical with those of Europe and the arctic regions. Considering how comparatively little the mountains of south temperate America are yet known, this is a very remarkable result, and it proves that the transmission of species must have gone on up to comparatively recent times. Yet, as only a few of these species are now found along the line of migration, we see that they only occupied such stations temporarily; and we may connect their disappearance with the passing-away of the last glacial period, which, by raising the snow-line, reduced the area on which alone they could exist, and exposed them to the competition of indigenous plants from the belt of country immediately below them.

Now, just as these numerous species and genera have undoubtedly passed along the great American range of mountains, although only now found at its two extremes, so others have doubtless passed on farther; and have found more suitable stations or less severe competition in the antarctic continent and islands, in New Zealand, in Tasmania, and even in Australia itself. The route by which they may have reached these countries is easily marked out. Immediately south of Cape Horn, at a distance of only 500 miles, are the South Shetland Islands and Graham's Land, whence the antarctic continent or a group of large islands probably extends across or around the south polar area to Victoria Land, and thence to Adélie Land. The outlying Young Island, 12,000 feet high, is about 750 miles south of the Macquarie Islands, which may be considered a southern outlier of the New Zealand group; and the Macquarie Islands are about the same distance from the 1000-fathom line, marking the probable southern extension of Tasmania. Other islands may have existed at intermediate points; but, even as it is, these distances are not greater than we know are traversed by plants both by flotation and by aerial currents, especially in such a stormy atmosphere as that of the antarctic regions. Now we may further assume that what we know occurred within the Arctic Circle also took place in the Antarctic—that is, that there have been alternations of climate during which some portion of what are now ice-clad lands became able to support a considerable amount of vegetation.[1] During such periods there would be a steady migration of plants from all southern circumpolar countries to people the comparatively unoccupied continent; and the southern extremity of America being considerably the nearest, and also being the best stocked with those

---

[1] The recent discovery of a rich flora on rocky peaks rising out of the continental ice of Greenland, as well as the abundant vegetation of the highest northern latitudes, renders it possible that even now the antarctic continent may not be wholly destitute of vegetation, although its climate and physical condition are far less favorable than those of the arctic lands.

northern types which have such great powers of migration and
colonization, such plants would form the bulk of the antarctic
vegetation, and during the continuance of the milder southern
climate would occupy the whole area.

When the cold returned and the land again became ice-clad,
these plants would be crowded towards the outer margins of
the antarctic land and its islands, and some of them would find
their way across the sea to such countries as offered on their
mountain summits suitable cool stations; and as this process of
alternately receiving plants from Chili and Fuegia and trans-
mitting them in all directions from the central antarctic land
may have been repeated several times during the Tertiary period, we have no difficulty in understanding the general community between the European and antarctic plants found in all
south temperate lands.   Kerguelen Land and the Crozets are
within about the same distance from the antarctic continent as
New Zealand and Tasmania, and we need not therefore be surprised at finding in each of these islands some Fuegian species
which have not reached the others.   Of course there will remain difficulties of detail, as there always must when we know
so imperfectly the past changes of the earth's surface and the
history of the particular plants concerned.   Sir Joseph Hooker
notes, for example, the curious fact that several Compositæ common to three such remote localities as the Auckland Islands,
Fuegia, and Kerguelen Land have no pappus or seed-down,
while such as have pappus are in no case common even to two
of these islands.   Without knowing the exact history and distribution of the genera to which these plants belong, it would
be useless to offer any conjecture, except that they are ancient
forms which may have survived great geographical changes, or
may have some peculiar and exceptional means of dispersion.

*Proofs of Migration by Way of the Himalayas and Southern
Asia.*—But although we may thus explain the presence of a
considerable portion of the European element in the floras of
New Zealand and Australia, we cannot account for the whole
of it by this means, because Australia itself contains a host of
European and Asiatic genera of which we find no trace in New
Zealand or South America, or any other antarctic land.   We

find, in fact, in Australia two distinct sets of European plants. First, we have a number of species identical with those of Northern Europe or Asia (of the most characteristic of which —thirty-eight in number—Sir Joseph Hooker gives a list); and, in the second place, a series of European genera, usually of a somewhat more southern character, mostly represented by very distinct species, and all absent from New Zealand; such as Clematis, Papaver, Cleome, Polygala, Lavatera, Ajuga, etc. Now of the first set—the North European species—about three fourths occur in some parts of America, and about half in south temperate America or New Zealand; whence we may conclude that most of these, as well as some others, have reached Australia by the route already indicated. The second set of Australo-European genera, however, and many others characteristic of the South European or the Himalayan flora, have probably reached Australia by way of the mountains of Southern Asia, Borneo, the Moluccas, and New Guinea, at a somewhat remote period when loftier ranges and some intermediate peaks may have existed, sufficient to carry on the migration by the aid of the alternate climatal changes which are known to have occurred. The long belt of Secondary and Palæozoic formations in East Australia from Tasmania to Cape York, continued by the lofty ranges of New Guinea, indicates the route of this immigration, and sufficiently explains how it is that these Northern types are almost wholly confined to this part of the Australian continent. Some of the earlier immigrants of this class no doubt passed over to New Zealand and now form a portion of the peculiar genera confined to these two countries; but most of them are of later date, and have thus remained in Australia only.

Proofs of Migration by Way of the African Highlands.—It is owing to this twofold current of vegetation flowing into Australia by widely different routes that we have in this distant land a better representation of the European flora, both as regards species and genera, than in any other part of the Southern Hemisphere; and, so far as I can judge of the facts, there is no general phenomenon — that is, nothing in the distribution of genera and other groups of plants as opposed to cases of individual species—that is not fairly accounted for by such an origin. It further receives support from the case of South Africa, which also contains a large and important representation of the northern flora. But here we see no indications (or very slight ones) of that southern influx which has given Australia such a community of vegetation with the antarctic lands. There are no less than sixty genera of strictly north temperate plants in South Africa, none of which occur in Australia; while very few of the species so characteristic of Australia, New Zealand, and Fuegia are found there. It is clear, therefore, that South Africa has received its European plants by the direct route through the Abyssinian highlands and the lofty equatorial mountains, and mostly at a distant period, when the conditions for migration were somewhat more favorable than they are now. The much greater directness of the route from Northern Europe to South Africa than to Australia, and the existence even now of lofty mountains and extensive highlands for a large portion of the distance, will explain (what Sir Joseph Hooker notes as "a very curious fact") why South Africa has more very northern European genera than Australia, while Australia has more identical species and a better representation, on the whole, of the European flora—this being clearly due to the large influx of species it has received from the antarctic islands, in addition to those which have entered it by way of Asia. The greater distance of South Africa even now from any of these islands, and the much deeper sea to the south of the African continent, than in the case of Tasmania and New Zealand, indicating a smaller recent extension southward, are all quite in harmony with the facts of distribution of the northern flora above referred to.

Supposed Connection of South Africa and Australia.—There remains, however, the small amount of direct affinity between the vegetation of South Africa and that of Australia, New Zealand, and temperate South America, consisting in all of fifteen genera, five of which are confined to Australia and South Africa, while several natural orders are better represented in these two countries than in any other part of the world. This resemblance has been supposed to imply some former land-connection of all the great southern lands, but it appears to me that any such supposition is wholly unnecessary. The differences between the faunas and floras of these countries are too great and too radical to render it possible that any such connection should have existed except at a very remote period. But if we have to go back so far for an explanation, a much simpler one presents itself, and one more in accordance with what we have learned of the general permanence of deep oceans and the radical changes that have taken place in the distribution of all forms of life. Just as we explain the presence of marsupials in Australia and America, and of Centetidæ in Madagascar and the Antilles, by the preservation in these localities of remnants of once wide-spread types, so we should prefer to consider the few genera common to Australia and South Africa as remnants of an ancient vegetation, once spread over the northern hemisphere, driven southward by the pressure of more specialized types, and now finding a refuge in these two widely separated southern lands. It is suggestive of such an explanation that these genera are either of very ancient groups, as Conifers and Cycads, or plants of low organization, as the Restiaceæ, or of world-wide distribution, as Melanthaceæ.

The Endemic Genera of Plants in New Zealand.—Returning now to the New Zealand flora, with which we are more especially concerned, there only remains to be considered the peculiar or endemic genera which characterize it. These are thirty-two in number, and are mostly very isolated. A few have affinities with arctic groups, others with Himalayan or Australian genera; several are tropical forms, but the majority appear to be altogether peculiar types of world-wide groups, as Leguminosæ, Saxifrageæ, Compositæ, Orchideæ, etc. We must evidently trace back these peculiar forms to the earliest immigrants, either from the north or from the south; and the great antiquity we are obliged to give to New Zealand—an antiquity supported by every feature in its fauna and flora, no less than by its geological structure and its extinct forms of life[1]—affords

---

[1] Dr. Hector notes the occurrence of the genus Dammara in Triassic deposits, while in the Jurassic period New Zealand produced the genera Palæozamia, Oleandrium, Alethopteris, Camptopteris, Cycadites, Echinostrobus, etc., all Indian forms of the same age. Neocomian beds contain a true dicotyledonous leaf with Dammara and Araucaria. The Cretaceous deposits have produced a rich flora of dicotyledonous

ample time for the changes in the general distribution of plants,
and for those due to isolation and modification under the influence of changed conditions, which are manifested by the extreme peculiarity of many of these interesting endemic forms.

*The Absence of Southern Types from the Northern Hemi-
sphere.*—We have now only to notice the singular want of rec-
iprocity in the migrations of northern and southern types of
vegetation. In return for the vast number of European plants
which have reached Australia, not one single Australian plant
has entered any part of the north temperate zone, and the same
may be said of the typical southern vegetation in general, whether developed in the antarctic lands, New Zealand, South America, or South Africa. The farthest northern outliers of the
southern flora are a few genera of antarctic type on the Bornean Alps; the genus Acæna, which has a species in California;
two representatives of the Australian flora—Casuarina and Stylidium—in the peninsula of India; while China and the Philippines have two strictly Australian genera of Orchideæ—Microtis
and Thelymitra—as well as a Restiaceous genus. Several distinct
causes appear to have combined to produce this curious inability
of the southern flora to make its way into the Northern Hemisphere. The primary cause is, no doubt, the totally different
distribution of land in the two hemispheres, so that in the south
there is the minimum of land in the colder parts of the temperate zone, and in the north the maximum. This is well shown
by the fact that on the parallel of lat. 50° N. we pass over 240°
of land or shallow sea, while on the same parallel of south latitude we have only 4° where we cross the southern part of Patagonia. Again, the three most important south temperate land-
areas—south temperate America, South Africa, and Australia
—are widely separated from each other, and have in all probability always been so; whereas the whole of the north temperate lands are practically continuous. It follows that, instead of

plants, many of which are of the same genera as the existing flora; while the Miocene and other Tertiary deposits produce plants apparently almost identical with
those now inhabiting the country (*Transactions of the New Zealand Institute*, Vol.
XI., 1879, p. 536). These facts agree well with the origin of the New Zealand flora
developed in the last chapter.

the enormous northern area in which highly organized and dominant groups of plants have been developed gifted with great colonizing and aggressive powers, we have in the south three comparatively small and detached areas, in which rich floras have been developed with special adaptations to soil, climate, and organic environment, but comparatively impotent and inferior beyond their own domain.

Another circumstance which makes the contest between the northern and southern forms still more unequal is the much greater hardiness of the former, from having been developed in a colder region, and one where alpine and arctic conditions extensively prevail; whereas the southern floras have been mainly developed in mild regions to which they have been altogether confined. While the northern plants have been driven north or south by each succeeding change of climate, the southern species have undergone comparatively slight changes of this nature, owing to the areas they occupy being unconnected with the ice-bearing antarctic continent. It follows that whereas the northern plants find in all these southern lands a milder and more equable climate than that to which they have been accustomed, and are thus often able to grow and flourish even more vigorously than in their native land, the southern plants would find in almost every part of Europe, North America, or Northern Asia a more severe and less equable climate, with winters that usually prove fatal to them even under cultivation. These causes, taken separately, are very powerful, but when combined they must, I think, be held to be amply sufficient to explain why examples of the typical southern vegetation are almost unknown in the north temperate zone, while a very few of them have extended so far as the northern tropic.[1]

---

[1] The fact stated in the last edition of the "Origin of Species" (p. 340) on the authority of Sir Joseph Hooker, that Australian plants are rapidly sowing themselves and becoming naturalized on the Neilgherry Mountains in the southern part of the Indian Peninsula, though an exception to the rule of the inability of Australian plants to become naturalized in the Northern Hemisphere, is yet quite in harmony with the hypothesis here advocated. For not only is the climate of the Neilgherries more favorable to Australian plants than any part of the north temperate zone, but the entire Indian Peninsula has existed for unknown ages as an *island*, and thus possesses the "insular" characteristic of a comparatively poor and less developed flora and

Concluding Remarks on the Last Two Chapters.—Our inquiry into the external relations and probable origin of the fauna and flora of New Zealand has thus led us on to a general theory as to the cause of the peculiar biological relations between the Northern and the Southern Hemisphere; and no better or more typical example could be found of the wide range and great interest of the study of the geographical distribution of animals and plants.

The solution which has here been given of one of the most difficult of this class of problems has been rendered possible solely by the knowledge very recently obtained of the form of the sea-bottom in the Southern Ocean, and of the geological structure of the great Australian continent. Without this knowledge we should have nothing but a series of guesses or probabilities on which to found our hypothetical explanation, which we have now been able to build up on a solid foundation of fact. The complete separation of East from West Australia during the Cretaceous period could never have been guessed till it was established by the laborious explorations of the Australian geologists; while the hypothesis of a comparatively shallow sea, uniting New Zealand by a long route with tropical Australia, while a profoundly deep ocean always separated it from temperate Australia, would have been rejected as too improbable a supposition for the foundation of even the most enticing theory. Yet it is mainly by means of these two facts that we are enabled to give an adequate explanation of the strange anomalies in the flora of Australia and its relation to that of New Zealand.

In the more general explanation of the relations of the various northern and southern floras, I have shown what an important aid to any such explanation is the theory of repeated changes of climate, not necessarily of great amount, given in our eighth chapter; while the whole discussion justifies the importance attached to the theory of the general permanence of continents

---

fauna as compared with the truly "continental" Malayan and Himalayan regions. Australian plants are thus enabled to compete with those of the Indian Peninsula highlands with a fair chance of success.

and oceans, as demonstrated in Chapter VI., since any rational
explanation based upon facts (as opposed to mere unsupported
conjecture) must take such general permanence as a starting-
point. The whole inquiry into the phenomena presented by
islands, which forms the main subject of the present volume,
has, I think, shown that this theory does afford a firm foundation for the discussion of questions of distribution and dispersal ;
and that by its aid, combined with a clear perception of the
wonderful powers of dispersion and modification in the organic
world when long periods are considered, the most difficult problems connected with this subject cease to be insoluble.

CHAPTER XXIV.

SUMMARY AND CONCLUSION.

The Present Volume is the Development and Application of a Theory.—Statement of the Biological and Physical Causes of Dispersal.—Investigation of the Facts of Dispersal.—Of the Means of Dispersal.—Of Geographical Changes Affecting Dispersal.—Of Climatal Changes Affecting Dispersal.—The Glacial Epoch and its Causes.—Alleged Ancient Glacial Epochs.—Warm Polar Climates and their Causes.—Conclusions as to Geological Climates.—How far Different from those of Mr. Croll.—Supposed Limitations of Geological Time.—Time Amply Sufficient both for Geological and Biological Development.—Insular Faunas and Floras.—The North Atlantic Islands.—The Galapagos.—St. Helena and the Sandwich Islands.—Great Britain as a Recent Continental Island.—Borneo and Java.—Japan and Formosa. —Madagascar as an Ancient Continental Island.—Celebes and New Zealand as Anomalous Islands.—The Flora of New Zealand and its Origin.—The European Element in the South Temperate Floras.—Concluding Remarks.

The present volume has gone over a very wide field both of facts and theories, and it will be well to recall these to the reader's attention, and point out their connection with each other, in a concluding chapter. I hope to be able to show that, although at first sight somewhat fragmentary and disconnected, this work is really the development of a clear and definite theory, and its application to the solution of a number of biological problems. That theory is, briefly, that the distribution of the various species and groups of living things over the earth's surface, and their aggregation in definite assemblages in certain areas, are the direct result and outcome of a complex set of causes, which may be grouped as "biological" and "physical." The biological causes are mainly of two kinds—firstly, the constant tendency of all organisms to increase in numbers and to occupy a wider area, and their various powers of dispersion and migration through which, when unchecked, they are enabled to spread widely over the globe; and, secondly, those laws of evolution and extinction which determine the manner in which groups of organisms arise and grow, reach their maximum, and then dwindle away, often breaking up into separate portions which long survive in very remote regions. The physical causes are also mainly of two kinds. We have, first, the geographical changes which at one time isolate a whole fauna and flora, at another time lead to their dispersal and intermixture with adjacent faunas and floras—and it was here important to ascertain and define the exact nature and extent of these changes, and to determine the question of the general stability or instability of continents and oceans; in the second place, it was necessary to determine the exact nature, extent, and frequency of the changes of climate which have occurred in various parts of the earth, because such changes are among the most powerful agents in causing the dispersal and extinction of plants and animals. Hence the importance attached to the question of geological climates and their causes, which have been here investigated at some length with the aid of the most recent researches of geologists, physicists, and explorers. These various inquiries led on to an investigation of the mode of formation of stratified deposits, with a view to fix within some limits their probable age; and also to an estimate of the probable rate of development of the organic world; and both these processes are shown to involve, in all probability, periods of time less vast than have generally been thought necessary.

The numerous facts and theories established in the First Part of the work are then applied to explain the phenomena presented by the floras and faunas of the chief islands of the globe, which are classified, in accordance with their physical origin, in three groups or classes, each of which is shown to exhibit certain well-marked biological features.

Having thus shown that the work is a connected whole, founded on the principle of tracing out the more recondite causes of the distribution of organisms, we will briefly indicate the scope and object of the several chapters by means of which this general conception has been carried out.

Beginning with simple and familiar facts relating to British and European quadrupeds and birds, I have defined and shown the exact character of " areas of distribution " as applied to species, genera, and families, and have illustrated the subject by maps showing the peculiarities of distribution of some well-known groups of birds. Taking, then, our British mammals and land birds, I follow them over the whole area they inhabit, and thus obtain a foundation for the establishment of "zoological regions," and a clear insight into their character as distinct from the usual geographical divisions of the globe.

The facts thus far established are then shown to be necessary results of the "law of evolution." The nature and amount of "variation" are exhibited by a number of curious examples; the origin, growth, and decay of species and genera are traced; and all the interesting phenomena of isolated groups and discontinuous generic and specific areas are shown to follow as logical consequences.

The next subject investigated is the means by which the various groups of animals are enabled to overcome the natural barriers which often seem to limit them to very restricted areas, how far those barriers are themselves liable to be altered or abolished, and what are the exact nature and amount of the changes of sea and land which our earth has undergone in past times. This latter part of the inquiry is shown to be the most important as it is the most fundamental; and as it is still a subject of controversy, and many erroneous views prevail in regard to it, it is discussed at some length. Several distinct classes of evidence are adduced to prove that the grand features of our globe —the position of the great oceans and the chief land-areas—have remained, on the whole, unchanged throughout geological time. Our continents are shown to be built up mainly of "shore-deposits;" and even the chalk, which is so often said to be the exact equivalent of the "globigerina ooze" now forming in mid-Atlantic, is shown to be a comparatively shallow-water deposit formed in inland seas, or in the immediate vicinity of land. The general stability of continents has, however, been accompanied by constant changes of form, and insular conditions have prevailed over every part in succession; and the effect of such changes on the distribution of organisms is pointed out.

We then approach the consideration of another set of changes —those of climate—which have probably been agents of the first importance in modifying the specific forms as well as the distribution of animals. Here, again, we find ourselves in the midst of fierce controversies. The occurrence of a recent glacial epoch of great severity in the Northern Hemisphere is now universally admitted, but the causes which brought it on are matter of dispute. But unless we can arrive at these causes, as well as at those which produced the equally well demonstrated mild climate in the arctic regions, we shall be quite unable to determine the nature and amount of the changes of climate which have occurred throughout past ages, and shall thus be left without a most important clew to the explanation of many of the anomalies in the distribution of animals and plants.

I have therefore devoted three chapters to a full investigation of this question. I have first given such a sketch of the most salient facts as to render the phenomena of the glacial epoch clear and intelligible. I then review the various suggested explanations, and, taking up the two which alone seem tenable, I endeavor to determine the true principles of each. While adopting generally Mr. Croll's views as to the causes of the "glacial epoch," I have introduced certain limitations and modifications. I have pointed out, with more precision than has, I believe, hitherto been done, the very different effects on climate of water in the liquid and in the solid state; and I have shown by a variety of evidence that without high land there can be no permanent snow and ice. From these facts and principles, the very important conclusion is reached that the alternate phases of precession—causing the winter of each hemisphere to be in aphelion and perihelion each 10,500 years—would produce a complete change of climate only where a country was *partially* snow-clad; while, whenever a large area became almost *wholly* buried in snow and ice, as was certainly the case with Northern Europe during the glacial epoch, then the glacial conditions would be continued, and perhaps even intensified, when the sun approached nearest to the earth in winter, instead of there being at that time, as Mr. Croll maintains, an almost perpetual spring. This important result is supported by reference to the existing differences between the climates of the Northern and Southern hemispheres, and by what is known to have occurred during the last glacial epoch; and it is shown to be in complete harmony with the geological evidence as to interglacial mild periods.

Discussing next the evidence for glacial epochs in earlier times, it is shown that Mr. Croll's views are opposed by a vast body of facts, and that the geological evidence leads irresistibly to the conclusion that during a large portion of the Secondary and Tertiary periods, uninterrupted warm climates prevailed in the north temperate zone, and so far ameliorated the climate of the arctic regions as to admit of the growth of a luxuriant vegetation in the highest latitudes yet explored. The geographical condition of the Northern Hemisphere at these periods is then investigated, and it is shown to have been such as to admit the warm tropical waters freely to penetrate the land, and to reach the arctic seas by several channels; and, adopting Mr. Croll's views as to the enormous quantity of heat that would thus be conveyed northward, it is maintained that the mild arctic climates are amply accounted for. With such favorable geographical conditions, it is shown that changes of eccentricity and of the phases of precession would have no other effect than to cause greater differences of temperature between summer and winter; but wherever there was a considerable extent of very lofty mountains the snow-line would be lowered, and, the snow-collecting area being thus largely increased, a considerable amount of glaciation might result. Thus may be explained the presence of enormous ice-borne rocks in Eocene and Miocene times in Central Europe, while at the very same period all the surrounding country enjoyed a tropical or subtropical climate.

The general conclusion is thus reached that geographical conditions are the primary causes of great changes of climate, and that the radically different distribution of land and sea in the Northern and Southern hemispheres has generally led to great diversity of climate in the arctic and antarctic regions. The form and arrangement of the continents are shown to be such as to favor the transfer of warm oceanic currents to the north far in excess of those which move towards the south; and whenever these currents had free passage *through* the northern land-masses to the polar area, a mild climate must have prevailed over the whole Northern Hemisphere. It is only in very recent times that the great northern continents have become so completely consolidated as they now are, thus shutting out the warm water from their interiors, and rendering possible a wide-spread and intense glacial epoch. But this great climatal change was actually brought about by the high eccentricity which occurred about 200,000 years ago ; and it is doubtful if a similar glaciation in equally low latitudes could be produced, by means of any such geographical combinations as actually occur, without the concurrence of a high eccentricity.

A survey of the present condition of the earth supports this view; for, though we have enormous mountain-ranges in every latitude, there is no glaciated country south of Greenland in N. lat. 61°. But directly we go back a very short period, we find the superficial evidences of glaciation to an enormous extent over three fourths of the globe. In the Alps and Pyrenees, in the British Isles and Scandinavia, in Spain and the Atlas, in the Caucasus and the Himalayas, in Eastern North America and west of the Rocky Mountains, in the Andes, in the mountains of Brazil, in South Africa, and in New Zealand, huge moraines and other unmistakable ice-marks attest the universal descent of the snow-line for several thousand feet below its present level. If we reject the influence of high eccentricity as the cause of this almost universal glaciation, we must postulate a general elevation of *all* these mountains about the same time; for the close similarity in the state of preservation of the ice-marks, and the known activity of denudation as a destroying agent, forbid the idea that they belong to widely separated epochs. It has, indeed, been suggested that denudation alone has lowered these mountains so much during the Quaternary epoch that they were previously of sufficient height to account for the glaciation of all of them, but this hardly needs refutation ; for it is clear that denudation could not at the same time have removed some thousands of feet of rock from many hundreds of square miles of lofty snow-collecting plateaus, and yet have left moraines and blocks, and even glacial striæ, undisturbed and uneffaced on the slopes and in the valleys of these same mountains.

The theory of geological climates set forth in this volume,

while founded on Mr. Croll's researches, differs from all that
have yet been made public, in clearly tracing out the comparative influence of geographical and astronomical revolutions,
showing that, while the former have been the chief, if not the
exclusive, causes of the long-continued mild climates of the arctic regions, the concurrence of the latter has been essential to the
production of glacial epochs in the temperate zones, as well as
of those local glaciations in low latitudes of which there is such
an abundance of evidence.

The next question discussed is that of geological time as
bearing on the development of the organic world. The periods
of time usually demanded by geologists have been very great,
and it was often assumed that there was no occasion to limit
them. But the theory of development demands far more; for
the earliest fossiliferous rocks prove the existence of many and
varied forms of life which require unrecorded ages for their
development—ages probably far longer than those which have
elapsed from that period to the present day. The physicists,
however, deny that any such indefinitely long periods are available. The sun is ever losing heat far more rapidly than it can
be renewed from any known or conceivable source. The earth
is a cooling body, and must once have been too hot to support
life; while the friction of the tides is checking the earth's rotation, and this cannot have gone on indefinitely without making
our day much longer than it is. A limit is therefore placed to
the age of the habitable earth, and it has been thought that
the time so allowed is not sufficient for the long processes of
geological change and organic development. It is therefore
important to inquire whether these processes are either of them
so excessively slow as has been supposed, and I devote a chapter
to the inquiry.

Geologists have measured with some accuracy the maximum
thickness of all the known sedimentary rocks. The rate of
denudation has also been recently measured by a method which,
if not precise, at all events gives results of the right order of
magnitude, and which err on the side of being too slow rather
than too fast. If, then, the *maximum* thickness of the *known*
sedimentary rocks is taken to represent the *average* thickness

of *all* the sedimentary rocks, and we also know the *amount* of sediment carried to the sea or lakes, and the *area* over which that sediment is spread, we have a means of calculating the *time* required for the building-up of all the sedimentary rocks of the geological system. I have here inquired how far the above suppositions are correct, or on which side they probably err ; and the conclusion arrived at is that the time required is very much less than has hitherto been supposed.

Another estimate is afforded by the date of the last glacial epoch as coincident with the last period of high eccentricity, while the Alpine glaciation of the Miocene period is assumed to have been caused by the next earlier phase of very high eccentricity. Taking these as data, the proportionate change of the species of mollusca affords a means of arriving at the whole lapse of time represented by the fossiliferous rocks ; and these two estimates agree in the *order* of their magnitudes.

It is then argued that the changes of climate every 10,500 years during the numerous periods of high eccentricity have acted as a motive power in hastening on both geological and biological change. By raising and lowering the snow-line in all mountain-ranges, it has caused increased denudation ; while the same changes have caused much migration and disturbance in the organic world, and have thus tended to the more rapid modification of species. The present epoch being a period of very low eccentricity, the earth is in a phase of *exceptional stability*, both physical and organic; and it is from this period of exceptional stability that our notions of the very slow rate of change have been derived.

The conclusion is, on the whole, that the periods allowed by physicists are not only far in excess of such as are required for geological and organic change, but that they allow ample margin for a lapse of time anterior to the deposit of the earliest fossiliferous rocks several times longer than the time which has elapsed since their deposit to the present day.

Having thus laid the foundation for a scientific interpretation of the phenomena of distribution, we proceed to the Second Part of our work—the discussion of a series of typical insular faunas and floras with a view to explain the interesting phenomena they present. Taking, first, two North Atlantic groups—the Azores and Bermuda—it is shown how important an agent in the dispersal of most animals and plants is a stormy atmosphere. Although 900 and 700 miles respectively from the nearest continents, their productions are very largely identical with those of Europe and America; and, what is more important, fresh arrivals of birds, insects, and plants are now taking place almost annually. These islands afford, therefore, test examples of the great dispersive powers of certain groups of organisms, and thus serve as a basis on which to found our explanations of many anomalies of distribution. Passing on to the Galapagos, we have a group less distant from a continent and of larger area, yet, owing to special conditions, of which the comparatively stormless equatorial atmosphere is the most important, exhibiting far more speciality in its productions than the more distant Azores. Still, however, its fauna and flora are as unmistakably derived from the American continent as those of the Azores are from the European.

We next take St. Helena and the Sandwich Islands, both wonderfully isolated in the midst of vast oceans, and no longer exhibiting in their productions an exclusive affinity to one continent. Here we have to recognize the results of immense antiquity, and of those changes of geography, of climate, and in the general distribution of organisms which we know have occurred in former geological epochs, and whose causes and consequences we have discussed in the First Part of our volume. This concludes our review of the oceanic islands.

Coming now to continental islands, we consider first those of most recent origin and offering the simplest phenomena; and begin with the British Isles as affording the best example of very recent and well-known continental islands. Reviewing the interesting past history of Britain, we show why it is comparatively poor in species, and why this poverty is still greater in Ireland. By a careful examination of its fauna and flora, it is then shown that the British Isles are not so completely identical, biologically, with the continent as has been supposed. A considerable amount of speciality is shown to exist, and that this speciality is real, and not apparent, is supported by the fact that small outlying islands, such as the Isle of Man, the Shetland Isles, Lundy

Island, and the Isle of Wight, all possess certain species or varieties not found elsewhere.

Borneo and Java are next taken, as illustrations of tropical islands which may be not more ancient than Britain, but which, owing to their much larger area, greater distance from the continent, and the extreme richness of the equatorial fauna and flora, possess a large proportion of peculiar species, though these are, in general, very closely allied to those of the adjacent parts of Asia. The preliminary studies we have made enable us to afford a simpler and more definite interpretation of the peculiar relations of Java to the continent and its differences from Borneo and Sumatra than was given in my former work, " The Geographical Distribution of Animals."

Japan and Formosa are next taken, as examples of islands which are decidedly somewhat more ancient than those previously considered, and which present a number of very interesting phenomena, especially in their relations to each other, and to remote rather than to adjacent parts of the Asiatic continent.

We now pass to the group of ancient continental islands, of which Madagascar is the most typical example. It is surrounded by a number of smaller islands which may be termed its satellites, since they partake of many of its peculiarities; though some of these, as the Comoros and Seychelles, may be considered continental; while others, as Bourbon, Mauritius, and Rodriguez, are decidedly oceanic. In order to understand the peculiarities of the Madagascar fauna, we have to consider the past history of the African and Asiatic continents, which it is shown are such as to account for all the main peculiarities of the fauna of these islands without having recourse to the hypothesis of a now-submerged Lemurian continent. Considerable evidence is further adduced to show that " Lemuria " is a myth, since not only is its existence unnecessary, but it can be proved that it would not explain the actual facts of distribution. The origin of the interesting Mascarene wingless birds is discussed, and the main peculiarities of the remarkable flora of Madagascar and the Mascarene Islands pointed out ; while it is shown that all these phenomena are to be explained on the general principles of the permanence of the great oceans and the comparatively slight fluctuations of the land area, and by taking account of established palæontological facts.

There remain two other islands, Celebes and New Zealand, which are classed as "anomalous"—the one because it is almost impossible to place it in any of the six zoological regions, or determine whether it has ever been actually joined to a continent; the other because it combines the characteristics of continental and oceanic islands.

The peculiarities of the Celebesian fauna have already been dwelt upon in several previous works, but they are so remarkable and so unique that they cannot be omitted in a treatise on "insular faunas;" and here, as in the case of Borneo and Java, fuller consideration and the application of the general principles laid down in our First Part lead to a solution of the problem at once more simple and more satisfactory than any which have been previously proposed. I now look upon Celebes as an outlying portion of the great Asiatic continent of Miocene times, which either by submergence or some other cause had lost the greater portion of its animal inhabitants, and since then has remained more or less completely isolated from every other land. It has thus preserved a fragment of a very ancient fauna along with a number of later types which have reached it from surrounding islands by the ordinary means of dispersal. This sufficiently explains all the peculiar *affinities* of its animals, though the peculiar and distinctive *characters* of some of them remain as mysterious as ever.

New Zealand is shown to be so completely continental in its geological structure, and its numerous wingless birds so clearly imply a former connection with some other land (as do its numerous lizards and its remarkable reptile, the Hatteria), that the total absence of indigenous land mammalia was hardly to be expected. Some attention is therefore given to the curious animal which has been seen but never captured, and this is shown to be probably identical with an animal referred to by Captain Cook. The more accurate knowledge which has recently been obtained of the sea-bottom around New Zealand enables us to determine that the former connection of that island with Australia was towards the north, and this is found to agree well with many of the peculiarities of its fauna.

The flora of New Zealand and that of Australia are now both so well known, and they present so many peculiarities and relations of so anomalous a character, as to present, in Sir Joseph Hooker's opinion, an almost insoluble problem. Much additional information on the physical and geological history of these two countries has, however, been obtained since the appearance of Sir Joseph Hooker's works, and I therefore determined to apply to them the same method of discussion and treatment which has been usually successful with similar problems in the case of animals. The fact above noted, that New Zealand was connected with Australia in its northern, tropical portion only, of itself affords a clew to one portion of the specialities of the New Zealand flora — the presence of an unusual number of tropical families and genera, while the temperate forms consist mainly of species either identical with those found in Australia or closely allied to them. But a still more important clew is obtained in the geological structure of Australia itself, which is shown to have been for long periods divided into an eastern and a western island, in the latter of which the highly peculiar flora of temperate Australia was developed. This is found to explain with great exactness the remarkable absence from New Zealand of all the most abundant and characteristic Australian genera, both of plants and of animals, since these existed at that time only in the *western* island; while New Zealand was in connection with the *eastern* island alone, and with the tropical portion of it. From these geological and physical facts, and the known powers of dispersal of plants, all the main features and many of the detailed peculiarities of the New Zealand flora are shown necessarily to result.

Our last chapter is devoted to a wider, and if possible more interesting, subject—the origin of the European element in the floras of New Zealand and Australia, and also in those of South America and South Africa. This is so especially a botanical question that it was with some diffidence I entered upon it; yet it arose so naturally from the study of the New Zealand and Australian floras, and seemed to have so much light thrown upon it by our preliminary studies as to changes of climate and the causes which have favored the distribution of plants, that I felt my work would be incomplete without a consideration of it. The subject will be so fresh in the reader's mind that a complete summary of it is unnecessary. I venture to think, however, that I have shown, not only the several routes by which the northern plants have reached the various southern lands, but have pointed out the special aids to their migration, and the motive power which has urged them on.

In this discussion, if nowhere else, will be found a complete justification of that lengthy investigation of the exact nature of past changes of climate which to some readers may have seemed unnecessary and unsuited to such a work as the present. Without the clear and definite conclusions arrived at by that discussion, and those equally important views as to the permanence of the great features of the earth's surface, and the wonderful dispersive powers of plants which have been so frequently brought before us in our studies of insular floras, I should not have ventured to attack the wide and difficult problem of the northern element in southern floras.

In concluding a work dealing with subjects which have occupied my attention for many years, I trust that the reader who has followed me throughout will be imbued with the conviction that ever presses upon myself, of the complete interdependence of organic and inorganic nature. Not only does the marvellous structure of each organized being involve the whole past history of the earth, but such apparently unimportant facts as the presence of certain types of plants or animals in one island rather than in another are now shown to be dependent on the long series of past geological changes; on those marvellous astronomical revolutions which cause a periodic variation of terrestrial climates; on the apparently fortuitous action of storms and currents in the conveyance of germs; and on the endlessly varied actions and reactions of organized beings on each other. And although these various causes are far too complex in their combined action to enable us to follow them out in the case of any one species, yet their broad results are clearly recognizable; and we are thus encouraged to study more completely every detail and every anomaly in the distribution of living things, in the firm conviction that by so doing we shall obtain a fuller and clearer insight into the course of nature, and with increased confidence that the "mighty maze" of Being we see everywhere around us is "not without a plan."

INDEX

INDEX.

THE END.

GREAT BOOKS IN PHILOSOPHY PAPERBACK SERIES

ETHICS

Aristotle—*The Nicomachean Ethics*	$8.95
Marcus Aurelius—*Meditations*	5.95
Jeremy Bentham—*The Principles of Morals and Legislation*	8.95
John Dewey—*The Moral Writings of John Dewey, Revised Edition*	
(edited by James Gouinlock)	10.95
Epictetus—*Enchiridion*	4.95
Immanuel Kant—*Fundamental Principles of the Metaphysic of Morals*	5.95
John Stuart Mill—*Utilitarianism*	5.95
George Edward Moore—*Principia Ethica*	8.95
Friedrich Nietzsche—*Beyond Good and Evil*	8.95
Plato—*Protagoras, Philebus, and Gorgias*	7.95
Bertrand Russell—*Bertrand Russell On Ethics, Sex, and Marriage*	
(edited by Al Seckel)	19.95
Arthur Schopenhauer—*The Wisdom of Life and Counsels and Maxims*	6.95
Benedict de Spinoza—*Ethics and The Improvement of the Understanding*	9.95

SOCIAL AND POLITICAL PHILOSOPHY

Aristotle—*The Politics*	7.95
Francis Bacon—*Essays*	6.95
Mikhail Bakunin—*The Basic Bakunin: Writings, 1869–1871*	
(translated and edited by Robert M. Cutler)	10.95
Edmund Burke—*Reflections on the Revolution in France*	7.95
John Dewey—*Freedom and Culture*	10.95
G. W. F. Hegel—*The Philosophy of History*	9.95
G. W. F. Hegel—*Philosophy of Right*	9.95
Thomas Hobbes—*The Leviathan*	7.95
Sidney Hook—*Paradoxes of Freedom*	9.95
Sidney Hook—*Reason, Social Myths, and Democracy*	11.95
John Locke—*Second Treatise on Civil Government*	5.95
Niccolo Machiavelli—*The Prince*	4.95
Karl Marx—*The Poverty of Philosophy*	7.95
Karl Marx/Frederick Engels—*The Economic and Philosophic Manuscripts of 1844*	
and *The Communist Manifesto*	6.95
John Stuart Mill—*Considerations on Representative Government*	6.95
John Stuart Mill—*On Liberty*	5.95
John Stuart Mill—*On Socialism*	7.95
John Stuart Mill—*The Subjection of Women*	5.95
Friedrich Nietzsche—*Thus Spake Zarathustra*	9.95
Thomas Paine—*Common Sense*	5.95
Thomas Paine—*Rights of Man*	7.95
Plato—*Lysis, Phaedrus, and Symposium*	6.95
Plato—*The Republic*	9.95
Jean-Jacques Rousseau—*The Social Contract*	5.95
Mary Wollstonecraft—*A Vindication of the Rights of Men*	5.95
Mary Wollstonecraft—*A Vindication of the Rights of Women*	6.95

METAPHYSICS/EPISTEMOLOGY

Aristotle—*De Anima*	6.95
Aristotle—*The Metaphysics*	9.95
George Berkeley—*Three Dialogues Between Hylas and Philonous*	5.95
René Descartes—*Discourse on Method* and *The Meditations*	6.95
John Dewey—*How We Think*	10.95
John Dewey—*The Influence of Darwin on Philosophy and Other Essays*	11.95
Epicurus—*The Essential Epicurus: Letters, Principal Doctrines,*	
Vatican Sayings, and Fragments	
(translated, and with an introduction, by Eugene O'Connor)	5.95
Sidney Hook—*The Quest for Being*	11.95
David Hume—*An Enquiry Concerning Human Understanding*	5.95
David Hume—*Treatise of Human Nature*	9.95
William James—*The Meaning of Truth*	11.95
William James—*Pragmatism*	7.95
Immanuel Kant—*Critique of Practical Reason*	7.95
Immanuel Kant—*Critique of Pure Reason*	9.95
Gottfried Wilhelm Leibniz—*Discourse on Method* and the *Monadology*	6.95
John Locke—*An Essay Concerning Human Understanding*	9.95
Plato—*The Euthyphro, Apology, Crito,* and *Phaedo*	5.95
Bertrand Russell—*The Problems of Philosophy*	8.95
Sextus Empiricus—*Outlines of Pyrrhonism*	8.95

PHILOSOPHY OF RELIGION

Marcus Tullius Cicero—*The Nature of the Gods* and *On Divination*	6.95
Ludwig Feuerbach—*The Essence of Christianity*	8.95
David Hume—*Dialogues Concerning Natural Religion*	5.95
John Locke—*A Letter Concerning Toleration*	5.95
Lucretius—*On the Nature of Things*	7.95
Thomas Paine—*The Age of Reason*	13.95
Bertrand Russell—*Bertrand Russell On God and Religion* (edited by Al Seckel)	19.95

ESTHETICS

Aristotle—*The Poetics*	5.95
Aristotle—*Treatise on Rhetoric*	7.95

GREAT MINDS PAPERBACK SERIES

ECONOMICS

Charlotte Perkins Gilman—*Women and Economics: A Study of the*	
Economic Relation between Women and Men	11.95
John Maynard Keynes—*The General Theory of Employment, Interest, and Money*	11.95
Alfred Marshall—*Principles of Economics*	11.95
David Ricardo—*Principles of Political Economy and Taxation*	10.95
Adam Smith—*Wealth of Nations*	9.95

RELIGION

Thomas Henry Huxley—*Agnosticism and Christianity and Other Essays*	10.95
Ernest Renan—*The Life of Jesus*	11.95
Voltaire—*A Treatise on Toleration and Other Essays*	8.95

SCIENCE

Nicolaus Copernicus—*On the Revolutions of Heavenly Spheres*	8.95
Charles Darwin—*The Descent of Man*	18.95
Charles Darwin—*The Origin of Species*	10.95
Albert Einstein—*Relativity*	8.95
Michael Faraday—*The Forces of Matter*	8.95
Galileo Galilei—*Dialogues Concerning Two New Sciences*	9.95
Ernst Haeckel—*The Riddle of the Universe*	10.95
William Harvey—*On the Motion of the Heart and Blood in Animals*	9.95
Julian Huxley—*Evolutionary Humanism*	10.95
Edward Jenner—*Vaccination against Smallpox*	5.95
Johannes Kepler—*Epitome of Copernican Astronomy* and *Harmonies of the World*	8.95
Isaac Newton—*The Principia*	14.95
Louis Pasteur and Joseph Lister—*Germ Theory and Its Application to Medicine* and *On the Antiseptic Principle of the Practice of Surgery*	7.95
Alfred Russel Wallace—*Island Life*	16.95

HISTORY

Edward Gibbon—*On Christianity*	9.95
Herodotus—*The History*	13.95
Andrew D. White—*A History of the Warfare of Science with Theology in Christendom*	19.95

SOCIOLOGY

Emile Durkheim—*Ethics and the Sociology of Morals* (translated with an introduction by Robert T. Hall)	8.95

CRITICAL ESSAYS

Desiderius Erasmus—*The Praise of Folly*	9.95
Jonathan Swift—*A Modest Proposal and Other Satires* (with an introduction by George R. Levine)	7.95
H. G. Wells—*The Conquest of Tme* (with an introduction by Martin Gardner)	7.95

(Prices subject to change without notice.)

ORDER FORM

Prometheus Books
59 John Glenn Drive • Amherst, New York 14228–2197
Telephone: (716) 691–0133

Phone Orders (24 hours):
Toll free (800) 421–0351 • FAX (716) 691–0137
Email: PBooks6205@aol.com

Ship to: _____

Address _____

_____

City _____

County (*N.Y. State Only*) _____

Telephone _____

Prometheus Acct. # _____

❑ Payment enclosed (or)

Charge to ❑ VISA ❑ MasterCard

A/C: ☐☐☐☐☐☐☐☐☐☐☐☐☐☐☐☐☐☐☐☐☐☐

Exp. Date _____ / _____

Signature _____